AF776645

Microbial Siderophores for Agricultural Productivity

Scrivener Publishing
100 Cummings Center, Suite 541J
Beverly, MA 01915-6106

Publishers at Scrivener
Martin Scrivener (martin@scrivenerpublishing.com)
Phillip Carmical (pcarmical@scrivenerpublishing.com)

Microbial Siderophores for Agricultural Productivity

Perspectives of Use in Agriculture

Edited by

Debasis Mitra

Ayush Madan

and

Anju Rani

Scrivener Publishing

WILEY

This edition first published 2026 by John Wiley & Sons, Inc., 111 River Street, Hoboken, NJ 07030, USA and Scrivener Publishing LLC, 100 Cummings Center, Suite 541J, Beverly, MA 01915, USA
© 2026 Scrivener Publishing LLC
For more information about Scrivener publications please visit www.scrivenerpublishing.com.

All rights reserved. No part of this publication may be reproduced, stored in a retrieval system, or transmitted, in any form or by any means, electronic, mechanical, photocopying, recording, or otherwise, except as permitted by law. Advice on how to obtain permission to reuse material from this title is available at http://www.wiley.com/go/permissions.

Wiley Global Headquarters
111 River Street, Hoboken, NJ 07030, USA

For details of our global editorial offices, customer services, and more information about Wiley products visit us at www.wiley.com.

The manufacturer's authorized representative according to the EU General Product Safety Regulation is Wiley-VCH GmbH, Boschstr. 12, 69469 Weinheim, Germany, e-mail: Product_Safety@wiley.com.

Limit of Liability/Disclaimer of Warranty
While the publisher and authors have used their best efforts in preparing this work, they make no representations or warranties with respect to the accuracy or completeness of the contents of this work and specifically disclaim all warranties, including without limitation any implied warranties of merchantability or fitness for a particular purpose. No warranty may be created or extended by sales representatives, written sales materials, or promotional statements for this work. The fact that an organization, website, or product is referred to in this work as a citation and/or potential source of further information does not mean that the publisher and authors endorse the information or services the organization, website, or product may provide or recommendations it may make. This work is sold with the understanding that the publisher is not engaged in rendering professional services. The advice and strategies contained herein may not be suitable for your situation. You should consult with a specialist where appropriate. Neither the publisher nor authors shall be liable for any loss of profit or any other commercial damages, including but not limited to special, incidental, consequential, or other damages. Further, readers should be aware that websites listed in this work may have changed or disappeared between when this work was written and when it is read.

Library of Congress Cataloging-in-Publication Data

ISBN 9781394384587

Cover image: Generated with AI using Adobe Firefly
Cover design by Russell Richardson

Set in size of 11pt and Minion Pro by Manila Typesetting Company, Makati, Philippines

Printed in the USA

10 9 8 7 6 5 4 3 2 1

Contents

Preface

This book comprehensively explored the role of siderophores in promoting plant growth and their potential applications in sustainable agriculture. It was structured into multiple sections, each dedicated to different facets of siderophores, from fundamental mechanisms to advanced applications in enhancing plant health and productivity. The first section covered the basics of siderophores, including their biosynthesis, regulation, and molecular mechanisms of action. It delved into the diverse types of siderophores produced by various microorganisms, their iron-chelating properties, and the intricate plant-microbe interactions that facilitated iron acquisition in plants. This section set the stage for understanding how these microbial iron chelators played a crucial role in nutrient cycling and plant growth promotion. Subsequent sections focused on the practical applications of siderophores in agriculture, including their use as biofertilizers, biocontrol agents, and biostimulants. Chapters explored case studies and field trials that demonstrated the efficacy of siderophores in improving crop yield, stress resistance, and overall plant health under various environmental conditions. Special attention was given to innovative strategies, such as the genetic engineering of microbes for enhanced siderophore production, integration with other beneficial microorganisms, and the development of formulations for commercial agricultural use. The book concluded with a forward-looking section on the challenges, opportunities, and future directions for the application of siderophores in agriculture. This included discussions on regulatory aspects, scalability, and the economic viability of siderophore-based products, as well as emerging research trends and potential breakthroughs. This content was tailored for researchers, academicians, and professionals in the fields of plant science, microbiology, biotechnology, and sustainable agriculture, providing them with in-depth knowledge and practical insights into harnessing the power of siderophores for sustainable crop production.

1

Overview of Siderophores: Types and Functions

Rahul Kumar[1], Ayush Madan[2,3], Edward Terhemen Akange[4], Rachan Karmakar[5] and Debasis Mitra[1*]

[1]Department of Microbiology, Graphic Era (Deemed to be University), Clement Town Dehradun, Uttarakhand, India
[2]Department of Biotechnology, School of Research & Technology (SORT), People's University, Bhanpur, Bhopal, Madhya Pradesh, India
[3]Institute of Tropical Aquaculture and Fisheries (AKUATROP), Universiti Malaysia Terengganu, Kuala Nerus, Terengganu, Malaysia
[4]Department of Fisheries and Aquaculture, Joseph Sarwuan Tarka University, Makurdi, Nigeria
[5]Department of Environmental Science, Graphic Era (Deemed to be University), Clement Town Dehradun, Uttarakhand, India

Abstract

Siderophores are small, high-affinity iron-chelating molecules produced by a wide range of organisms, including bacteria, fungi, and plants, to acquire their iron from the environment. Iron is an essential element in many biological processes, yet its bioavailability is also frequently limited in aerobic and host-associated environments. Indeed siderophores provide one of the primary mechanisms for overcoming this limitation by binding iron tightly and then facilitating transport into cells by receptor-mediated processes. This chapter offers a general overview of siderophores, focusing on categorization by chemical structure and origin—from catecholates, hydroxamates, and carboxylates to mixed types. We explore the biosynthetic pathways involved in siderophore production, highlighting genetic and enzymatic mechanisms controlling siderophore synthesis. Siderophore functions extend well beyond iron acquisition; they modulate ecological interactions, microbial competition, antibiotic resistance, and pathogenicity within host–pathogen dynamics. This review chapter covers the diverse applications of siderophores in bioremediation, agriculture, and medicine, an overview of recent breakthroughs

**Corresponding author*: debasismitra3@gmail.com

Debasis Mitra, Ayush Madan and Anju Rani (eds.) Microbial Siderophores for Agricultural Productivity: Perspectives of Use in Agriculture, (1–30) © 2026 Scrivener Publishing LLC

on the use of siderophores for drug delivery and treatment of iron-related disorders. Understanding the multifunctional roles of siderophores is necessary in developing novel therapeutic strategies as well as new biotechnological applications. The scope of summarizing the current knowledge on siderophore biology and their general significance in both natural and applied contexts is what this chapter is aimed to achieve.

Keywords: Siderophores, microbial competition, pathogenicity, bioremediation

1.1 Introduction to Siderophores

Siderophores are small, high-affinity iron-chelating compounds secreted by microorganisms and plants in response to iron scarcity [16]. These compounds play a pivotal role in iron acquisition, a process crucial for the survival and growth of organisms in environments where bioavailable iron is limited [6]. The name "siderophore" originates from the Greek words "sideros", meaning iron, and "phore", meaning bearer, reflecting their primary function as iron transporters. Iron is an essential element for virtually all living organisms [41]. Despite its abundance in the Earth's crust, iron is often unavailable due to its tendency to form insoluble hydroxides and oxides under aerobic and neutral pH conditions [31]. Consequently, many organisms have evolved sophisticated mechanisms, including the production of siderophores, to overcome this challenge and acquire the necessary iron for their metabolic and physiological processes [85]. This document explores the definition of siderophores, the importance of iron in biological systems, the role of siderophores in iron acquisition, and an overview of siderophore-producing organisms [96]. Siderophores are organic compounds with high specificity and affinity for binding ferric iron (Fe^{3+}). These molecules are typically synthesized and secreted by microorganisms such as bacteria and fungi, as well as by some plants [1]. Siderophores facilitate iron uptake by forming soluble iron–siderophore complexes, which are recognized and transported into the cells *via* specific receptor proteins [4]. Siderophores exhibit diverse chemical structures, broadly classified into three main categories based on their iron-chelating functional groups. Other less common classes include mixed-type siderophores, which combine different functional groups, and amphiphilic siderophores that possess both hydrophilic and hydrophobic regions [72] (Figure 1.1).

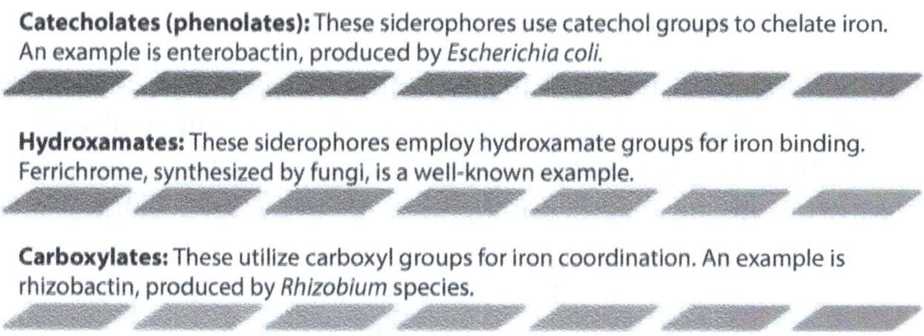

Figure 1.1 Chemical structure of siderophores.

1.1.1 Importance of Iron in Biological Systems

Iron is a critical micronutrient involved in a wide array of biological processes. It acts as a cofactor in enzymes, participates in electron transfer reactions, and is essential for DNA synthesis, respiration, and other metabolic activities [10]. However, excess iron can be toxic due to its ability to catalyze the production of reactive oxygen species (ROS) *via* Fenton reactions, leading to oxidative stress and cellular damage [55]. This dual nature of iron necessitates the tight regulation of its homeostasis in biological systems (Figure 1.2).

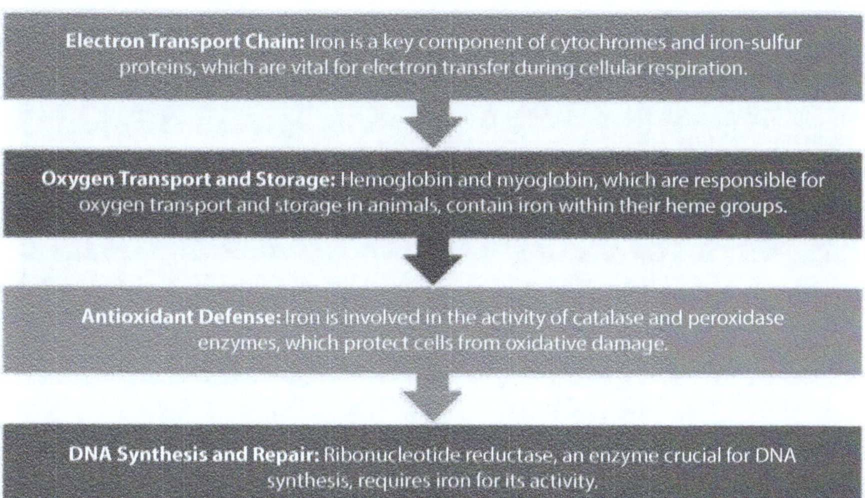

Figure 1.2 Importance of iron in biological systems.

1.1.2 Role of Siderophores in Iron Acquisition

Siderophores are indispensable in environments where iron availability is restricted [82]. The role of siderophores in iron acquisition involves several key steps: This efficient system allows organisms to thrive in iron-depleted environments, giving them a competitive advantage in ecological niches (Figure 1.3).

1.1.3 Overview of Siderophore-Producing Organisms' Ecological and Biotechnological Implications

Siderophores are essential microorganisms that bacteria use to scavenge iron from various environments [4]. Pathogenic bacteria, such as *Escherichia coli*, *Pseudomonas aeruginosa*, and *Mycobacterium tuberculosis*, and fungi rely on siderophores for iron acquisition, contributing to their virulence [35]. Fungi produce hydroxamate-type siderophores, while plants, like rice and barley, secrete phytosiderophores to enhance iron uptake in calcareous soils [56]. Graminaceous plants release mugineic acids to chelate iron and transport it into root cells. Siderophores provide a competitive edge in iron-limited environments, influencing microbial interactions and community dynamics [29]. They also have significant biotechnological applications, such as promoting plant growth by enhancing

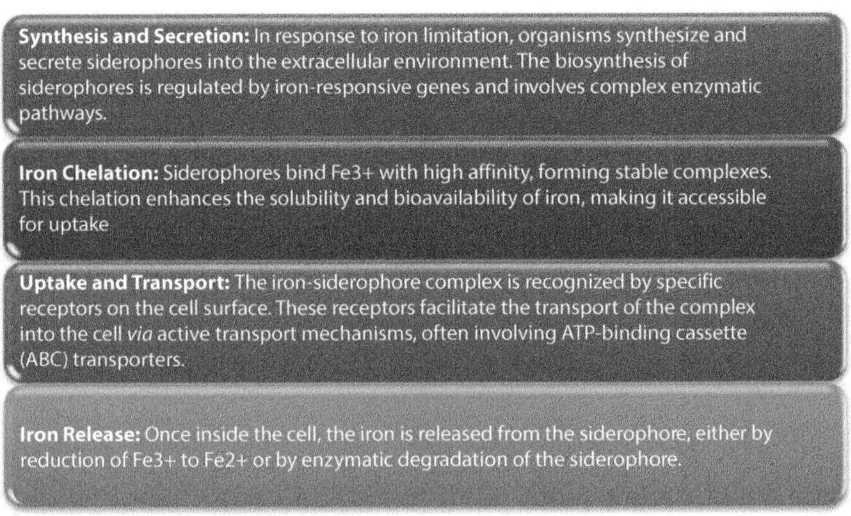

Figure 1.3 Role of siderophores.

iron uptake and suppressing pathogens. Targeting siderophore-mediated iron acquisition is a promising strategy for developing antimicrobial therapies [85]. Additionally, siderophores can mobilize heavy metals in contaminated soils and waters, facilitating bioremediation efforts.

1.2 Types of Siderophores and Classification Based on Chemical Structure

Siderophores are specialized molecules produced by microorganisms, plants, and fungi to capture iron from their environment [89]. Iron is crucial for cellular functions but is often insoluble under aerobic conditions, making it difficult for organisms to acquire enough. Siderophores chelate ferric iron (Fe^{3+}) from the environment and facilitate its transport into the organism [75]. They are essential for microbial growth and have found applications in medicine, agriculture, and biotechnology. Siderophores are classified based on their chemical structure and the type of ligands they use to bind iron (Table 1.1). They possess a high affinity for iron, typically in the range of 10^{18} to 10^{22} M^{-1}, enabling them to compete effectively with other iron-binding molecules in the environment [27]. They bind to ferric iron through donor atoms such as oxygen, nitrogen, and sometimes sulfur. Different siderophores have varying levels of specificity for iron and different metal ions, although their primary target is Fe^{3+} [9]. The classification of siderophores can be done based on the structural categories: hydroxamates, catecholate, and carboxylates. Hydroxamates are common and well-studied, containing hydroxamic acid groups that bind ferric iron *via* the nitrogen atom in the hydroxamate group. Catecholate siderophores, characterized by the presence of a catechol group, form a bidentate complex with iron, creating a stable complex [98]. Carboxylate siderophores contain carboxyl groups as the primary ligands that bind to ferric iron, playing a significant role in iron acquisition. Siderophores are molecules that can be classified based on their size and complexity. Monomeric siderophores are small molecules with one or two chelating groups produced by bacteria and fungi [57]. They are highly efficient in scavenging iron, such as enterobactin (catecholate) and ferrichrome (hydroxamate). Polymeric siderophores are larger molecules with repeated units of smaller siderophore subunits, often having more than one binding site [88]. They are involved in more complex iron uptake systems, particularly in competitive and harsh environments. Some siderophores

Table 1.1 Types of siderophore and their classification.

Types of siderophore	Classes of chemical	Structural features	Examples
Phenolate–catecholate hybrids	Phenolic and catecholic groups	Hybrid of phenolate and catecholate for iron sequestration	Entrobactin derivatives
Polyamines	Polyamine backbone	Utilize polyamine structures for binding	Staphyloferrin A, B
Hydroxy carboxylates	Hydroxy acid derivatives	Combine hydroxy and carboxylic acid groups	Citrate, rhizoferrin
Mixed-type siderophores	Combination of functional groups	Contain both hydroxamate and catecholate features	Azotobactin, serratiochelin
Carboxylates	Carboxylic acid derivatives	Possess carboxylate groups for iron coordination	Rhizobactin, rhodotorulic acid
Hydroxamates	Hydroxamic acid derivatives	Feature Hydroxamate functional groups for Fe (III) binding	Ferrichrome, desferrioxamine
Catecholates (phenolates)	Phenolic hydroxyl groups	Contain catechol or phenol groups for Fr (III) chelation	Enterobactin, agrobactin

are peptides or contain peptide-like components, while others are purely organic (non-peptide) [37]. Peptide siderophores contain amino acid sequences and are synthesized through non-ribosomal peptide synthesis pathways [89]. Non-peptide siderophores are small organic molecules without peptide components and often contain hydroxamate, catecholate, or carboxylate groups as their iron-binding ligands [22]. Organizations produce siderophores differently based on their type. Bacteria are prolific producers of siderophores, with a vast diversity in siderophore types [96]. Fungal siderophores, often with more complex structures, are key to their survival in environments where iron is limiting. Plant siderophores, on the other hand, assist plants in acquiring iron from the soil, such as mugineic acid produced by grasses like *Zea mays* [40].

1.3 Classification Based on Origin

Siderophores are specialized compounds synthesized by various organisms to scavenge iron, an essential element required for a range of biological processes [12]. Because iron is often in a limited and insoluble form in the environment primarily as Fe^{3+}, siderophores play a critical role in facilitating the acquisition of this nutrient by organisms [50]. These molecules have evolved in diverse biological systems, including bacteria, fungi, plants, and even some marine organisms, each developing unique mechanisms and chemical structures to capture iron effectively [42]. A common way to classify siderophores is based on their origin, as different types are produced by various organisms to suit their specific ecological niches and survival strategies. Broadly, siderophores can be categorized into four major groups based on their origin. Each of these origins presents a different biochemical landscape and strategy for iron acquisition, and the diversity of siderophores reflects the wide-ranging evolutionary adaptations across these organisms [80].

1.3.1 Bacterial Siderophores

Bacteria are the most studied producers of siderophores, which are essential for survival in nutrient-limited environments like soil, human body, and aquatic ecosystems [13]. These siderophores are classified based on their chemical structure, such as hydroxamates, catecholates, and carboxylates. They are diverse and competitive in iron-limited environments, outperforming other microbes for available iron [44]. Some species secrete siderophores that can also bind iron from competitors, known as stealth iron acquisition. Siderophores play a significant role in ecological functions, allowing bacteria to colonize various habitats and aiding in biogeochemical cycles by mobilizing iron in natural environments [15].

The types of siderophores include hydroxamates, which are produced by many pathogenic bacteria to acquire iron from their hosts; catecholates [89], which are produced by Gram-negative bacteria; and carboxylates, which are less common but produced by some species. Understanding the origin of these siderophores is crucial for understanding their ecological roles.

1.3.2 Fungal Siderophores

Fungi, like bacteria, face iron limitations in their environments, particularly in soil and plant-associated habitats. To address this, they produce siderophores, which are larger and more complex than bacterial ones, often containing peptide or polyketide components. Hydroxamate siderophores, like ferrichrome and fusigen, are structurally like bacterial siderophores but have additional functional groups or a more complex backbone. Peptidic siderophores, like coprogens and mugineic acid, have peptide backbones that help anchor siderophores to cell walls or facilitate iron transport into fungal hyphae [19]. Fungal siderophores are crucial in nutrient-poor soils, where fungi act as decomposers, recycling organic matter. They also enable fungi to access iron from decaying organic material for their metabolic processes [17]. Some fungi form symbiotic relationships with plants, exchanging nutrients, including iron, for carbohydrates. In pathogenic fungi, siderophores play a role in virulence, allowing them to scavenge iron from host tissues, promoting disease [56].

1.3.3 Plant Siderophores

Plants, like bacteria and fungi, are constantly involved in iron acquisition, particularly from the soil, where iron is less bioavailable [66]. Plant roots secrete siderophores to mobilize iron and other essential nutrients from the soil. These plant-derived siderophores, which are diverse in structure but share some similarities with fungal siderophores, play a critical role in ensuring that plants can grow in iron-deficient environments [49]. Plant siderophores, like those from bacteria and fungi, chelate iron, making it more soluble and easier for plants to absorb [4]. They are classified into non-proteinaceous and proteinaceous types, which form complexes with iron and are transported into root cells. Plant siderophores are crucial in iron uptake from the soil, especially in environments with poor iron availability. They play a vital role in plant growth, health, and development, managing soil fertility and promoting iron bioavailability [36]. They can also help create symbiotic relationships with bacteria, aiding in nitrogen fixation and nutrient exchange. Plants have developed sophisticated mechanisms to regulate the production of siderophores in response to iron stress, optimizing iron uptake and conserving energy when iron is abundant [32].

1.3.4 Marine and Other Siderophores

Siderophores, produced by marine microorganisms, play a crucial role in the iron cycle, particularly in oceanic regions where iron is a limiting nutrient [8]. These microorganisms, such as certain *Pseudomonas* species, produce siderophores to help them acquire iron from the iron-limited oceanic environments. These siderophores are specialized to function in the specific conditions of seawater, where iron exists in extremely low concentrations [24]. Cyclo-peptide siderophores, based on cyclopeptide structures, are more complex and can bind iron effectively in seawater. Abiotic siderophores, found in non-living marine environments like ocean sediments and particulate matter, are thought to assist in iron cycling in marine biogeochemical cycles [61]. Marine siderophores play a critical role in the iron cycle, enabling marine bacteria to thrive in nutrient-poor environments and maintaining the ecological balance of marine ecosystems [11].

1.4 Biosynthesis of Siderophores

Siderophores are small, high-affinity iron-chelating molecules produced by microorganisms to acquire iron from their environment [74]. Iron is essential for the survival of all living organisms, but its bioavailability is limited due to its poor solubility under aerobic conditions. Siderophores bind ferric iron (Fe^{3+}) with high affinity and facilitate its transport into the cell, enabling the organism to thrive in iron-limited environments [5]. The biosynthesis of siderophores is a complex, multi-step process involving several enzymes and cofactors, with biosynthetic pathways varying between different organisms and depending on the siderophore's chemical structure. The two most common classes of siderophores are hydroxamate, catecholate, and carboxylate-based, and their biosynthesis typically involves precursor formation, assembly of siderophore components, and iron binding [91]. Siderophores can be broadly classified based on their chemical structure, with each class following distinct biosynthetic pathways. Some of the most common types of siderophores include hydroxamate siderophores, catecholate siderophores, and carboxylate siderophores [89]. The biosynthesis of these siderophores is crucial for the survival of living organisms in iron-limited environments.

1.4.1 Hydroxamate Siderophores

Hydroxamate siderophores are ligands that bind iron, with hydroxamic acid groups (-C = NOH) acting as the primary ligands [26]. They are produced by actinomycetes and fungi and begin with the amino acid L-ornithine or L-lysine as the initial precursor. Non-ribosomal peptide synthetases (NRPS) play a crucial role in the assembly of the siderophore backbone [46]. The side chains of the precursor molecules are modified by hydroxylation reactions, catalyzed by enzymes like hydroxylase, forming the hydroxamic acid functional group key for binding iron. The final step is the formation of a cyclic structure that enhances stability, typically mediated by cyclization enzymes. For example, in desferrioxamine B biosynthesis, L-ornithine is converted into a peptide, which undergoes hydroxylation and cyclization to form a stable hydroxamate iron-binding ligand [74].

1.4.2 Catecholate Siderophores

Catecholate siderophores, produced by *Escherichia coli* and other enteric bacteria, contain a phenolic compound called the catechol group, which can bind iron through its two hydroxyl groups [47]. The biosynthesis of these siderophores typically begins with the amino acid tyrosine or anthranilic acid, which is converted into anthranilic acid and undergoes enzymatic transformations to form the catecholate precursor [78]. The (enterobactin) biosynthetic pathway is used to synthesize enterobactin, a classic catecholate siderophore. The amino acid L-serine is converted into anthranilic acid, and enzymatic transformations are performed to form the catecholate precursor. The resulting cyclic structure, consisting of three catechol groups linked by amide bonds, can tightly bind iron *via* bidentate coordination with the ferric iron ion (Fe^{3+}) [65]. The resulting enterobactin–Fe^{3+} complex is recognized by bacterial transport systems, facilitating its uptake into the cell.

1.4.3 Carboxylate Siderophores

Carboxylate siderophores are essential in certain environmental niches, containing carboxyl groups that chelate iron [48]. They are less common than hydroxamates and catecholates but are important. Carboxylate siderophores are synthesized using amino acids glycine or serine, which form the backbone of the siderophore. The *Pseudomonas* genus produces pyoverdine, a carboxylate siderophore, through a biosynthetic pathway that incorporates peptides and polyketides [89]. The pathway involves activating precursor amino acids by non-ribosomal peptide synthetases (NRPS) and polyketide synthases (PKS). The iron-binding motif is often completed with a peptide or cyclic structure. The siderophore binds iron with the carboxylate groups, forming a stable Fe^{3+} complex that can be transported into the bacterial cell. Carboxylate siderophores are more common in *Pseudomonas* species, and their biosynthesis involves a combination of peptide and polyketide synthesis [63].

1.4.4 Regulation of Siderophore Biosynthesis

Siderophores are produced by organisms in response to iron availability, with transcriptional control often being used to maximize iron uptake [39]. In *Escherichia coli*, the Fur protein regulates iron availability, affecting gene expression. Under low iron conditions, Fur's repression is lifted, leading to increased production of siderophores like enterobactin. Other regulatory systems, like RegB-RegA in *Pseudomonas* species, also modulate siderophore production in response to environmental iron levels.

1.5 Role of Siderophores in Various Fields

Siderophores are small, high-affinity iron-chelating molecules produced by microorganisms, plants, fungi, and even some animals to acquire iron from their environment [24]. Iron is a crucial element for many biological processes, including respiration, DNA synthesis, and enzymatic functions. However, iron is often insoluble or limited in many environments, making it challenging for organisms to access. Siderophores play a vital role in solving this problem by binding to ferric iron (Fe^{3+}), forming a stable complex and facilitating its transport into the cell. The significance of siderophores extends beyond microbial survival; they have applications in various fields, including medicine, agriculture, biotechnology, and environmental science [20]. This comprehensive exploration will cover the diverse roles of siderophores in these domains.

1.5.1 Role of Siderophores in Medicine

Siderophores are essential in the virulence of pathogenic microorganisms, particularly bacteria and fungi, as they help them scavenge iron from their host. These siderophores are also being explored for therapeutic applications and drug delivery systems [33]. Pathogens face iron limitation within the host, which the human immune system sequesters in proteins like transferrin and lactoferrin. To overcome this, pathogens synthesize siderophores to capture the limited iron available. Examples of siderophores include enterobacter, ferrichrome, desferrioxamine B, and pyoverdine, which help pathogens extract iron from the host and aid in virulence. Siderophores are considered potential therapeutic targets due to their crucial role in microbial infections [74]. Inhibiting siderophore synthesis or blocking siderophore-iron binding can limit the pathogen's ability to cause disease.

1.5.2 Siderophore-Based Drug Delivery

Siderophores are used for drug delivery, especially antibiotics, due to their high specificity for bacterial iron transporters. They can be used as vectors to deliver therapeutic agents directly to bacterial cells, improving drug efficiency and targeting, reducing side effects, and preventing drug resistance [45]. Researchers are exploring siderophore–antibiotic conjugates, where an antibiotic molecule is linked to a siderophore, enhancing the bactericidal activity of antibiotics and potentially overcoming challenges with multidrug-resistant pathogens [88].

1.5.3 Siderophores in Cancer Treatment

Recent research has explored the use of siderophores in cancer therapy, as cancer cells require large amounts of iron for growth and division. Researchers have developed strategies to deliver cytotoxic agents specifically to cancer cells, utilizing their high-affinity iron-binding properties [7]. Siderophores can bind to overexpressed iron transporters on cancer cell surfaces, allowing targeted drug delivery and minimizing damage to healthy tissues. Some siderophore-based drug delivery systems reduce side effects associated with conventional chemotherapy [69].

1.5.4 Role of Siderophores in Agriculture

Siderophores are crucial in agriculture, particularly in iron-deficient soils, as they aid in plant growth by acquiring iron from the soil, enhancing nutrient uptake, and improving plant health [54]. They also help suppress plant pathogens, offering an environmentally friendly alternative to chemical fertilizers and pesticides. Iron is an essential micronutrient for plants, but its availability can be limited in soil, especially in alkaline or waterlogged conditions. To address this, plants secrete phytosiderophores, such as mugineic acid, which chelate ferric iron and make it more available for plant roots [97]. This ensures an adequate supply of iron for vital physiological processes like chlorophyll synthesis and photosynthesis. Siderophores produced by plant-associated bacteria and fungi also improve soil fertility, enhancing the availability of iron and other nutrients in the rhizosphere, thereby promoting plant growth and crop yield. Siderophores are essential in controlling plant pathogens, as they require iron for growth. Some beneficial microorganisms, like plant growth-promoting rhizobacteria (PGPR) like *Pseudomonas fluorescens* and *Bacillus* species, produce siderophores that compete with pathogens for iron, inhibiting their growth [94]. This competition, known as siderophore-mediated antagonism can also induce systemic resistance in plants, providing an eco-friendly and sustainable alternative to chemical pesticides. In sustainable agriculture, siderophores offer the potential to reduce the reliance on synthetic chemical fertilizers. By promoting efficient iron acquisition and enhancing plant health, siderophore-producing microbes can improve soil fertility and reduce the need for chemical inputs [76]. This can lead to more environmentally friendly farming practices with reduced costs and less environmental impact.

1.5.5 Role of Siderophores in Biotechnology

Siderophores are versatile materials with diverse applications in biotechnology, particularly in bioremediation, microbial fuel cells, and the development of novel materials [83]. They can bind metals other than iron, such as aluminum, chromium, and copper, making them useful in environmental and industrial biotechnology. Siderophores enhance the solubility and bioavailability of heavy metals, allowing microorganisms to access and detoxify them [77]. They can sequester toxic metals like chromium, aluminum, and cobalt, neutralizing their toxicity or facilitating their uptake by microorganisms. This makes bioremediation techniques more effective in cleaning up contaminated soil and water, offering a sustainable alternative to traditional chemical remediation methods. Microbial fuel cells (MFCs) are devices that convert chemical energy from organic compounds into electrical energy using microorganisms [53]. Siderophores facilitate electron transfer, improving microbial metabolism and electron flow and making MFC technology more viable for sustainable energy production. Siderophores are also being explored in nanotechnology and materials science, as they can synthesize nanoparticles with high precision [72]. For example, siderophore–iron complexes have been used to create iron oxide nanoparticles, which have potential applications in medicine, electronics, and environmental science.

1.5.6 Role of Siderophores in Environmental Science

Siderophores are crucial in environmental science, influencing the cycling of nutrients and metals in ecosystems [24]. They affect the bioavailability of iron and other trace metals in soil, water, and marine environments, affecting microbial and plant communities. Iron availability in natural environments can limit microbial growth, but siderophores solubilize ferric iron, making it accessible to microorganisms. This impacts primary productivity in aquatic ecosystems and soil microbial activity. Siderophores in marine bacteria support marine food webs by utilizing phytoplankton for iron utilization [86]. In terrestrial ecosystems, siderophores facilitate nutrient cycling and contribute to soil fertility. Soil microbes produce siderophores, which improve soil health by increasing essential nutrients for plants and suppressing plant pathogens. This has significant implications for agriculture, especially in low-fertility soils, where siderophores can promote sustainable crop production [64].

1.6 Iron Transport Systems in Microorganisms through Siderophores

Iron is a crucial element for all living organisms, playing a vital role in cellular processes like respiration, DNA synthesis, and electron transfer [93]. However, iron is often limited in many environments, especially in aerobic conditions, where it oxidizes to Fe^{3+} and precipitates out of solution. Microorganisms have developed highly efficient and specialized systems to acquire iron from their surroundings, particularly through the production and utilization of siderophores. These small, high-affinity iron-chelating molecules are crucial for overcoming iron scarcity. The process of acquiring iron through siderophores involves several key steps: biosynthesis of the siderophore, binding to ferric iron, transport to the cell surface, recognition by iron uptake systems, and internalization of the siderophore–iron complex [38]. There are two primary types of siderophore-mediated iron transport systems: TonB-dependent receptors (TBDRs) and ABC (ATP-binding cassette) transporters. TonB-dependent receptors (TBDRs) are large outer membrane proteins that facilitate the active transport of iron–siderophore complexes across the outer membrane of Gram-negative bacteria. The TonB–ExbB–ExbD complex plays a crucial role in energizing these transport systems. The siderophore–Fe^{3+} complex first binds to the specific TBDR on the bacterial cell surface, providing the energy needed for the translocation of the iron–siderophore complex across the outer membrane. Inner membrane proteins, often linked to the TonB complex, facilitate the passage of the complex through the inner membrane into the periplasmic space [30].

1.6.1 ABC Transporters

ABC transporters are a family of transporters that use ATP hydrolysis to transport various substrates, including iron–siderophore complexes, across Gram-negative bacteria's inner and outer membranes [3]. They are found in Gram-negative and Gram-positive bacteria, fungi, and plants. The siderophore–iron complex binds to periplasmic-binding proteins or membrane-bound receptors, which deliver it to the transporter. The energy for transporting the siderophore–iron complex is provided by ATP hydrolysis. Once inside the cell, the iron is released by molecular reductase or hydrolytic enzymes, allowing the microorganism to use it for cellular functions [14].

1.6.2 Siderophore–Fe^{3+} Complex Internalization

Siderophore–iron complexes in bacteria are essential for various cellular processes, including respiration, DNA synthesis, and enzyme activation [74]. They are regulated by the Fur system in Gram-negative bacteria, which acts as a repressor of siderophore biosynthesis under iron-replete conditions and activates under iron-limited conditions. When intracellular iron levels are high, Fur binds to iron and represses the transcription of siderophore biosynthetic genes [59]. However, when the iron levels drop, Fur undergoes a conformational change, allowing the production of siderophores and the activation of transporters. Other regulatory systems, such as the PfeR–PfeB system in *Pseudomonas* and SodB in *Escherichia coli*, also control siderophore production in response to iron availability. In Gram-positive bacteria, iron regulation is often mediated by the DtxR protein or IdeR (iron-dependent regulator) in certain species like *Mycobacterium tuberculosis*. These proteins function similarly to Fur in Gram-negative bacteria, repressing siderophore production in the presence of sufficient iron [34].

1.6.3 Siderophore-Mediated Iron Transport in Fungi and Plants

Fungi and plants produce siderophores to acquire iron, like those in bacteria but with differences in proteins and mechanisms [56]. Fungi like *Aspergillus* and *Fusarium* produce siderophores that bind to ferric iron and use high-affinity receptors for iron uptake. Plants secrete phytosiderophores, like mugineic acid, to bind Fe^{3+} in the soil and form soluble complexes that are absorbed by roots. This process is essential for iron uptake in many plants, especially in alkaline soils [68]. Siderophore-mediated iron transport systems have evolved independently in different microbial and eukaryotic lineages but share common features like high-affinity binding and the need for specific transporters to internalize siderophore–iron complexes. The variation in siderophore types across organisms reflects the adaptive evolution of these systems in response to different environmental conditions [39].

1.7 Uses of Siderophores in Various Sectors

Siderophores are iron-chelating compounds produced by microorganisms, fungi, and plants to scavenge iron from their environment, particularly when iron is in limited supply [16]. These molecules bind to ferric iron (Fe^{3+}) with high affinity and form stable complexes that are then transported back into the producing organism for use in various biochemical processes. While siderophores are primarily studied for their biological role in microbial iron acquisition, their applications span a broad spectrum of industries and research fields [92]. The unique properties of siderophores, such as their specificity for metal ions and their ability to facilitate targeted transport, make them useful in medicine, agriculture, biotechnology, environmental science, and materials engineering. This article explores the diverse uses of siderophores in these sectors.

1.7.1 Medical Applications of Siderophores

Siderophores play a crucial role in microbial pathogenesis and drug delivery systems. They enable pathogenic bacteria, fungi, and parasites to acquire iron from the host during infection, which is essential for microbial growth and survival [35]. Iron availability is limited by host defense mechanisms, such as transferrin and lactoferrin. Pathogens produce siderophores to capture iron, allowing them to proliferate and establish infections. *Pseudomonas aeruginosa* produces pyoverdine, which enhances virulence, while enterobactin binds iron with high affinity. Understanding siderophores' role in pathogenicity has led to new strategies for combating infections. Inhibiting siderophore production or blocking siderophore–iron uptake systems can reduce the pathogens' ability to cause disease [14]. Research into siderophore-based antibiotics aims to target the pathogens' iron acquisition systems, potentially leading to new antimicrobial therapies.

1.7.2 Siderophore-Mediated Drug Delivery

Siderophores, which are highly specific to bacterial iron transporters, are being used in drug delivery systems to deliver drugs, particularly antibiotics, directly to bacterial cells [58]. This enhances drug efficacy and reduces side effects of non-targeted therapies. Key applications include siderophore–antibiotic conjugates, where antibiotics like gentamicin or vancomycin are recognized by siderophore receptors, leading to specific uptake into bacterial cells. This targeted delivery can help overcome antibiotic resistance. In cancer treatment, siderophores are being used to target cancer cells, as many overexpress iron transporters, allowing for the selective targeting of cancer cells, improving treatment outcomes and minimizing damage to healthy cells [21].

1.7.3 Agricultural Applications of Siderophores

Siderophores are crucial in agriculture for improving plant health, nutrient uptake, and disease management [87]. They facilitate iron acquisition and play a role in biocontrol. Iron is a vital micronutrient for plants, but its availability can be limited in alkaline or waterlogged soils. Plants secrete phytosiderophores, which solubilize iron and make it available for root uptake, enhancing plant growth and productivity. Phytosiderophores in cereals like wheat and barley release mugineic acid, a type of phytosiderophore, to mobilize iron from the soil. Microbial siderophores, produced by soil bacteria like *Pseudomonas* and *Azospirillum*, help plants acquire iron, phosphorus, and nitrogen, forming beneficial relationships with plants, promoting growth, and improving crop yield [90].

1.7.4 Biocontrol of Plant Pathogens

Siderophores are bacteria that suppress plant diseases through competitive exclusion, competently competing for iron in soil. This mechanism is called siderophore-mediated biocontrol [15]. Siderophores are used as biocontrol agents, such as *Pseudomonas fluorescens*, to inhibit pathogens like *Fusarium* and *Rhizoctonia*. They are also used in integrated pest management (IPM) strategies, reducing the need for chemical pesticides and promoting sustainable agricultural practices. This bio-based approach reduces the need for chemical pesticides [73].

1.7.5 Biotechnology and Industrial Applications

Siderophores are essential in biotechnological applications, particularly in bioremediation, metal recovery, and microbial fuel cells [84]. They can bind and solubilize toxic metals, making them easier for microorganisms to metabolize or immobilize, making them valuable for remediating polluted environments contaminated with heavy metals. Siderophores can also chelate toxic metals, making them less toxic or facilitating their uptake and conversion by metal-reducing bacteria [2]. They can also be used in microbial cleanup, removing pollutants from contaminated water and soils, providing an environmentally friendly alternative to chemical treatments. In industrial processes, siderophores can be used for metal recovery, extracting trace metals from low-grade ores and facilitating metal separation in metal recycling industries [95]. They also enhance the efficiency of microbial fuel cells (MFCs), which generate electricity through the metabolic processes of microorganisms. Their high-affinity iron-binding capacity assists in electron transfer between microorganisms and electrodes, improving the power output of MFCs [70]. Additionally, siderophores act as electron shuttles, facilitating the transfer of electrons from microbial cells to the anode, enhancing the performance of MFCs, making them useful in sustainable energy production, waste treatment, and environmental monitoring.

1.7.6 Environmental Applications of Siderophores

Siderophores are microorganisms that play a crucial role in environmental science, particularly in nutrient cycling, bioremediation, and ecosystem health [24]. They play a significant role in the biogeochemical cycling of iron in natural ecosystems, helping solubilize and mobilize iron, ensuring that it remains bioavailable for plants and algae. Siderophores also influence soil fertility by making essential nutrients more accessible to plants, especially in iron-deficient soils. In marine ecosystems, siderophores produced by marine bacteria help marine phytoplankton acquire iron, sustaining marine food webs [71]. Siderophores are also used in environmental cleanup to treat pollutants, including heavy metals and organic contaminants. They can bind to and detoxify heavy metals, making them less toxic to the environment. Additionally, siderophore-producing microorganisms aid in the breakdown of organic pollutants, such as petroleum hydrocarbons, in polluted environments.

1.7.7 Materials Science and Nanotechnology

Siderophores, due to their metal-binding properties, have gained significant use in nanotechnology and materials science [43]. They are used in the synthesis of metal oxide nanoparticles like iron oxide (magnetite), which have applications in drug delivery, environmental monitoring, medical imaging, drug delivery, and targeted therapies. Siderophores also stabilize nanoparticles, preventing aggregation and enhancing their stability in various applications [72].

1.8 Siderophores in Human Health and Disease

Siderophores are iron-chelating molecules produced by various organisms, including bacteria, fungi, and plants, to scavenge iron from their environment [16]. Iron is crucial for cellular processes like respiration, DNA synthesis, and enzyme activity, making it essential for both pathogens and host organisms. Iron is often limited in biological systems, particularly in human hosts, where it is tightly regulated to prevent pathogenic overgrowth [18].

Siderophores are produced by many pathogenic microorganisms as part of their strategy to acquire iron from the host during infection. Understanding and manipulating siderophores offer promising therapeutic avenues for treating infections, managing iron overload diseases, and advancing drug delivery systems. One of the most significant roles of siderophores in human health is their involvement in the pathogenesis of bacterial infections. Iron is a critical nutrient for most microorganisms, but in the human body, iron is tightly controlled to limit its availability to invading pathogens [51]. Many pathogens have evolved sophisticated strategies to overcome these limitations by producing and utilizing siderophores. For example, *Escherichia coli* produces enterobactin, a potent siderophore that binds iron with high affinity, enabling the bacteria to acquire iron from the host's tissues during infection. *Pseudomonas aeruginosa* produces pyoverdine, a siderophore that helps the bacterium acquire iron from the host during infections, particularly in cystic fibrosis patients [23]. *Staphylococcus aureus* produces staphyloferrin A and B, siderophores that aid in the uptake of iron from the host.

1.8.1 Siderophore-Dependent Virulence Mechanisms

Siderophores play a crucial role in microbial virulence, enabling pathogens to invade tissues, evade the immune system, and cause disease [60]. They are often linked to the expression of other virulence factors, such as toxins and adhesins, which contribute to the pathogen's ability to invade tissues and cause disease. For example, *Yersinia pestis* uses the siderophore yersiniabactin to acquire iron from the host, while *Helicobacter pylori* produces ferric enterobactin to scavenge iron from the host stomach mucosa. In some cases, pathogens can hijack host iron-binding proteins, such as transferrin or ferritin, to gain access to iron. The immune system controls iron availability during infection, with iron sequestration being part of the host's innate immune response [25]. The body's first line of defense against bacterial infection involves the nutritional immunity response, which includes iron sequestration by host proteins like transferrin, ferritin, and lactoferrin. This reduces the free iron available to pathogens, limiting their ability to grow and replicate [67]. However, some pathogens have evolved strategies to overcome iron limitation, such as the production of siderophores that can strip iron away from these host proteins. The immune system can respond to the presence of siderophores during infection by directly activating immune cells or serving as antigens recognized by the host's immune system. Siderophores have been studied as potential vaccine candidates, with some pathogens' siderophores, such as enterobactin from *E. coli*, used as targets for vaccine development to promote protective immunity [28]. In some cases, pathogens can use siderophores not only for iron acquisition but also as a mechanism for immune evasion.

1.9 Recent Advances and Future Directions in Siderophore Research

Siderophores, small iron-chelating molecules produced by microorganisms, fungi, and plants, are crucial for acquiring iron from the environment, especially in iron-limited conditions [14]. They play a significant role in infection, immunity, and therapeutic interventions. Recent advancements in siderophore biology and chemistry have opened new avenues for therapeutic and biotechnological applications, transforming microbial pathogenesis, antimicrobial therapies, environmental remediation, and agricultural practices. Recent advances in molecular genetics, biochemistry, and structural biology have provided new insights into the complex pathways that microorganisms use to produce and regulate siderophore synthesis [96]. This has implications for drug development, environmental applications, and biocontrol strategies. New classes of siderophores with novel chemical scaffolds and structures include azotobactins, produced by certain nitrogen-fixing bacteria, and tridentate siderophores, which coordinate iron using three binding sites. Researchers are making significant strides in understanding the structural basis for the high affinity of siderophores for iron. X-ray crystallography, cryo-electron microscopy, and advanced spectroscopy techniques are being used to elucidate the three-dimensional structures of siderophores and their receptor complexes, providing deeper insights into the molecular mechanisms of iron uptake and enabling the rational design of siderophore-based therapeutic agents.

1.9.1 Genetic and Metabolic Engineering of Siderophore Production

Genetic engineering and synthetic biology have enabled the manipulation of microbial metabolic pathways to enhance siderophore production, increasing the yield of siderophores for industrial or therapeutic applications [62]. Metabolic engineering of microorganisms like *Escherichia coli* and *Pseudomonas fluorescens* has led to the overproduction of high-value siderophores, which can be utilized for medical and agricultural purposes. Genetic modifications can be made to enhance the production of siderophores with specific properties, such as increased stability in harsh environments or the ability to chelate a broader range of metals. This opens the door to scalable production of siderophores for bioremediation, metal recovery, and targeted drug delivery systems. The role of siderophores in bacterial virulence and pathogenesis is a focal point of research, especially in the context of antimicrobial resistance (AMR) and the development of novel therapeutic strategies [85]. Recent advancements in understanding how pathogens use siderophores to acquire iron from the host and evade immune responses are paving the way for innovative approaches to prevent and treat infections. Siderophores are often part of larger, complex iron-regulated gene clusters that control the expression of virulence factors in response to iron availability.

1.9.2 Targeting Siderophore Pathways in Antimicrobial Therapy

Siderophores are essential in pathogenesis, making them attractive targets for antimicrobial drug development [52]. Siderophore-based antibiotics and siderophore-antibiotic conjugates are promising approaches, as they use siderophores' high affinity for bacterial iron transport systems to deliver antibiotics directly to bacterial cells. These conjugates have shown efficacy in targeting iron-dependent bacterial pathogens like *E. coli* and *Klebsiella pneumoniae*. Siderophore-based vaccines are being explored for pathogens like *Neisseria meningitidis* and *Haemophilus influenzae*. Targeting siderophore receptors or disrupting their production can diminish bacterial virulence, providing a promising strategy for vaccine development. Siderophores are also playing a crucial role in environmental science, particularly in bioremediation, metal recovery, and soil health management [81]. They can mobilize and transport metal ions, making them ideal candidates for applications in environmental cleanup and resource recovery. Siderophores can selectively bind to specific metals and facilitate their recovery through biotechnological processes, such as bioleaching [79]. In agriculture, siderophores can enhance crop growth, improve nutrient uptake, and combat plant diseases. Recent research in agricultural microbiology has highlighted the role of siderophore-producing bacteria in promoting plant health and yield. Using siderophore-producing bacteria as part of integrated pest management strategies could help reduce reliance on chemical pesticides, promoting sustainable farming practices.

1.10 Future Directions in Siderophore Research

Siderophore research holds promise for therapeutic siderophores, particularly siderophore–antibiotic conjugates, which could revolutionize bacterial infection treatment. Advances in synthetic biology will enable the large-scale production of customized siderophores for industrial applications like drug delivery and metal recovery. Agricultural applications, such as soil fertility enhancement and disease protection, are expected to grow due to the growing demand for sustainable farming solutions.

AI disclosure statement: The author acknowledges the use of AI-based tools for minor language editing or grammatical correction. However, all intellectual contributions, analyses, and conclusions presented in this chapter are entirely the author's own.

References

1. Abdul Hamid, N.W. and Nadarajah, K., Microbe related chemical signalling and its application in agriculture. *Int. J. Mol. Sci.*, 23, 16, 8998, 2022.
2. Adhikary, S., Saha, J., Dutta, P., Pal, A., Bacterial Homeostasis and Tolerance to Potentially Toxic Metals and Metalloids through Diverse Transporters: Metal-Specific Insights. *Geomicrobiol. J.*, 1–23, 2024.
3. Akhtar, A.A. and Turner, D.P., The role of bacterial ATP-binding cassette (ABC) transporters in pathogenesis and virulence: Therapeutic and vaccine potential. *Microb. Pathogen.*, 171, 105734, 2022.
4. Albelda-Berenguer, M., Monachon, M., Joseph, E., Siderophores: From natural roles to potential applications. *Adv. Appl. Microbiol.*, 106, 193–225, 2019.
5. Årstøl, E. and Hohmann-Marriott, M.F., Cyanobacterial siderophores—physiology, structure, biosynthesis, and applications. *Mar. Drugs*, 17, 5, 281, 2019.
6. Baek, G., Kim, J., Lee, C., A review of the effects of iron compounds on methanogenesis in anaerobic environments. *Renewable Sustainable Energy Rev.*, 113, 109282, 2019.
7. Bialasek, M., Kubiak, M., Górczak, M., Braniewska, A., Kucharzewska-Siembieda, P., Król, M., Taciak, B., Exploiting iron-binding proteins for drug delivery. *J. Physiol. Pharmacol.*, 70, 5, 2019.

8. Boiteau, R.M., Mende, D.R., Hawco, N.J., McIlvin, M.R., Fitzsimmons, J.N., Saito, M.A., Repeta, D.J., Siderophore-based microbial adaptations to iron scarcity across the eastern Pacific Ocean. *Proc. Natl. Acad. Sci.*, 113, 50, 14237–14242, 2016.
9. Brick, M.B., Hussein, M.H., Mowafy, A.M., Hamouda, R.A., Ayyad, A.M., Refaay, D.A., Significance of siderophore-producing cyanobacteria on enhancing iron uptake potentiality of maize plants grown under iron-deficiency. *Microb. Cell Fact.*, 24, 1, 3, 2025.
10. Cakmak, I., Brown, P., Colmenero-Flores, J.M., Husted, S., Kutman, B.Y., Nikolic, M., Zhao, F.J., Micronutrients, in: *Marschner's Mineral Nutrition of Plants*, pp. 283–384, Academic Press, London, United Kingdom, 2023.
11. Calvanese, M., D'Angelo, C., Tutino, M.L., Lauro, C., Whole-Cell Biosensor for Iron Monitoring as a Potential Tool for Safeguarding Biodiversity in Polar Marine Environments. *Mar. Drugs*, 22, 7, 299, 2024.
12. Chakraborty, S., Verma, E., Singh, S.S., Cyanobacterial siderophores: ecological and biotechnological significance, in: *Cyanobacteria*, pp. 383–384, Academic Press, London, United Kingdom, 2019.
13. Chevallereau, A., Pons, B.J., van Houte, S., Westra, E.R., Interactions between bacterial and phage communities in natural environments. *Nat. Rev. Microbiol.*, 20, 1, 49–62, 2022.
14. Cui, F., Fan, R., Wang, D., Li, J., Li, T., Research progress on iron uptake pathways and mechanisms of foodborne microorganisms and their application in the food sector. *Crit. Rev. Food Sci. Nutr.*, 64, 24, 8892–8910, 2024.
15. Deb, C.R. and Tatung, M., Siderophore producing bacteria as biocontrol agent against phytopathogens for a better environment: A review. *S. Afr. J. Bot.*, 165, 153–162, 2024.
16. Dhuldhaj, U. and Pandya, U., Diversity, function, and application of fungal iron chelators (Siderophores) for integrated disease management, in: *Role of microbial communities for sustainability*, pp. 259–288, 2021.
17. Dong, H., Zeng, Q., Sheng, Y., Chen, C., Yu, G., Kappler, A., Coupled iron cycling and organic matter transformation across redox interfaces. *Nat. Rev. Earth Environ.*, 4, 9, 659–673, 2023.
18. Donker, A.E., van der Staaij, H., Swinkels, D.W., The critical roles of iron during the journey from fetus to adolescent: Developmental aspects of iron homeostasis. *Blood Rev.*, 50, 100866, 2021.
19. Drechsel, H. and Winkelmann, G., Iron chelation and siderophores, in: *Transition metals in microbial metabolism*, pp. 1–49, CRC Press, Boca Raton, Florida, USA, 2022.
20. Eid, A.M., Fouda, A., Abdel-Rahman, M.A., Salem, S.S., Elsaied, A., Oelmüller, R., Hassan, S.E.D., Harnessing bacterial endophytes for promotion of plant growth and biotechnological applications: an overview. *Plants*, 10, 5, 935, 2021.
21. Fan, D. and Fang, Q., Siderophores for medical applications: imaging, sensors, and therapeutics. *Int. J. Pharm.*, 597, 120306, 2021.

22. Ghosh, S.K., Trichoderma and Its Products From Laboratory to Patient Bedside in Medical Science: An Emerging Aspect, in: *Advances in Trichoderma Biology for Agricultural Applications*, pp. 499–544, Springer International Publishing, Cham, 2022.
23. Ghssein, G. and Ezzeddine, Z., A review of Pseudomonas aeruginosa metallophores: Pyoverdine, pyochelin and pseudopaline. *Biology*, 11, 12, 1711, 2022.
24. Gomes, A.F., Almeida, M.C., Sousa, E., Resende, D.I., Siderophores and metallophores: Metal complexation weapons to fight environmental pollution. *Sci. Total Environ.*, 173044, 2024.
25. Haschka, D., Hoffmann, A., Weiss, G., Iron in immune cell function and host defense. *Semin. Cell Dev. Biol.*, 115, 27–36, 2021, July.
26. Huber, A.L., Holbein, B.E., Kidby, D.K., Metal uptake by synthetic and biosynthetic chemicals, in: *Biosorption of Heavy Metals*, pp. 249–292, 1990.
27. Ianiro, G., Rosa, L., Bonaccorsi di Patti, M.C., Valenti, P., Musci, G., Cutone, A., Lactoferrin: From the structure to the functional orchestration of iron homeostasis. *Biometals*, 36, 3, 391–416, 2023.
28. Ismail, C. M. K. H. B., Shabani, N.R.M., Chuah, C., Hassan, Z., Majeed, A.B.A., Leow, C.H., Leow, C.Y., Shigella iron-binding proteins: An insight into molecular physiology, pathogenesis, and potential target vaccine development. *Vaccine*, 40, 30, 3991–3998, 2022.
29. Joshi, H. and Khan, A., Competition-driven phenotypic plasticity in Iron acquisition and aromatic utilization confers a fitness advantage to Pseudomonas putida in an Iron-limited rhizospheric environment. *World J. Microbiol. Biotechnol.*, 40, 12, 386, 2024.
30. Josts, I., Veith, K., Tidow, H., Ternary structure of the outer membrane transporter FoxA with resolved signalling domain provides insights into TonB-mediated siderophore uptake. *Elife*, 8, e48528, 2019.
31. Kappler, A., Emerson, D., Gralnick, J.A., Roden, E.E., Muehe, E.M., Geomicrobiology of iron, in: *Ehrlich's Geomicrobiology*, vol. 6, p. 635, 2015.
32. Kermeur, N., Pédrot, M., Cabello-Hurtado, F., Iron availability and homeostasis in plants: a review of responses, adaptive mechanisms, and signaling, in: *Plant Abiotic Stress Signaling*, pp. 49–81, 2023.
33. Khairnar, A. and Goyal, A.K., Iron Harvesters: Exploring Microbial Siderophores and Their Diverse Applications in Biomedicine. *Biomed. Mater. Devices*, 1–11, 2024.
34. Khare, G., Nangpal, P., Tyagi, A.K., An Insight into the Importance of Ferritins in the Physiology of Mycobacterium tuberculosis: Unique Structural and Functional Properties, in: *Frontiers in Protein Structure, Function, and Dynamics*, pp. 425–452, 2020.
35. Khasheii, B., Mahmoodi, P., Mohammadzadeh, A., Siderophores: Importance in bacterial pathogenesis and applications in medicine and industry. *Microbiol. Res.*, 250, 126790, 2021.

36. Khatoon, Z., Huang, S., Rafique, M., Fakhar, A., Kamran, M.A., Santoyo, G., Unlocking the potential of plant growth-promoting rhizobacteria on soil health and the sustainability of agricultural systems. *J. Environ. Manage.*, 273, 111118, 2020.
37. Kitson, S.L., Squaryl Molecular Metaphors Application to Rational Drug Design and Imaging Agents. *J. Diagn. Imaging Ther.*, 4, 1, 35–75, 2017.
38. Klebba, P.E., Newton, S.M., Six, D.A., Kumar, A., Yang, T., Nairn, B.L., Chakravorty, S., Iron acquisition systems of gram-negative bacterial pathogens define TonB-dependent pathways to novel antibiotics. *Chem. Rev.*, 121, 9, 5193–5239, 2021.
39. Kramer, J., Özkaya, O., Kümmerli, R., Bacterial siderophores in community and host interactions. *Nat. Rev. Microbiol.*, 18, 3, 152–163, 2020.
40. Kumar, U., Priyanka, Malik, R., Prexha, Yogita, Malik, K., Fe chelation and zinc solubilization: A promising approach for cereals biofortification, in: *Soil Microbiomes for Sustainable Agriculture: Functional Annotation*, pp. 149–174, 2021.
41. Kümmerli, R., Iron acquisition strategies in pseudomonads: mechanisms, ecology, and evolution. *Biometals*, 36, 4, 777–797, 2023.
42. Kurth, C., Kage, H., Nett, M., Siderophores as molecular tools in medical and environmental applications. *Org. Biomol. Chem.*, 14, 35, 8212–8227, 2016.
43. Laisney, J., Chevallet, M., Fauquant, C., Sageot, C., Moreau, Y., Predoi, D., Michaud-Soret, I., Ligand-Promoted Surface Solubilization of TiO2 Nanoparticles by the Enterobactin Siderophore in Biological Medium. *Biomolecules*, 12, 10, 1516, 2022.
44. Lauderdale, J.M., Braakman, R., Forget, G., Dutkiewicz, S., Follows, M.J., Microbial feedbacks optimize ocean iron availability. *Proc. Natl. Acad. Sci.*, 117, 9, 4842–4849, 2020.
45. Lin, D., Feng, X., Mai, B., Li, X., Wang, F., Liu, J., Wang, X., Bacterial-based cancer therapy: An emerging toolbox for targeted drug/gene delivery. *Biomaterials*, 277, 121124, 2021.
46. Liras, P. and Martín, J.F., Interconnected Set of Enzymes Provide Lysine Biosynthetic Intermediates and Ornithine Derivatives as Key Precursors for the Biosynthesis of Bioactive Secondary Metabolites. *Antibiotics*, 12, 1, 159, 2023.
47. Maier, G.P. and Butler, A., Siderophores and mussel foot proteins: the role of catechol, cations, and metal coordination in surface adhesion. *JBIC J. Biol. Inorg. Chem.*, 22, 739–749, 2017.
48. Maret, W., The Extracellular Metallometabolome: Metallophores, Metal Ionophores, and Other Chelating Agents as Natural Products. *Nat. Prod. Commun.*, 19, 8, 1934578X241271701, 2024.
49. Molnár, Z., Solomon, W., Mutum, L., Janda, T., Understanding the mechanisms of Fe deficiency in the rhizosphere to promote plant resilience. *Plants*, 12, 10, 1945, 2023.

50. Monika, G., Kim, S.R.M., Kumar, P.S., Gayathri, K.V., Rangasamy, G., Saravanan, A., Biofortification: A long-term solution to improve global health-a review. *Chemosphere*, 314, 137713, 2023.
51. Murdoch, C.C. and Skaar, E.P., Nutritional immunity: The battle for nutrient metals at the host–pathogen interface. *Nat. Rev. Microbiol.*, 20, 11, 657–670, 2022.
52. Negash, K.H., Norris, J.K., Hodgkinson, J.T., Siderophore–antibiotic conjugate design: New drugs for bad bugs? *Molecules*, 24, 18, 3314, 2019.
53. Obileke, K., Onyeaka, H., Meyer, E.L., Nwokolo, N., Microbial fuel cells, a renewable energy technology for bio-electricity generation: A mini-review. *Electrochem. Commun.*, 125, 107003, 2021.
54. Pattnaik, S., Mohapatra, B., Gupta, A., Plant growth-promoting microbe mediated uptake of essential nutrients (Fe, P, K) for crop stress management: Microbe–soil–plant continuum. *Front. Agron.*, 3, 689972, 2021.
55. Paunovic, J., Vucevic, D., Radosavljevic, T., Mandić-Rajčević, S., Pantic, I., Iron-based nanoparticles and their potential toxicity: Focus on oxidative stress and apoptosis. *Chem. Biol. Interact.*, 316, 108935, 2020.
56. Pecoraro, L., Wang, X., Shah, D., Song, X., Kumar, V., Shakoor, A., Rani, R., Biosynthesis pathways, transport mechanisms and biotechnological applications of fungal siderophores. *J. Fungi*, 8, 1, 21, 2021.
57. Petrik, M., Zhai, C., Haas, H., Decristoforo, C., Siderophores for molecular imaging applications. *Clin. Transl. Imaging*, 5, 15–27, 2017.
58. Pham, T.N., Loupias, P., Dassonville-Klimpt, A., Sonnet, P., Drug delivery systems designed to overcome antimicrobial resistance. *Med. Res. Rev.*, 39, 6, 2343–2396, 2019.
59. Pi, H. and Helmann, J.D., Sequential induction of Fur-regulated genes in response to iron limitation in Bacillus subtilis. *Proc. Natl. Acad. Sci.*, 114, 48, 12785–12790, 2017.
60. Pita-Grisanti, V., Chasser, K., Sobol, T., Cruz-Monserrate, Z., Understanding the potential and risk of bacterial siderophores in cancer. *Front. Oncol.*, 12, 867271, 2022.
61. Pontoni, L., La Vecchia, C., Boguta, P., Sirakov, M., D'Aniello, E., Fabbricino, M., Locascio, A., Natural organic matter controls metal speciation and toxicity for marine organisms: A review. *Environ. Chem. Lett.*, 20, 1, 797–812, 2022.
62. Puja, H., Mislin, G.L., Rigouin, C., Engineering Siderophore biosynthesis and regulation pathways to increase diversity and availability. *Biomolecules*, 13, 6, 959, 2023.
63. Quadri, L.E., Assembly of aryl-capped siderophores by modular peptide synthetases and polyketide synthases. *Mol. Microbiol.*, 37, 1, 1–12, 2000.
64. Racioppo, A., d'Amelio, A., De Santis, A., Bevilacqua, A., Corbo, M.R., Sinigaglia, M., Potential Use of Plant Growth-Promoting Bacteria to Enhance Growth and Soil Fertility in Marginal Areas: Focus on the Apulia Region, Italy. *Agronomy*, 13, 12, 2983, 2023.

65. Raju, L. and Rajkumar, E., Coordination compounds of iron, ruthenium and osmium, in: *Photochemistry and Photophysics of Coordination Compounds*, pp. 135–203, Elsevier, Amsterdam, 2023.
66. Rashid, M.I., Mujawar, L.H., Shahzad, T., Almeelbi, T., Ismail, I.M., Oves, M., Bacteria and fungi can contribute to nutrients bioavailability and aggregate formation in degraded soils. *Microbiol. Res.*, 183, 26–41, 2016.
67. Ratledge, C. and Dover, L.G., Iron metabolism in pathogenic bacteria. *Annu. Rev. Microbiol.*, 54, 1, 881–941, 2000.
68. Riaz, M.U., Ayub, M.A., Khalid, H., ul Haq, M.A., Rasul, A., ur Rehman, M.Z., Ali, S., Fate of micronutrients in alkaline soils, in: *Resources use Efficiency in Agriculture*, pp. 577–613, 2020.
69. Ribeiro, M., Sousa, C.A., Simões, M., Harnessing microbial iron chelators to develop innovative therapeutic agents. *J. Adv. Res.*, 39, 89–101, 2022.
70. Ruacho, A., Richon, C., Whitby, H., Bundy, R.M., Sources, sinks, and cycling of dissolved organic copper binding ligands in the ocean. *Commun. Earth Environ.*, 3, 1, 263, 2022.
71. Sanchez, N., Brown, E.A., Olsen, Y., Vadstein, O., Iriarte, J.L., Gonzalez, H.E., Ardelan, M.V., Effect of siderophore on iron availability in a diatom and a dinoflagellate species: contrasting response in associated bacteria. *Front. Mar. Sci.*, 5, 118, 2018.
72. Sarvepalli, M., Velidandi, A., Ramachandravarapu, A.K., Korrapati, N., Marine Actinomycetes Siderophores: Types, High Throughput Characterization Techniques, Applications, and Their Association with Nanotechnology: A Comprehensive Review. *Nanoworld J.*, 10, 1, 1–21, 2024.
73. Sarwar, S., Advancing Sustainable Agriculture: A Comprehensive Analysis of Integrated Pest Management Strategies in Global Rice Production. *Int. J. Agric. Sustainable Dev.*, 6, 1, 1–14, 2024.
74. Schalk, I.J., Bacterial siderophores: diversity, uptake pathways and applications. *Nat. Rev. Microbiol.*, 1–17, 2024.
75. Secli, V., Michetti, E., Pacello, F., Iacovelli, F., Falconi, M., Astolfi, M.L., Battistoni, A., Investigation of Zur-regulated metal transport systems reveals an unexpected role of pyochelin in zinc homeostasis. *Mbio*, 15, 10, e02395-24, 2024.
76. Sharma, D., Gahtyari, N.C., Chhabra, R., Kumar, D., Role of microbes in improving plant growth and soil health for sustainable agriculture, in: *Advances in Plant Microbiome and Sustainable Agriculture: Diversity and Biotechnological Applications*, pp. 207–256, 2020.
77. Sharma, P. and Kumar, S., Bioremediation of heavy metals from industrial effluents by endophytes and their metabolic activity: Recent advances. *Bioresour. Technol.*, 339, 125589, 2021.
78. Shende, V.V., Bauman, K.D., Moore, B.S., The shikimate pathway: Gateway to metabolic diversity. *Nat. Prod. Rep.*, 41, 4, 604–648, 2024.
79. Sieber, A., Spiess, S., Rassy, W.Y., Schild, D., Rieß, T., Singh, S., Guebitz, G.M., Fundamentals of bio-based technologies for selective metal recovery

from bio-leachates and liquid waste streams. *Front. Bioeng. Biotechnol.*, 12, 1528992, 2025.
80. Sierra, M.A., Ryon, K.A., Tierney, B.T., Foox, J., Bhattacharya, C., Afshin, E., Tighe, S.W., Microbiome and metagenomic analysis of Lake Hillier Australia reveals pigment-rich polyextremophiles and wide-ranging metabolic adaptations. *Environ. Microbiome*, 17, 1, 60, 2022.
81. Singh, P., Chauhan, P.K., Upadhyay, S.K., Singh, R.K., Dwivedi, P., Wang, J., Jiang, M., Mechanistic insights and potential use of siderophores producing microbes in rhizosphere for mitigation of stress in plants grown in degraded land. *Front. Microbiol.*, 13, 898979, 2022.
82. Kramer, J., Özkaya, Ö., Kümmerli, R., Bacterial siderophores in community and host interactions. *Nat. Rev. Microbiol.*, 18, 3, 152–163, 2020.
83. Soares, E.V., Perspective on the biotechnological production of bacterial siderophores and their use. *App. Microbiol. Biotechnol.*, 106, 11, 3985–4004, 2022.
84. Sreedevi, P.R., Suresh, K., Jiang, G., Bacterial bioremediation of heavy metals in wastewater: A review of processes and applications. *J. Water Process Eng.*, 48, 102884, 2022.
85. Stelitano, G., Cocorullo, M., Mori, M., Villa, S., Meneghetti, F., Chiarelli, L.R., Iron acquisition and metabolism as a promising target for antimicrobials (bottlenecks and opportunities): where do we stand? *Int. J. Mol. Sci.*, 24, 7, 6181, 2023.
86. Sutak, R., Camadro, J.M., Lesuisse, E., Iron uptake mechanisms in marine phytoplankton. *Front. Microbiol.*, 11, 566691.3, 2020.
87. Swarnalatha, G.V., Goudar, V., Reddy, E. C. R. G. S., Al Tawaha, A.R.M., Sayyed, R.Z., Siderophores and their applications in sustainable management of plant diseases, in: *Secondary Metabolites and Volatiles of PGPR in Plant-Growth Promotion*, pp. 289–302, 2022.
88. Terreni, M., Taccani, M., Pregnolato, M., New antibiotics for multidrug-resistant bacterial strains: latest research developments and future perspectives. *Molecules*, 26, 9, 2671, 2021.
89. Timofeeva, A.M., Galyamova, M.R., Sedykh, S.E., Bacterial siderophores: Classification, biosynthesis, perspectives of use in agriculture. *Plants*, 11, 22, 3065, 2022.
90. Timofeeva, A.M., Galyamova, M.R., Sedykh, S.E., Plant growth-promoting soil bacteria: nitrogen fixation, phosphate solubilization, siderophore production, and other biological activities. *Plants*, 12, 24, 4074, 2023.
91. Tyśkiewicz, R., Nowak, A., Ozimek, E., Jaroszuk-Ściseł, J., Trichoderma: The current status of its application in agriculture for the biocontrol of fungal phytopathogens and stimulation of plant growth. *Int. J. Mol. Sci.*, 23, 4, 2329, 2022.
92. Vijay, K., Shibasini, M., Sivasakthivelan, P., Kavitha, T., Microbial siderophores as molecular shuttles for metal cations: sources, sinks and application perspectives. *Arch. Microbiol.*, 205, 9, 322, 2023.

93. Vogt, A.C.S., Arsiwala, T., Mohsen, M., Vogel, M., Manolova, V., Bachmann, M.F., On iron metabolism and its regulation. *Int. J. Mol. Sci.*, 22, 9, 4591, 2021.
94. Wang, H., Liu, R., You, M.P., Barbetti, M.J., Chen, Y., Pathogen biocontrol using plant growth-promoting bacteria (PGPR): Role of bacterial diversity. *Microorganisms*, 9, 9, 1988, 2021.
95. Williamson, A.J., Folens, K., Matthijs, S., Cortes, Y.P., Varia, J., Du Laing, G., Hennebel, T., Selective metal extraction by biologically produced siderophores during bioleaching from low-grade primary and secondary mineral resources. *Miner. Eng.*, 163, 106774, 2021.
96. Xie, B., Wei, X., Wan, C., Zhao, W., Song, R., Xin, S., Song, K., Exploring the Biological Pathways of Siderophores and Their Multidisciplinary Applications: A Comprehensive Review. *Molecules*, 29, 10, 2318, 2024.
97. Yoneyama, T., Iron delivery to the growing leaves associated with leaf chlorosis in mugineic acid family phytosiderophores-generating graminaceous crops. *Soil Sci. Plant Nutr.*, 67, 4, 415–426, 2021.
98. Zhang, Y., Sen, S., Giedroc, D.P., Iron Acquisition by Bacterial Pathogens: Beyond tris-Catecholate Complexes. *ChemBioChem*, 21, 14, 1955–1967, 2020.

2

Microbial Siderophores and Their Role in Plant–Microbe Interactions

K. Kamesh Krishnamoorthy

Department of Plant Pathology, Amrita Vishwa Vidyapeetham, Amrita School of Agricultural Sciences, Coimbatore, India

Abstract

Siderophores are specialized iron-chelating molecules synthesized by various microorganisms, including both beneficial and pathogenic species. They play a pivotal role in enabling microorganisms to acquire iron in iron-deficient environments, thereby supporting their survival and activity. Their diverse roles extend beyond iron acquisition, as they also have the capacity to chelate other essential elements, contributing to their versatility. In plant systems, siderophores significantly influence plant–microbe interactions, including stress mitigation, nutrient uptake, and disease management. They also enhance plant growth promotion, improve soil fertility, and aid in bioremediation processes, showcasing their multifaceted utility in sustainable agriculture. By limiting iron availability to plant pathogens while simultaneously promoting plant health, they act as effective biocontrol agents. The production of siderophores is regulated by several environmental and physiological factors, which influence their activity and effectiveness. Moreover, their high iron-binding capacity and multifunctionality have established siderophores as valuable tools in biomanagement and biocontrol applications.

***Keywords*:** Siderophores, bioremediation, plant disease management, biocontrol agents, nutrient uptake, soil fertility

Email: kamukrishagri@gmail.com

Debasis Mitra, Ayush Madan and Anju Rani (eds.) Microbial Siderophores for Agricultural Productivity: Perspectives of Use in Agriculture, (31–48) © 2026 Scrivener Publishing LLC

2.1 Introduction

Iron is a vital element involved in various biological processes, including the synthesis of essential compounds such as vitamins, antibiotics, and nucleic acids. By facilitating iron uptake, siderophores ensure that microorganisms can sustain these critical metabolic activities even in challenging conditions [11, 28, 46]. Besides this, iron is an essential element required for several fundamental biological processes, making it indispensable for the proper functioning of living organisms. Despite the abundance of iron in the earth crust it bioavailability is low due to its presence in a ferric state. In order to overcome this deficiency microorganism have evolved with a mechanism to produce siderophores which have the capacity to scavenge iron [9, 22]. Microbial siderophores are defined as specialized molecules synthesized by microorganisms when they encounter iron-limiting conditions in their environment. These molecules play a crucial role in enabling microorganisms to acquire iron that is otherwise inaccessible due to its low solubility in many natural settings [39]. Siderophores are produced by both bacteria and fungi and they possess a low molecular weight ranging from 400 to 1,000 kDa [50, 51].

2.2 Diverse Roles of Siderophores

Siderophores are widely recognized for their multitude of functions among which a few notable ones include plant growth promotion, bioremediation, soil fertility improvement, iron binding capacity, biomanagement, and biotechnology applications [68]. Chemically siderophores are water-soluble low molecular weight compounds and primarily exist in oxidized Fe^{3+} form [56, 61]. In addition to scavenging iron, siderophores produced by bacteria also have the capacity to chelate other elements. The most notable among these include desferroxamines which chelate nickel, aluminum, and vanadium, while the siderophore coelichelin chelates nickel and cadmium [21, 52]. Though siderophores aid in obtaining iron from soil environment, their presence in an increased concentration has an inverse relationship with siderophore production by microorganisms, highlighting their production in iron-deficient environments [55]. Siderophores not only play a vital role in providing essential nutrients to plants, particularly iron, but also contribute significantly to plant protection by inhibiting the growth of phytopathogens. These compounds achieve this by sequestering iron, making it unavailable to harmful microbes that rely on it for their survival and proliferation. Due to this dual functionality, siderophores are widely recognized as effective biocontrol agents, offering an eco-friendly and sustainable approach to

Table 2.1 List of recently reported siderophores produced by different microorganisms.

Producer organism	Name of siderophore	References
Streptomyces spp.	Streptobactin and tribenarthin	[3]
Pseudomonas aeruginosa	Pyoverdin and pyochelin	[36]
Chryseobacterium sp.	Chryseochelin A	[57]
Pseudomonas sp. NCIMB 10586	Mupirochelin and triabactin	[30]
Pseudomonas fluorescens C7R12	Pyoverdine	[6]

managing plant diseases while promoting plant health [42]. Siderophores play a significant role in enhancing plant nutrition, even in hydroponically grown systems, where plants rely entirely on nutrient solutions for their growth. These iron-chelating molecules help plants access iron, a crucial micronutrient, by binding to it and making it more bioavailable. This ability to support iron acquisition in a soilless environment underscores the versatility of siderophores [18]. Table 2.1 shows the list of microorganisms and the siderophores they produce. The table highlights different microbial species such as *Streptomyces* spp., which produce siderophores like Streptobactin and Tribenarthin. Species of *Pseudomonas aeruginosa* are shown to produce pyoverdin and pyochelin, while *Chryseobacterium* sp. is known for producing Chryseochelin A. It also includes *Pseudomonas* sp. NCIMB 10586, which produces Mupirochelin and Triabactin, and *Pseudomonas fluorescens* C7R12, a producer of Pyoverdine. This compilation illustrates the diversity of microbial siderophore producers and reflects their important roles in environmental and agricultural processes, such as iron acquisition and promotion of plant growth.

2.3 Classification of Siderophore Types Based on Chemical Composition

Siderophores are classified into three categories, namely, catecholate, hydroxymate, and mixed types based on their chelating groups [47]. The siderophores produced by fungal microorganism mainly fall under the category of hydroxymate siderophores [33, 71].

2.3.1 Catecholate-Type Siderophores

These are also known as phenolate type of siderophores. These group of siderophores are derived from dihydroxybenzoic acid or salicylate and have been reported only from bacteria. However, the marine fungi *Penicillium bilaii* is an exception shown to be producing the catecholate siderophore called pistillasin [12, 16]. Catechol siderophores are mainly assayed using Arnow test, which is a quantitative assay [5]. A few bacteria like *Bacillus subtilis*, *Rhizobium radiobacter*, *Pantoea vagans*, etc., are known to produce the catecholate type of siderophores [24, 35, 53]. A few notable microorganisms producing the catecholate type of siderophores include *Pantoea dispersa* MPJ19, *Pseudomonas putida* MPJ2 [54], *Bacillus subtilis* [41], and *Rhizobium* sp. strain R1 [34] under conditions encompassing iron-limiting situations, pot experiments, and drought-like situations, respectively.

2.3.2 Hydroxamate-Type Siderophores

The hydroxamate siderophores are represented by the structure C(=O) N-(OH)R, where R represents an amino acid or its derivative containing two oxygen atoms. The oxygen atoms form a bidentate ligand. When a hydroxamate siderophore interacts with a Fe^{3+} ion, the functional group is deprived of a proton from the hydroxylamine group (-NOH), leading to the creation of a bidentate ligand [25, 70]. *Bacillus megaterium* ATC 19213, *Rhizobium leguminosarium* IARI 917, and *Rhizobium meliloti* are examples of a few organisms capable of producing hydroxamate type of siderophores [10, 62, 65]. Csaky test is the most popular method for the detection of hydroxamate type of siderophore [17]. A few important bacteria producing the hydroxamate type of siderophores include *Bacillus megaterium*, *Pantoea allii* [26], *Penicillium chrysogenum* [14], *Pseudomonas fluroscens* SBW 25 [8], and fluorescent *Pseudomonas* [19].

2.3.3 Carboxylate-Type Siderophores

The carboxylate type of siderophores are exceptional in the way that they bind to iron using hydroxyl carboxylate and carboxylate groups. These have been identified in bacteria as well as in fungi [58].

2.4 Classification of Siderophores Based on Producer Organism

2.4.1 Bacterial Siderophores

The bacteria producing siderophores have been accounted in nearly 20 genera, some of which are also known to have other beneficial properties like atmospheric nitrogen fixation and phosphate solubilization [69]. A *Streptomyces* spp. collected from the rhizosphere of medicinal plants from Iran was shown produce two siderophores, namely, streptobactin and tribenarthin, when assayed using metabolomic and proteomic approaches. This was also shown to be the first report of a streptobactin siderophore produced by a terrestrial *Streptomyces* spp. [3]. Most of the siderophore-producing bacteria fall under the category of plant-growth-promoting rhizobacteria (PGPR) known for their antagonistic effects on harmful bacteria and their indigenous nature [38]. Bacterial siderophores were known to significantly influence plant growth and development particularly during stressful conditions like increased concentration of heavy metals in soil [29]. Among five bacterial isolates, viz., *Azotobacter vinelandii, Bacillus megaterium, Bacillus subtilis, Pantoea alli*, and *Rhizobium radiobacter*, evaluated for their siderophore production efficiency using CAS test, it was found that most of the bacteria reached their peak siderophore production capacity within a time frame of 24–48 h, with *A. vinelandii* being an exception. The catechol-type siderophores were found to be produced by *B. subtilis* and *R. radiobacter*, whereas *B. megaterium* and *P. alli* produced hydroxamate type of siderophores and *A. vinelandii* displayed production of both types [26]. The ideal temperature for siderophore production has been found to be often coinciding with the optimal or suboptimal conditions for bacterial growth. Most of the described bacterial siderophores are of mesophilic nature with an affinity to grow in the temperature range of 25–37°C [15, 27, 63].

2.4.2 Fungal Siderophores

The capacity to chelate iron in fungal siderophores is lower when compared to bacterial siderophores even when both are belonging to the same functional group [32]. The fungal siderophores show a great diversity, with nearly 55 types of siderophores produced by species of *Aspergillus* alone. In addition to that, many fungi are capable of producing more than one type of siderophore [37]. Siderophore production by endophytic fungi isolated from the orchid *Cymbidium aloifolium* was examined. Among them, the fungi *Penicillium crysogeum* was capable of producing the hydroxamate type of siderophores. A comparison of siderophore production efficiency among the various fungi revealed that the highest percentage of siderophore production was demonstrated by *Peniciullium chrysogenum* (CAL1), followed by *Aspergillus sydowii* (CAR 12) and *Aspergillus terreus*, respectively [14]. The siderophore synthesized by fungal phylum Ascomycota and Basidiomycota have been shown to belong to the hydroxamate type of siderophores [31]. The fungal pathogen *Colletortrichum graminicola* synthesizes an intracellular and three other siderophores which are utilized by the pathogen for growth under conditions of iron deficiency, conidiation process, and tolerance to reactive oxygen species [2].

2.5 Role of Siderophores in Plant–Microbe Interactions

The ability to produce siderophores is well exemplified by plant-growth-promoting rhizobacteria (PGPR) which have demonstrated a remarkable ability to colonize the rhizosphere, the soil region closely associated with plant roots. These beneficial microorganisms play a crucial role in enhancing iron uptake by plants, thereby influencing the bioavailability of iron in the soil [56]. Discussed in the following subsections are the roles of siderophores in plant microbe interactions. Figure 2.1 shows evidence of how siderophores, in conjunction with plant-microbe interactions, contribute significantly to the overall health and well-being of plants. These interactions bring

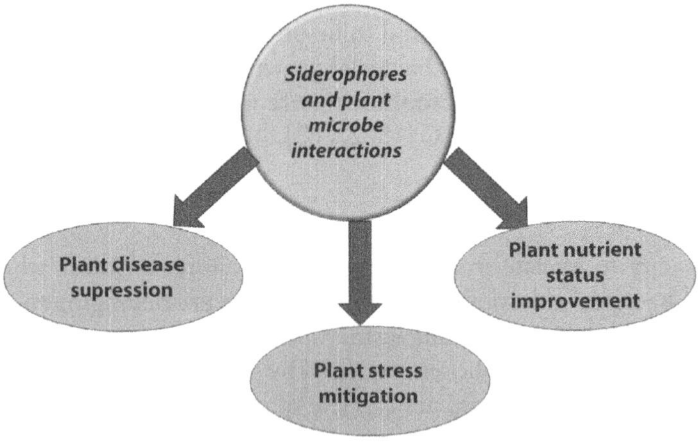

Figure 2.1 Siderophores and plant–microbe interactions.

about a range of valuable benefits, including the suppression of various plant diseases, which plays a vital role in protecting plants from harmful pathogens. In addition to disease resistance, these interactions also help mitigate stress factors affecting plants, thereby enhancing their ability to withstand adverse environmental conditions. Moreover, such interactions promote better nutrient status in plants by facilitating improved uptake and utilization of essential nutrients. Together, these siderophore-mediated processes play a crucial role in promoting plant vitality, growth, and productivity.

2.5.1 Plant Disease Suppression

The mechanism of plant pathogen suppression by siderophore-producing microorganisms is mainly through competition for iron in the rhizosphere [49]. Strains of *Pseudomonas* spp. demonstrated significant effectiveness in inhibiting the growth of the plant pathogen *Ralstonia solanacearum*. However, the degree of pathogen inhibition varied depending on the availability of iron in the environment. Notably, most *Pseudomonas* strains exhibited enhanced pathogen suppression under iron-limiting conditions, suggesting that their antagonistic activity is influenced by iron-dependent mechanisms [59]. The *Pseudomonas fluorescens* strain ATCC 17400 demonstrated a unique form of antagonistic activity against the oomycete pathogen *Pythium*, which was notably influenced by the availability of iron in its environment. Specifically, the activity was inhibited under iron-sufficient conditions, indicating that it is an iron-limited or iron-regulated mechanism. This suggests that the strain relies on iron scarcity as a trigger for activating certain biological processes that suppress the growth of *Pythium*. Among the two types of siderophores synthesized by *Pseudomonas fluorescens* ATCC 17400, the siderophore known as TQB was identified as playing a major role in this antagonistic interaction. TQB likely contributes significantly to the strain's ability to inhibit the growth and proliferation of *Pythium* [45]. The bacteria *Pseudomonas monteilii* was able to produce siderophore conveniently at an optimum pH of 7 under *in vitro* conditions. In addition to this, the bacteria also exhibited antagonism against the fungal plant pathogens *Fusarium oxysporum*, *Aspergillus calidoustus*, and *Telaromyces* spp., ranging from 40% to 65% [64]. The siderophores streptobactin and tribenarthin obtained from *Streptomyces* spp. isolated from the rhizosphere of medicinal plants was shown to display an antagonistic activity against gram-positive and gram-negative bacteria [3]. Some siderophores like those produced by *Pseudomonas aeruginosa* FP6 have an effect on fungal growth and were effective in supressing the growth of fungal plant pathogen *Colletotrichum gleiosporoides* in pepper ecosystems [20]. Some of the major soil-borne plant pathogens inhibited by siderophores include the oomycete pathogens *Pythium* and *Phytophthora* and the ascomycete pathogen *Sclerotinia* [74]. The siderophores produced by the endophytic fungi from *Cymbidium aloifolium* were capable of inhibiting the growth of bacterial pathogens, namely, *Ralstonia solanacearum* causing wilt disease in groundnut and *Xanthomonas oryzae* pv. *oryzae* causing bacterial blight disease in paddy [14]. Agriculture in high-altitude areas is often prone to low temperature stress and susceptibility to temperate plant pathogens capable of causing destructive plant diseases. The *Pseudomonas* strain AMRO1 showed an exceptional potential of siderophore production in the Himachal region at temperatures as low as 10°C, making it effective as a psychrotrophic bacteria with siderophore production efficiency and further highlighting its role in plant nutrition and protection against plant diseases. Evidence also suggested using this strain as a bioinoculant to protect plants against diseases existing in the Indian Himalayan region [60].

The bacteria isolated from paddy field soil samples from Northern Thailand were tested for siderophore production and it was observed that around 23% of isolated bacteria displayed siderophore production and showed inhibitory activity against the fungal plant pathogens *Sclerotium* sp., *Pyricularia oryzae*, *Alternaria*, and *Fusarium oxysporum* with inhibitory activity ranging from 10% to 38%. These pathogens infect major economically important crops [13]. Biocontrol-based strategies were investigated for the management of the oomycete pathogen *Globisporangium ultimum*, a fungal pathogen which is capable of infecting the juvenile tissues of plants. The siderophore mupirochelin produced by the bacterium *Pseudomonas* sp. NCIMB 10586 was found to be active against the pathogen and the only source of inhibition. Besides this, the bacterium also produced the antibiotic mupirocin which was not known to have any inhibitory activity against the fungus. The siderophore mupirochelin was identified as a non-ribosomal peptide synthestase [30].

2.5.2 Plant Stress Mitigation

Siderophores have been known to contribute to resilience in plants under challenging and adverse conditions by various means like reduction of toxicity stress in contaminated soils and facilitating nutrient delivery to plants. They also promote seedling development activity in important field and horticultural crops in addition to optimizing their early growth and performance. The siderophore bacillibactin produced by *Bacillus subtilis* was found to be effective against the plant pathogens *Alternaria porri* and *Fusarium equisetti*. The application of the siderophore on the infected leaves resulted in the reduction of plant stress by inhibition of ROS accumulation and necrosis. In addition to siderophore production, *Bacillus subtilis* also demonstrated direct antagonistic effects against the pathogens like cellular deformities and membrane damage which were confirmed by compound and scanning electron microscopy [40]. The endophytic siderophores produced by bacterium *Bacillus subtilis* augment the survival and growth of wheat plants under drought conditions [43].

2.5.3 Plant Nutrient Status Improvement

Siderophore-producing microorganisms play an important role in the extraction of heavy metals in contaminated environments and transporting the same in an accessible form to the plant roots in a controlled manner. In this way, it enhances the phytoextraction process and thereby alleviates the effects of heavy metal toxicity in plants [20]. The presence of siderophore-producing bacteria in the rhizosphere zone is highly essential to provide iron supply to plants, and the knowledge of such siderophore-producing organisms can also be utilized to apply them externally as a bio-inoculant for crop fields and tree-based ecosystems to enhance yield and soil fertility [64]. Siderophores enhance the nutritional status of plants by converting unavailable nutrients in the soil into forms that are more readily accessible and absorbable by plant roots. Siderophores were responsible for increased shoot and root length development in crops like chickpea, groundnut, and pigeonpea and increased cob length and grain number with an increased amount of iron content in plant parts in the case of maize [66]. A mixed-effect model meta-analysis study comparing the plant growth in the presence and absence of siderophore-producing bacteria revealed that siderophore-producing bacteria significantly increased plant growth up to 30% where the main growth indices stimulated included plant height and germination rate. Such an enhanced growth increase was reported in 10 families of plants which consisted of economically important crop families like Leguminosae, Graminae, Musaceae, and Solanaceae, and the mode of plant growth modification by siderophores differed with each plant family [75]. Iron deficiency poses problems in calcareous soils, leading to reduced iron content in the plant and seed of legumes. Three different pyoverdines siderophores from *Pseudomonas* bacteria were tested on the iron status of an iron-deficient chlorosis-tolerant and susceptible cultivar of pea. It was observed that the tolerant cultivar of pea accumulated an enhanced concentration of iron when compared to the susceptible one. This also displayed the role of ferripyoverdines in the optimization of internal nutrient distribution. In addition, siderophores isolated from a specific cultivar conferred maximum positive effect to that specific cultivar demonstrating host-specific interactions [44].

The two PGPR bacteria, namely, *Gluconacetobacter diazotrophicus* PAL5 and *Azospirillumm brasilense* REC 3, are known to produce siderophores which greatly contributed to iron nutrition in hydroponically grown strawberry plants. Besides this, growth metrics like growth index, root area, and greenness index were also enhanced [18]. Among four salt-tolerant PGPR bacteria isolated from the paddy rhizosphere, the bacteria *Bacillus aryabhattai* MS3 displayed maximum efficiency of siderophore production. This increased siderophore production capacity in the bacterium was attributed to the presence of ent D gene which plays a crucial role in siderophore biosynthesis, thereby improving the uptake of iron in plants under conditions of salinity [67]. The use of synthetic iron chelates have led to concerns of environmental impact. This has led to focus on siderophore-based fertilizers. The bacterium *Rhizobium radiobacter* was capable of producing the bio-based iron chelate fertilizer. This fertilizer was successful in the alleviation of iron chlorosis in soybean plants grown in calcareous soils and stabilization of chlorophyll content in plants grown during the experimental trials [4]. *Arabidopis thaliana* plants treated with pyoverdine siderophore from the bacterium *Pseudomonas fluorescens* C7R12 were found to have increased iron concentration in the roots [6]. The endophytic bacteria *Enterobacter quasihormaechei* (NBRIFY32) isolated from iron-deficient spinach plants exhibited enhanced siderophore production capacity and phytase and oxalate degradation ability. The plants inoculated with the bacterial strain showed improved vegetative growth parameters and elevated levels of essential plant nutrients which was irrespective of the iron concentration in the soil. The bacteria also displayed good colonization ability in the root and shoots of spinach plants, making it an ideal candidate displaying multiple plant-growth-promoting activities [48].

2.6 Factors Influencing Siderophore Production by Microbes

Siderophores are known to be highly resistant to degradation, which is attributed to their chemical modification which protect them from various kind of degradation. In addition to this, the process of iron chelation helps form a complex which insulates the siderophores, making them more resilient [73]. Siderophores, naturally synthesized primarily in the rhizospheric region of plants, tend to accumulate at their highest concentrations within this zone. The rhizosphere, being the narrow region of soil directly influenced by root activity and secretions, serves as a hotspot for siderophore production. Beyond this zone, their concentration progressively diminishes, and outside the rhizospheric region the siderophore concentration has been documented to be as low as 10 nM [1]. Besides this, sunlight exposure triggers the degradation of siderophores [7]. Certain farming practices like intercropping help bacteria increase their efficiency of siderophore production. This is achieved by microbiome convergence between two different plant species. Intercropping maize and groundnut helped *Pseudomonas* spp. to get transferred from maize to groundnut rhizosphere, thus enabling them to increase the bioavailability of iron by secretion of the siderophore pyoverdine which, in turn, helped improve the iron deficiency chlorosis observed in groundnut leaves [72].

2.7 Conclusion

Siderophores hold immense potential in agriculture due to their multifaceted role in promoting plant health. They contribute significantly to maintaining optimal plant nutrient status, enhancing resistance to diseases, and mitigating environmental stressors. By facilitating efficient iron acquisition, siderophores support robust plant growth, even under challenging conditions such as nutrient-deficient or alkaline soils. To harness these benefits sustainably, there is a pressing need to focus on utilizing microorganisms capable of producing siderophores as environmentally friendly iron chelators. These biological agents not only offer a greener alternative to synthetic chelates but also align with the principles of sustainable agriculture. Furthermore, it is essential to ensure that these microbial solutions are cost-effective, making them accessible for widespread application in agricultural practices.

AI disclosure statement: The author acknowledges the use of AI-based tools for minor language editing or grammatical correction. However, all intellectual contributions, analyses, and conclusions presented in this chapter are entirely the author's own.

Bibliography

1. Ahmed, E. and Holmström, S.J.M., The Effect of Soil Horizon and Mineral Type on the Distribution of Siderophores in Soil. *Geochim. Cosmochim. Acta*, 131, 184–195, 2014.

2. Albarouki, E., Schafferer, L., Ye, F., von Wirén, N., Haas, H., Deising, H.B., Biotrophy-specific downregulation of siderophore biosynthesis in C olletotrichum graminicola is required for modulation of immune responses of maize. *Mol. Microbiol., 92*, 2, 338–355, 2014.
3. Alinejad, F., Khodakaramian, G., Chandramowli, D., Devreese, B., Catechol-type siderophore producer Streptomyces spp. with quorum-quenching activity isolated from wild medicinal plants rhizosphere. *Microbe, 2*, 100033, 2024.
4. Arcas, A., Sadeghi, S.M., Lucena, J.J., Vindeirinho, J.M., Soares, E.V., Soares, H.M., López-Rayo, S., Alleviation of Fe-induced chlorosis of soybean plants grown in calcareous soil by a freeze-dried iron fertilizer containing siderophores produced by Rhizobium radiobacter. *J. Agric. Food Res., 18*, 101504, 2024.
5. Arnow, L.E., Colorimetric determination of the components of 3,4-dihydroxyphenylalanine-tyrosine mixtures. *J. Biol. Chem.*, 118, 531–537, 1937.
6. Avoscan, L., Lurthy, T., Lherminier, J., Arnould, C., Loria, P.M., Wu, T.D., Guerquin-Kern, J.L., Pivato, B., Lemaître, J.P., Lemanceau, P., Mazurier, S., Iron status and root cell morphology of Arabidopsis thaliana as modified by a bacterial ferri-siderophore. *Physiol. Plant., 176*, 1, e14223, 2024.
7. Barbeau, K., Rue, E.L., Trick, C.G., Bruland, K.W., Butler, A., Photochemical Reactivity of Siderophores Produced by Marine Heterotrophic Bacteria and Cyanobacteria Based on Characteristic Fe(III) Binding Groups. *Limnol. Oceanogr.*, 2003.
8. Boiteau, R.M., Markillie, L.M., Hoyt, D.W., Hu, D., Chu, R.K., Mitchell, H.D., *et al.*, Metabolic interactions between Brachypodiumand Pseudomonas fluorescens under controlled iron-limited conditions. *mSystem*, 6, e00580–20, 2021, doi: 10.1128/mSystems.00580-20.
9. Boyd, P.W. and Ellwood, M.J., The biogeochemical cycle of iron in the ocean. *Nat. Geosci.*, 3, 675–682, 2010.
10. Byers, B.R., Powell, M.V., Lankford, C.E., Iron-Chelating Hydroxamic Acid (Schizokinen) Active in Initiation of Cell Division in Bacillus Megaterium. *J. Bacteriol.*, 93, 286–294, 1967.
11. Cai, Y., Wang, R., An, M.M., Bei-Bei, L., Iron-depletion prevents biofilm formation in Pseudomonas aeruginosa through twitching motility and quorum sensing. *Braz. J. Microbiol.*, 41, 1, 37–41, 2010.
12. Capon, R.J., Stewart, M., Ratnayake, R., Lacey, E., Gill, J.H., Citromycetins and bilains A-C: new aromatic polyketides and diketopiperazines from Australian marine-derived and terrestrial, 2007.
13. Chaiharn, M., Chunhaleuchanon, S., Lumyong, S., Screening siderophore producing bacteria as potential biological control agent for fungal rice pathogens in Thailand. *World J. Microbiol. Biotechnol., 25*, 1919–1928, 2009.
14. Chowdappa, S., Jagannath, S., Konappa, N., Udayashankar, A.C., Jogaiah, S., Detection and characterization of antibacterial siderophores secreted by endophytic fungi from Cymbidium aloifolium. *Biomolecules, 10*, 10, 1412, 2020.

15. Colquhoun, D.J. and Sorum, H., Temperature dependent siderophore production in Vibrio salmonicida. *Microb. Pathog.*, *31*, 213–9, 2001.
16. Cox, G., Lowe, P., Winter, M., Private rights and public responsibilities: the prospects for agricultural and environmental controls. *J. Rural Stud.*, 4, 4, 323–337, 1988.
17. Csáky, T.Z., Hassel, O., Rosenberg, T., Lång, S., Turunen, E., Tuhkanen, A., On the Estimation of Bound Hydroxylamine in Biological Materials. *Acta Chem. Scand.*, 2, 450–454, 1948.
18. Delaporte-Quintana, P., Lovaisa, N.C., Rapisarda, V.A., Pedraza, R.O., The plant growth promoting bacteria Gluconacetobacter diazotrophicus and Azospirillum brasilense contribute to the iron nutrition of strawberry plants through siderophores production. *Plant Growth Regul.*, *91*, 185–199, 2020.
19. Deori, M., Jayamohan, N.S., Kumudini, B.S., Production, characterization and iron binding affinity of hydroxamate siderophores from rhizosphere associated fluorescent Pseudomonas. *J. Plant Product. Res.*, 58, 36–43, 2018, doi: 10.24425/119116.
20. Dimkpa, C.O., Merten, D., Svatoš, A., Büchel, G., Kothe, E., Metal-induced oxidative stress impacting plantgrowth in contaminated soil is alleviated by microbial siderophores. *Soil Biol. Biochem.*, 41, 154–162, 2009.
21. Dimkpa, C.O., Svatoš, A., Dabrowska, P., Schmidt, A., *et al.*, Involvement of siderophores in the reduction of metal-induced inhibition of auxin synthesis in *Streptomyces* spp. *Chemosphere*, 74, 19–25, 2008.
22. Emerson, D., Roden, E., Twining, B.S., The microbial ferrous wheel: iron cycling in terrestrial, freshwater, and marine environments. *Front. Microbiol.*, 3, 2012.
23. Eng-Wilmot, D.L. and Van der Helm, D., Molecular and Crystal Structure of the Linear Tricatechol Siderophore, Agrobactin. *J. Am. Chem. Soc.*, 102, 7719–7725, 1980.
24. Feistner, G. and Ishimaru, C., Proferrioxamine Profiles of Erwinia Herbicola and Related Bacteria. *Biometals*, 9, 337–344, 1996.
25. Feistner, G.J., Stahl, D.C., Gabrik, A.H., Proferrioxamine Siderophores OfErwinia Amylovora. A Capillary Liquid Chromatographic/Electrospray Tandem Mass Spectrometric Study. *Org. Mass Spectrom.*, 28, 163–175, 1993.
26. Ferreira, C.M., Vilas-Boas, A., Sousa, C.A., Soares, H.M., Soares, E.V., Comparison of five bacterial strains producing siderophores with ability to chelate iron under alkaline conditions. *AMB Express*, *9*, 1, 78, 2019.
27. Garner, B.L., Arceneaux, J.E.L., Byers, B.R., Temperature control of a 3,4-dihydroxybenzoate (protocatechuate)-based siderophore in *Bacillus anthracis*. *Curr. Microbiol.*, 49, 89–94, 2004.
28. Glick, R., Gilmour, C., Tremblay, J., Satanower, S., Avidan, O., Deziel, E., Greenberg, E.P., Poole, K., Banin, E., Increase in rhamnolipid synthesis under iron-limiting conditions influences surface motility and biofilm formation in Pseudomonas aeruginosa. *J. Bacteriol.*, 192, 12, 2973–2980, 2010.

29. Grobelak, A. and Hiller, J., Bacterial siderophores promote plant growth: Screening of catechol and hydroxamate siderophores. *Int. J. Phytorem.*, *19*, 9, 825–833, 2017.
30. Grosse, C., Brandt, N., Van Antwerpen, P., Wintjens, R., Matthijs, S., Two new siderophores produced by Pseudomonas sp. NCIMB 10586: The anti-oomycete non-ribosomal peptide synthetase-dependent mupirochelin and the NRPS-independent triabactin. *Front. Microbiol.*, *14*, 1143861, 2023.
31. Haas, H., Eisendle, M., Turgeon, B.G., Siderophores in fungal physiology and virulence. *Annu. Rev. Phytopathol.*, 46, 149–187, 2008.
32. He, X., Zhang, Q., Li, C., Li, M.G., Xu, S.T., Chen, S.Q., Yang, P.W., Screening and Identification of Secretory Siderophore from Plant Pathogen. *J. Yunnan Agric. Univ.*, 35, 235–242, 2020.
33. Holinsworth, B. and Martin, J., Siderophore production by marine-derived fungi. *BioMetals*, 22, 625–632, 2009.
34. Igiehon, N.O., Babalola, O.O., Aremu, B.R., Genomic insights into plant growth promoting rhizobia capable of enhancing soybean germination under drought stress. *BMC Microbiol.*, 19, 159, 2019.
35. Ito, T., Enzymatic Determination of Itoic Acid, a Bacillus Subtilis Siderophore, and 2,3-Dihydroxybenzoic Acid. *Appl. Environ. Microbiol.*, 59, 2343–2345, 1993.
36. Jeong, G.J., Khan, F., Tabassum, N., Jo, D.M., Jung, W.K., Kim, Y.M., Roles of Pseudomonas aeruginosa siderophores in interaction with prokaryotic and eukaryotic organisms. *Res. Microbiol.*, 104211, 2024.
37. Khan, A., Singh, P., Srivastava, A., Synthesis, nature and utility of universal iron chelator—Siderophore: A review. *Microbiol. Res.*, 212–213, 103–111, 2018.
38. Kong, W.L., Zhou, M., Wu, X.Q., Characteristics of siderophores production by Rahnella aquatilis JZ-GX1 and its antagonism against forest pathogens. *Microbiology*, 46, 3278–3285, 2019.
39. Kramer, J., Özkaya, O., Kümmerli, R., Bacterial siderophores in community and host interactions. *Nat. Rev. Microbiol.*, *18*, 3, 152–163, 2020.
40. Kumar, R., Singh, A., Shukla, E., Singh, P., Khan, A., Singh, N.K., Srivastava, A., Siderophore of plant growth promoting rhizobacterium origin reduces reactive oxygen species mediated injury in Solanum spp. caused by fungal pathogens. *J. Appl. Microbiol.*, *135*, 2, lxae036, 2024.
41. Kumari, S., Kiran, S., Kumari, S., Kumar, P., Singh, A., Optimization of siderophore production by Bacillus Subtilis DR2 and its effect on growth of Coriandrum Sativum. *Res. Squ.*, 1–25, 2021, doi: 10.21203/rs.3.rs-567897/v1.
42. Lahlali, R., Ezrari, S., Radouane, N., Kenfaoui, J., Esmaeel, Q., El Hamss, H., *et al.*, Biological control of plant pathogens: a global perspective. *Microorganisms*, 10, 3, 596, 2022.
43. Lastochkina, O., Garshina, D., Pusenkova, L., Effect of endophytic Bacillus subtilis on drought stress tolerance of Triticum aestivum plants of steppe volga and forest-steppe west siberian agroecological groups, in: *2nd International*

Conference Plants and Microbes: The Future of Biotechnology (Saratov), pp. 5–9, 2020.

44. Lurthy, T., Cantat, C., Jeudy, C., Declerck, P., Gallardo, K., Barraud, C., Leroy, F., Ourry, A., Lemanceau, P., Salon, C., Mazurier, S., Impact of bacterial siderophores on iron status and ionome in pea. *Front. Plant Sci.*, *11*, 730, 2020.
45. Matthijs, S., Tehrani, K.A., Laus, G., Jackson, R.W., Cooper, R.M., Cornelis, P., Thioquinolobactin, a Pseudomonas siderophore with antifungal and anti-Pythium activity. *Environ. Microbiol.*, *9*, 2, 425–434, 2007.
46. Messenger, A.J. and Barclay, R., Bacteria, iron and pathogenicity. *Biochem. Educ.*, 11, 2, 54–63, 1983.
47. Miethke, M. and Marahiel, M.A., Siderophore-based iron acquisition and pathogen control. *Microbiol. Mol. Biol. Rev.*, 71, 413–451, 2007.
48. Misra, S., Semwal, P., Pandey, D.D., Mishra, S.K., Chauhan, P.S., Siderophore-producing Spinacia Oleracea bacterial endophytes enhance nutrient status and vegetative growth under iron-deficit conditions. *J. Plant Growth Regul.*, *43*, 5, 1317–1330, 2024.
49. Murthy, K.N., Soumya, K., Udayashankar, A.C., Srinivas, C., Jogaiah, S., Biocontrol potential of plant growth-promoting rhizobacteria (PGPR) against Ralstonia solanacearum: current and prospects, in: *Biocontrol Agents and Secondary Metabolites*, S. Jogaiah (Ed.), pp. 153–180, Woodhead Publishing, Sawston, 2021, doi: 10.1016/B978-0-12-822919-4.00007-7.
50. Neilands, J.B., Microbial iron compounds. *Annu. Rev. Biochem.*, 50, 715–731, 1981.
51. Neilands, J.B., Microbial envelope proteins related to iron. *Annu. Rev. Microbiol.*, 36, 285–309, 1982.
52. Neu, M.P., Matonic, J.H., Ruggiero, C.E., Scott, B.L., Structural characterization of a plutonium(iv) siderophore complex: Single-crystal structure of Pu-Desferrioxamine E. *Angew. Chem. Int. Edit.*, 39, 1442–1444, 2000.
53. Ong, S.A., Peterson, T., Neilands, J.B., Agrobactin, a Siderophore from Agrobacterium Tumefaciens. *J. Biol. Chem.*, 254, 1860–1865, 1979.
54. Patel, T.M. and Minocheherhomji, F.P., Isolation and characterization of several siderophore producing bacteria from cotton plant. *Inter. J. Inn. Res. Sci. Eng. Tech.*, 7, 159–166, 2018, doi: 10.15680/IJIRSET.2018.0701026.
55. Rachid, D. and Ahmed, B., Effect of iron and growth inhibitors on siderophores production by *Pseudomonas fluorescens*. *Afr. J. Biotechnol.*, 4, 7, 697–702, 2005.
56. Raymond, K.N., Dertz, E.A., Kim, S.S., Enterobactin: An Archetype for Microbial Iron Transport. *Proc. Natl. Acad. Sci. U.S.A.*, 100, 3584–3588, 2003.
57. Rehm, K., Vollenweider, V., Gu, S., Friman, V.P., Kümmerli, R., Wei, Z., Bigler, L., Chryseochelins—structural characterization of novel citrate-based siderophores produced by plant protecting Chryseobacterium spp. *Metallomics*, *15*, 3, 008, 2023.
58. Schwyn, B. and Neilands, J., Universal chemical assay for the detection and determination of siderophores. *Anal. Biochem.*, *160*, 1, 47–56, 1987.

59. Shao, Z., Gu, S., Zhang, X., Xue, J., Yan, T., Guo, S., Pommier, T., Jousset, A., Yang, T., Xu, Y., Shen, Q., Siderophore interactions drive the ability of Pseudomonas spp. consortia to protect tomato against Ralstonia solanacearum. *Hortic. Res.*, *11*, 9, 186, 2024.
60. Sharma, P., Kaushal, G., Borker, S.S., Lepcha, A., Kumar, A., Kumar, R., Physiological and Genomic Elucidation of Cold-Resilient Rhizobacteria Reveals Plant Growth Promotion by Siderophore Optimization and Enhanced Biocontrol Potential Against Fungal Pathogens. *J. Plant Growth Regul.*, 1–14, 2024.
61. Sheng, Q., Li, L., Dong, W., He, L., Sheng, X., Impact of the Iron-Response Regulator Genes on the Release of Iron and Aluminum from Biotite by Rhizobium pusense S41. *Geomicrobiol. J.*, 41, 298–307, 2024.
62. Smith, M.J. and Neilands, J.B., Rhizobactin, a Siderophore from Rhizobium Meliloti. *J. Plant Nutr.*, 7, 449–458, 1984.
63. Soares, E.V., Perspective on the biotechnological production of bacterial siderophores and their use. *Appl. Microbiol. Biotechnol.*, 106, 3985–4004, 2022.
64. Srivastava, P., Sahgal, M., Sharma, K., Enshasy, H.A.E., Gafur, A., Alfarraj, S., Ansari, M.J., Sayyed, R.Z., Optimization and identification of siderophores produced by Pseudomonas monteilii strain MN759447 and its antagonism toward fungi associated with mortality in Dalbergia sissoo plantation forests. *Front. Plant Sci.*, *13*, 984522, 2022.
65. Storey, E.P., Boghozian, R., Little, J.L., Lowman, D.W., Chakraborty, R., Characterization of 'Schizokinen'; a Dihydroxamate-Type Siderophore Produced by Rhizobium Leguminosarum IARI 917. *BioMetals*, 19, 637–649, 2006.
66. Sulochana, M., Jayachandra, S., Kumar, S.A., Parameshwar, A., Reddy, K.M., Dayanand, A., Siderophore as a potential plant growth-promoting agent produced by Pseudomonas aeruginosa JAS-25. *Appl. Biochem. Biotechnol.*, 174, 1, 297e308, 2014.
67. Sultana, S., Alamb, S., Karimc, M.M., Screening of siderophore-producing salt-tolerant rhizobacteria suitable for supporting plant growth in saline soils with iron limitation. *J. Agri. Food Res.*, 4, 100150, 2021.
68. Swarnalatha, G.V., Goudar, V., Reddy, E.C.R.G.S., Al Tawaha, A.R.M., Sayyed, R.Z., Siderophores and their applications in sustainable management of plant diseases, in: *Secondary metabolites and volatiles of PGPR in plant-growth promotion*, pp. 289–302, 2022.
69. Timofeeva, A.M., Galyamova, M.R., Sedykh, S.E., Bacterial Siderophores: Classification, Biosynthesis, Perspectives of Use in Agriculture. *Plants*, 11, 3065, 2022.
70. Ustiatik, R., Nuraini, Y., Suharjono, S., Handayanto, E., Siderophore Production of the Hg-Resistant Endophytic Bacteria Isolated from Local Grass in the Hg-Contaminated Soil. *J. Ecol. Eng.*, 22, 129–138, 2021.
71. Van der Helm, D. and Winkelmann, G., Hydroxamates and polycarboxylates as iron transport agents (siderophores) in fungi, in: *Metal Ions in Fungi*, vol.

11, G. Winkelmann, and D.R. Winge, (Eds.), pp. 39–98, Marcel Dekker, Inc., NewYork, 1994.

72. Wang, N., Wang, T., Chen, Y., Wang, M., Lu, Q., Wang, K., Dou, Z., Chi, Z., Qiu, W., Dai, J., Niu, L., Microbiome convergence enables siderophore-secreting-rhizobacteria to improve iron nutrition and yield of peanut intercropped with maize. *Nat. Commun.*, *15*, 1, 839, 2024.
73. Winkelmann, G., Busch, B., Hartmann, A., Kirchhof, G., Süßmuth, R., Jung, G., Degradation of desferrioxamines by Azospirillum irakense: assignment of metabolites by HPLC/electrospray mass spectrometry. *Biometals*, *12*, 255–264, 1999.
74. Yu, X.M., Ai, C.X., Xin, L., Zhou, G.F., The siderophore-producing bacterium, Bacillus subtilis CAS15, has a biocontrol effect on Fusarium wilt and promotes the growth of pepper. *Eur. J. Soil Biol.*, 2011.
75. Zhang, S., Deng, Z., Borham, A., Ma, Y., Wang, Y., Hu, J., Wang, J., Bohu, T., Significance of soil siderophore-producing bacteria in evaluation and elevation of crop yield. *Horticulturae*, *9*, 3, 370, 2023.

3

Siderophores in the Rhizosphere: Enhancing Plant Nutrition and Growth

Beatriz Elena Guerra Sierra* and Alvaro Jose Hernández Tasco

Facultad de Ciencias Naturales y Agropecuarias, Grupo de investigación Biotecnología Agroambiente y Salud-Microbiota-Universidad de Santander, Bucaramanga, Colombia

Abstract

Siderophores, high-affinity iron-binding molecules produced by microorganisms, play a crucial role in plant nutrition by facilitating the uptake of iron and other essential nutrients. Several studies have shown that microbial siderophore production enhances iron solubilization, thereby promoting plant growth and resilience under adverse conditions. Other microorganisms, such as endophytic bacteria, have also been proven effective in mobilizing nutrients within plants. Research has identified microbial strains efficient in siderophore production and explored their application in sustainable agricultural systems. However, large-scale production and application pose challenges, including siderophore stability across diverse soil conditions and formulation costs. Future studies should focus on optimizing microbial strains, developing effective formulations, and evaluating environmental impacts. Integrating siderophores into sustainable agricultural practices can reduce the dependency on chemical fertilizers, thereby improving soil health and crop productivity.

***Keywords*:** Plant–microbe interaction, nutrient metabolism, siderophores, rhizosphere, plant nutrition

**Corresponding author*: bguerra@udes.edu.co

Debasis Mitra, Ayush Madan and Anju Rani (eds.) Microbial Siderophores for Agricultural Productivity: Perspectives of Use in Agriculture, (49–68) © 2026 Scrivener Publishing LLC

3.1 Introduction

Plants are considered holobiont organisms that simultaneously interact across three main ecological sectors: the phyllosphere (aerial parts), the endosphere (internal tissues), and the rhizosphere (soil contact area) [23]. Among these, the rhizosphere stands out as a highly dynamic ecological niche where roots establish complex relationships with soil microbiota to ensure nutrient acquisition and promote plant growth. This rhizospheric microbiota encompasses a vast diversity of bacteria, fungi, archaea, oomycetes, viruses, and other eukaryotes [59], whose presence and activity are strongly influenced by root exudates. Through processes such as mineralization, mineral solubilization, growth promotion, increased stress tolerance, and pathogen suppression, rhizospheric microorganisms play an essential role in plant nutrition [62, 65, 66].

Within this microbial community, plant growth-promoting rhizobacteria (PGPR) and arbuscular mycorrhizal fungi (AMF) are particularly noteworthy for their roles in enhancing nutrient uptake and plant resilience against biotic and abiotic stresses. The composition and structure of these communities are shaped by constantly changing biotic and abiotic factors [1, 60]. Moreover, the activity of these microorganisms is regulated by various chemical signals derived from root exudates, such as secondary metabolites, volatile organic compounds, phytohormones, and siderophores. These exudates critically determine both the organization of the rhizospheric microbiome and the availability of essential nutrients for plant growth [11, 59, 67].

In addition to influencing microbial community structure and functionality, root exudates regulate plant–microbe interactions. By modulating microbial diversity and activity, plants can encourage beneficial populations while simultaneously restricting pathogen proliferation through competitive and exclusionary mechanisms [14, 38]. Among these compounds, siderophores are particularly important for iron acquisition, a micronutrient essential for plant development and nutrition. Iron plays a vital role in oxygen transport and the electron transport chain, participating in physiological processes such as photosynthesis and chlorophyll biosynthesis. These chelating molecules, synthesized by both plants and microorganisms, improve iron bioavailability in low-solubility soils, facilitating root absorption [34]. Additionally, siderophores contribute to suppressing soil pathogens by competing for iron, thereby favoring beneficial microorganisms and strengthening plant immune responses [67, 68].

In this context, the current chapter aims to delve deeper into the role of siderophores in plant nutrition and growth, emphasizing their interaction with the rhizosphere microbiota and their influence on the absorption of iron and other essential micronutrients. It will discuss the mechanisms enabling iron acquisition in environments with limited availability, their impact on rhizospheric microbiome structure, and potential applications in sustainable agriculture. A detailed understanding of siderophore-mediated microbial interactions in the rhizosphere and their impact on plant development and nutrition will facilitate innovative strategies to address nutritional deficiencies in crops. Ultimately, this will contribute to reducing reliance on chemical fertilizers and promoting regenerative and sustainable agriculture.

3.2 Iron: Essential Micronutrient and Its Soil Chemistry Challenges

Iron (Fe) is an essential micronutrient for plant growth as it participates in redox reactions, electron transfer, and nutrient assimilation through enzymes and proteins containing iron cofactors. Numerous biochemical pathways, particularly in mitochondria and chloroplasts, require iron cofactors, making the soluble concentration of this element crucial for crop development and health [29].

Iron deficiency is the primary nutritional disorder in plants, leading to reduced productivity and considerable economic losses [9]. This issue is becoming increasingly relevant under the current climate change conditions, as elevated atmospheric CO_2 concentrations decrease iron content in the edible parts of various staple crops, including rice, wheat, legumes, and maize [36]. Additionally, although iron is critical for nutrient assimilation, its high reactivity in the ferrous form (Fe^{2+}) can lead to the generation of reactive oxygen species (ROS), negatively affecting plant physiology [44].

In field conditions, iron predominantly occurs as Fe^{2+}; however, weathering under aerobic environments transforms it into highly insoluble Fe^{3+} oxides and hydroxides. This transformation results in various compositions of iron oxides and hydroxides with different degrees of crystallization, impacting their solubility. Furthermore, iron availability sharply decreases when soil pH approaches neutrality or becomes slightly alkaline (pH 7–8.5) [31]. Indeed Fe^{3+} solubility decreases by a factor of 1,000 for each unit increase in pH [30]. To maintain iron in solution and ensure its efficient availability, this metal must be stabilized by chelating agents or ligands. Among the factors contributing to this stabilization are siderophores, compounds produced by soil and rhizosphere microorganisms capable of solubilizing mineral phosphates and increasing iron bioavailability for plants.

3.3 Siderophores as Mediators of Nutrient Availability

Siderophores are key components in the rhizosphere due to their ability to mobilize, chelate, and transport essential metals, including Fe, zinc (Zn), manganese (Mn), molybdenum (Mo), and copper (Cu) [4]. Iron, in particular, is one of Earth's most abundant elements and is essential for numerous physiological processes, acting as a cofactor in enzymes involved in respiration, DNA synthesis, and photosynthesis. However, in oxygenated soils, iron exhibits low solubility, which limits its availability to plants. To overcome this constraint, organisms have developed iron acquisition strategies based on acidolysis and chelation. On one hand, microbial secondary metabolites such as citrate and gluconate, which are organic acids, acidify the environment, dissolve minerals, and release nutrients. On the other hand, microorganisms produce siderophores—low molecular weight organic chelating compounds—that sequester metals into more soluble forms [4, 8, 37]. These siderophores, secreted by rhizospheric bacteria, fungi, and even plants, exhibit a high affinity for Fe^{3+}, converting it into bioavailable forms readily absorbed by roots [27, 53]. Additionally, siderophores can modulate the availability of other micronutrients and, in some cases, stimulate root exudation of organic compounds that further enhances nutrient solubilization in the soil [3].

In the rhizosphere, the competition for iron between beneficial microorganisms and pathogens significantly shapes the structure and regulation of the soil microbiome. Plant-growth-promoting rhizobacteria produce siderophores that not only facilitate iron uptake by plants but also restrict pathogen access to this micronutrient, thereby reducing pathogen proliferation and virulence [12, 67]. Furthermore, siderophores can modulate plant immune responses by activating defense mechanisms, including induced systemic resistance (ISR), thereby enhancing plant resilience against pathogens and other stressors [35]. Recent studies also indicate that siderophores influence hormonal regulation in plants by promoting auxin and cytokinin synthesis, thus positively impacting root development and nutrient uptake efficiency [53].

Although the role of siderophores in iron availability has been extensively studied, their contribution to plant nutrition extends to other micronutrients as well. Due to their variable structures, siderophores can form complexes with various metals and metalloids, making them effective not only in chelating essential micronutrients but also in sequestering heavy metals from the environment [35]. The affinity of siderophores for different metals depends on their ligand characteristics, the coordination environment, and the specific metal ion involved. Due to this property, siderophores can facilitate the detoxification of soils contaminated with toxic elements such as cadmium (Cd), lead (Pb), or mercury (Hg), aiding in bioremediation efforts and reducing toxicity to plants [19, 47]. Indeed research has shown that applying siderophores to crops grown in contaminated soils promotes tolerance to heavy metal stress by stimulating exclusion mechanisms and the differential accumulation of these metals [63].

The stability of the metal–siderophore complex is critical for its effectiveness. When affinity is excessively high, metals may remain immobilized in the soil and become unavailable for root absorption. Complex formation and dissociation depend on factors such as soil pH, mineral composition, and root exudate activity. The use of siderophore-producing bacteria in biofertilizers has been proposed as an effective method to enhance nutrient absorption in various agricultural systems [3, 15]. However, it is essential to select bacterial strains compatible with specific soil conditions to ensure long-term effectiveness. Additionally, integrating siderophores into comprehensive fertilization programs could amplify synergistic effects when combined with organic amendments, thereby improving soil fertility and crop yields.

3.4 Impact of Siderophores on Plant Growth and Health

Siderophores are chelating compounds, primarily targeting iron, playing a crucial role in plant nutrition and protection. Their importance lies in their ability to facilitate the uptake of Fe and other essential trace metals, regulate the toxicity of heavy metals in contaminated soils, and strengthen plant defenses against both biotic and abiotic stress [22, 69]. In alkaline or calcareous soils, the low solubility of Fe limits its absorption, potentially triggering deficiencies that affect vital processes such as respiration, DNA synthesis, and photosynthesis. To counteract these limitations, rhizospheric microorganisms such as bacteria and fungi secrete siderophores, which enhance the solubility of Fe^{3+} and facilitate its root absorption. Once chelated, Fe^{3+} can be assimilated after its reduction to Fe^{2+} in the apoplast or *via* ligand exchange with phytosiderophores, depending on complex stability and soil physicochemical conditions [69]. Additionally, these siderophores capture other trace metals like Zn, Cu, and Mn, which are essential for various enzymatic processes [48].

Increased nutrient availability promotes more vigorous root development, further supported by the synthesis of phytohormones such as auxins, cytokinins, and gibberellins, produced both by plants and PGPR [7, 58]. Furthermore, enhanced uptake of trace elements through siderophores contributes to crop biofortification, improving nutritional quality and reducing dependency on chemical fertilizers [70]. This leads to optimal plant growth and higher-value food products, thus supporting more sustainable agricultural practices.

Plants interact with numerous microorganisms, not all beneficial. Pathogenic microorganisms causing diseases present significant threats, necessitating robust plant defenses for survival. Beneficial microorganisms employ various strategies, including siderophore production, since Fe and other trace elements are also essential for pathogens. By capturing Fe more efficiently, siderophores secreted by beneficial microbes (e.g., *Pseudomonas* spp.) reduce Fe availability for pathogens such as *Fusarium oxysporum* and *Aspergillus* spp., thus limiting their proliferation and pathogenicity [15, 17]. Iron competition in the rhizosphere significantly influences soil microbiome structure. Additionally, various siderophores can activate induced systemic resistance (ISR), strengthening plant defenses against a wide range of phytopathogens [22]. This effect, combined with antimicrobial compound production by rhizobacteria, enhances plant protection and root system health [58]. In Figure 3.1, the main benefits (1–13) provided by siderophores to promote plant productivity and health are summarized.

Plant stress is not limited to biotic factors; it also includes abiotic conditions such as heavy metal toxicity (Cd, Pb, Zn, Hg) or salinity [16, 28]. Due to their strong chelating capabilities, siderophores sequester or immobilize harmful metals, reducing their bioavailability and associated oxidative damage [71]. This action prevents the excessive formation of reactive oxygen species, which are damaging to plants, thereby promoting cellular homeostasis. Additionally, bacteria like *Gluconacetobacter diazotrophicus* and *Azospirillum brasilense* produce siderophores that enhance Fe absorption under saline or other stress conditions, enabling cultivation in marginal soils without relying on transgenic varieties [51]. Overall, siderophores

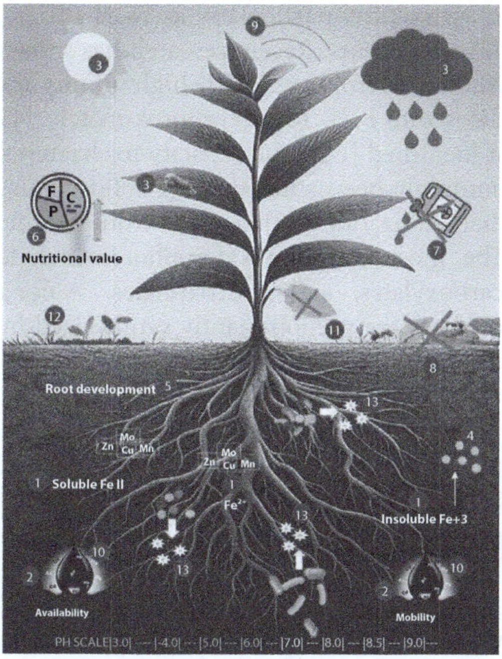

Figure 3.1 Main benefits provided by siderophores to enhance plant productivity and health. The numbered points highlight key functions such as facilitating the uptake of iron and other micronutrients, regulating heavy metal toxicity, strengthening plant defenses against pathogens, activating systemic resistance responses, and contributing to crop biofortification.

exert multidimensional effects on plant growth and health. They enhance Fe and other micronutrient uptake, promoting robust root development and phytohormone production. Simultaneously, they compete with pathogens for Fe, activate systemic defense responses, and help mitigate abiotic stress in saline or heavy-metal-contaminated soils. Moreover, siderophores contribute to the biofortification of crops essential for human nutrition, encouraging more sustainable and efficient agricultural practices.

3.5 Microbial Siderophores and Their Importance to Plants

Numerous studies have demonstrated the capability of various microbial groups to synthesize siderophores, highlighting their significant potential in regenerative and sustainable agriculture, particularly as biofertilizers, biocontrol agents, and bioremediation agents. Recent research has shown that bacteria and fungi can produce siderophores with high affinity for iron, forming highly stable chelates with Fe^{3+}. To date, approximately 945 distinct siderophores have been identified [24]. The capacity for bacterial siderophore production has commonly been studied using solid culture medium assays, enabling the characterization of this trait within specific microbial populations [41]. The most common siderophores include hydroxamates, catecholates, carboxylates, and hydroximates. Among these, hydroxamate-type siderophores are predominantly synthesized by fungi [40], whereas bacteria frequently produce hydroximates, catecholates, and carboxylates [61].

3.6 Research on Bacterial Siderophore Production (PGPR and Endophytes)

The genus *Pseudomonas* is notable for secreting pyoverdine, a siderophore with a high affinity for iron. A study evaluating siderophore production at alkaline pH demonstrated the biostimulant capability of the bacterium through the secretion of organic acids such as indole-3-acetic acid and glutamic acid. This study identified *Pseudomonas monsensis*, revealing genes associated with biofertilization, plant growth promotion, phosphorus solubilization, phytohormone production, and biological control [33].

Additionally, Traxler *et al.* [55] observed that certain actinomycetes inhibited aerial hyphae formation in adjacent *Streptomyces* colonies affected by iron availability and the presence of amichelin, a siderophore that limited their development. This finding underscores the importance of siderophores in regulating microbial interactions and promoting sustainable agricultural practices. Similarly, subsequent research demonstrated reduced disease incidence caused by *Ralstonia solanacearum* in tomato when plants were inoculated with siderophore-producing microbial consortia that the pathogen could not exploit for its benefit [20].

Nitrogen-fixing bacteria also produce siderophores to acquire iron. A study conducted in cucumber rhizospheres compared wild-type strains and mutants differing in siderophore production capacities. The results showed that strains with high siderophore production significantly increased plant biomass, enhanced soil enzymatic activity, and reduced bacterial community diversity compared to siderophore-deficient strains [64].

Correspondingly, other studies have shown improved primary productivity in croplands and pastures due to interactions between nitrogen-fixing bacteria and mycorrhizal fungi, whereas pathogens negatively affect crop yields. This highlights the critical role of soil biodiversity and health in sustaining productivity across different land uses [45].

Siderophore production has also been studied in endophytic microorganisms, which reside within plant tissues without causing harm. In acidic soils, phosphorus is frequently immobilized by iron, negatively impacting plant nutrition and development. Research involving the endophyte *Streptomyces* sp. CoT10, isolated from *Camellia oleifera*, demonstrated its ability to mobilize insoluble phosphorus bound to iron through siderophore production. This capability benefits both the soil microbial community and the phosphorus cycle, increasing phosphorus availability and promoting plant growth. Additionally, an increase in microbial diversity and soil interactions was observed, confirming the value of siderophores as a tool to address phosphorus scarcity in iron-rich soils [13].

3.7 Mycorrhizal Fungi and Their Role in Siderophore Production

Arbuscular mycorrhizal fungi represent the oldest and possibly most widespread symbiosis between plants and fungi, predominant in most terrestrial ecosystems [39]. Over 80% of terrestrial plants—including most agricultural crops in both temperate and tropical regions—form symbiotic relationships with these fungi. Approximately 355 AMF species across 44 genera have been described to date [5].

In general terms, AMF associated with various plant species enhance nutrient acquisition by expanding the root exploration surface area and capturing essential nutrients such as nitrogen (N), phosphorus (P), and various micronutrients. Additionally, this symbiosis improves plant resilience against both abiotic and biotic stresses. The extensive hyphal networks established between fungi and plants connect with other microorganisms, improving soil structure and increasing soil porosity and water retention through the extracellular protein glomalin, which also contributes to carbon storage, stress tolerance, and soil aggregate stabilization [50].

Mycorrhizal fungi—including those forming ericoid, orchidaceous, and ectomycorrhizal symbioses—secrete hydroxamate-type siderophores.

Specifically, AMF, known for nutrient exchange symbioses, produce siderophores characteristic of the Zygomycota phylum [42]. Under iron deficiency, these fungi release siderophores to chelate and transport iron to their host plants [53]. A study conducted on sorghum revealed increased total and bioavailable iron and zinc levels in grains treated with AMF, alongside a significant reduction in phytate content, an antinutrient compound [18].

AMF also interact closely *via* their mycelium and spores with a broad bacterial community, playing crucial roles in plant growth promotion, nutrition, and health [56]. An analysis of taxa associated with AMF spores has identified bacterial orders such as Actinomycetales, Bacillales, Burkholderiales, Pseudomonadales, and Rhizobiales [2, 32, 46]. These microbial interactions offer new possibilities for developing multifunctional consortia aimed at optimizing plant nutrition and promoting sustainable agricultural systems. Thus, AMF holds significant potential to alleviate micronutrient deficiencies in important crops.

It is essential to emphasize that iron accessibility is critical for plant development. The ability of siderophores to capture iron depends on factors influencing its solubilization and availability, including soil type and quality, pH, and the chemical form of soil iron. Moreover, siderophore production varies among different microbial species, microbial consortia, and microbial networks in the rhizosphere, further influencing the efficacy of these mechanisms. Most microorganisms actively produce and release compounds for diverse purposes such as communication, resource acquisition, motility, and defense against other organisms.

3.8 Application of Siderophore-Producing Microorganisms in Agriculture: A Compendium of Recent Case Studies

In recent years, the search for sustainable alternatives to enhance crop yield and resilience has gained increasing relevance. Facing rising food demands and impacts from climate change, agricultural producers and the scientific community have focused their attention on biological strategies to promote plant development and strengthen plant health. In this context, siderophore-producing rhizospheric microorganisms have generated significant interest due to their essential role in acquiring and transporting iron and other trace elements, considered essential micronutrients for plant metabolism. Besides facilitating greater mineral availability, siderophore production stimulates robust root development, enhances pathogen resistance, and helps mitigate various types of environmental stress.

The following compilation highlights recent case studies evaluating the impact of inoculating siderophore-producing microorganisms in agricultural crops. These investigations illustrate how their application benefits nutrient absorption, boosts plant growth and yield, and can be integrated with other management strategies to enhance sustainability and productivity across various cropping systems.

1. In *Zea mays*, inoculation with *Synechococcus mundulus* significantly enhances growth under iron-deficient conditions, particularly in stressed seedlings [10].
2. In *Oryza sativa*, the use of *Bacillus aryabhattai* increases yield under both normal and saline conditions, showing improvements of 60% and 43%, respectively [69].
3. In *Triticum aestivum*, the endophytic strain *Acinetobacter guillouiae* EU-B2RT.R1 produces siderophores and other beneficial metabolites (N fixation, P solubilization, phytohormone production), enhancing nutrient absorption, and tolerance to biotic and abiotic stresses, resulting in improved growth and yield under greenhouse conditions [43].
4. In *Vigna radiata*, *Bacillus amyloliquefaciens* D5 stands out for its siderophore production (≈75% of units) and exopolysaccharides, as well as for controlling *Cercospora canescens* (71.4% fungal inhibition). This reduces disease severity, improves root colonization, and increases biomass [25]. The same strain also activates enzymatic defenses (POD, PPO, chitinase, phenols) and increases the expression of immune-related genes (PR10, catalase, epoxide hydrolase, calmodulin), leading to reduced *Cercospora* incidence and increased pathogen tolerance [26].
5. In *Pisum sativum*, strains isolated from cow manure with high siderophore production (up to 98.42%), phosphate solubilization, and antifungal activity have shown improvements in plant height (up to 14.76 cm), germination (≈91.77%), and pod numbers per plant [49].
6. For *Solanum lycopersicum*, inoculation with *Pseudomonas fluorescens* A506 and LMG strains mitigates lead stress, increases shoot and root length, improves physiological and

enzymatic parameters, and reduces Pb-induced damage in plants [6].

7. Finally, in various crops belonging to the Solanaceae and Cucurbitaceae families, strains *Bacillus megaterium* and *B. altitudinis* demonstrate high siderophore production, phosphate solubilization, indoleacetic acid (IAA) synthesis, and antagonistic effects against *Fusarium oxysporum* and *Alternaria alternata*. This set of traits enhances beneficial microbial diversity and improves crop health and yield, particularly in organic or reduced-tillage systems [54].

Collectively, these studies underline the importance of siderophores and their microbial producers as promising tools for sustainable agriculture. Their application not only enhances iron uptake but also strengthens pathogen resistance and helps crops adapt to adverse conditions, thereby supporting agroecosystem productivity and health.

3.9 Limitations and Future Perspectives

The large-scale production and application of siderophores in agriculture, as well as their uniform distribution across crops, face significant challenges due to diverse soil conditions in agricultural fields. These differences complicate the long-term stability and effectiveness of siderophores, thus requiring further studies in this area. Additionally, the production and formulation costs of siderophore-based biofertilizers may be prohibitively high for many farmers. Although promising advances have been made in genetic engineering and the development of microbial strains capable of producing siderophores more efficiently, technical and regulatory obstacles still persist on a global scale. It is crucial to ensure that such microorganisms do not negatively impact ecosystems [57].

Biotechnological strategies offer pathways to overcome these limitations—for instance, through the development of more efficient microbial strains or the formulation of siderophore mixtures tailored to specific soil types and crops. Combining siderophores with other biofertilizers and biological control agents could enhance their overall effectiveness. Thus, siderophores emerge as tools that can be integrated into regenerative agricultural systems, reducing dependency on chemical fertilizers and improving soil health. Their adoption holds the potential to foster more sustainable and resilient farming practices, contributing to food security and environmental conservation.

In a recent study, Gu *et al.* [21] developed bioinformatics tools to predict how siderophores secreted by different bacteria influence the growth of other members within the microbial community. Specifically, by analyzing *Pseudomonas* genomes and designing a co-evolution algorithm, researchers mapped siderophore-mediated interactions, revealing variations based on the microorganisms' habitats and lifestyles. These interactions were denser and more competitive in natural, non-pathogenic environments, becoming more fragmented in pathogen-associated and human-influenced settings. This approach provides critical insights for understanding and specifically manipulating microbial communities, with potential applications in ecology, medicine, and biotechnology.

3.10 Conclusions

Agriculture is increasingly shifting toward regenerative and conservation models that promote soil health and enhance crop productivity. This approach represents an evolution of sustainable agriculture toward methods that not only maintain but also improve soil quality and agricultural production efficiency. Within this context, siderophores play a crucial role, facilitating the uptake of iron and other essential nutrients for plants, thereby promoting their growth and strengthening their resilience against adverse conditions. To maximize these benefits, research must focus on optimizing microbial strains that produce siderophores, tailored to various soil types and environments, as well as on developing stable formulations and effective application methods to ensure uniform distribution. Additionally, it is essential to evaluate the environmental impacts of their use, ensuring no negative effects on biodiversity or ecosystem health. Finally, integrating siderophores into regenerative and organic agricultural systems, in combination with other sustainable practices, can reduce reliance on chemical inputs and enhance soil health, thus contributing to a more resilient and productive agriculture.

AI disclosure statement: The author confirms that no artificial intelligence (AI)-generated content or AI-assisted writing tools were used in the preparation of this chapter (including summaries and the abstract). All text, analysis, and conclusions are the result of the author's original work.

Bibliography

1. Adeleke, B.S., Chaudhary, P., Ayilara, M.S., Ojo, F.M., Erinoso, S.M., Upadhayay, V.K., Adeyemo, A.I., Akinola, S.A., Rhizosphere Microbiomes Mediating Abiotic Stress Mitigation for Improved Plant Nutrition. *Ecologies*, 5, 3, 375–401, 2024, https://doi.org/10.3390/ecologies5030024.
2. Agnolucci, M., Avio, L., Pepe, A., Turrini, A., Cristani, C., Bonini, P., Cirino, V., Colosimo, F., Ruzzi, M., Giovannetti, M., Bacteria Associated with a Commercial Mycorrhizal Inoculum: Community Composition and Multifunctional Activity as Assessed by Illumina Sequencing and Culture-Dependent Tools. *Front. Plant Sci.*, 9, 1956, 2019, https://doi.org/10.3389/fpls.2019.01956.
3. Aguiar, D., Melo, V.F., Nogueira, M.A., Correa, R.S., El papel de los mecanismos microbianos en la disponibilidad de fósforo del suelo a partir de fracciones minerales fijas y estructurales. *J. Soil Sci. Plant Nutr.*, 24, 8192–8207, 2024, https://doi.org/10.1007/s42729-024-02106-z.
4. Alzate Zuluaga, M.Y., Fattorini, R., Cesco, S., Pii, Y., Plant–microbe interactions in the rhizosphere for smarter and more sustainable crop fertilization: the case of PGPR-based biofertilizers. *Front. Microbiol.*, 15, 1440978, 2024, https://doi.org/10.3389/fmicb.2024.1440978.
5. AMF-Phylogeny, Glomeromycota Phylogeny, Disponible en: http://www.amf-phylogeny.com/, accedido el 25 de febrero de 2025.
6. Anjum, S., Sarwar, M., Mukhtar, A., Hussain, T., Liaquat, M., Batool, A., Alam, M.W., Ali, Q., Ashraf, M.I., Impact of rhizobacterial inoculation in plant growth medium to mitigate lead stress in Tomato (*Solanum lycopersicum* L.). *Sci. Rep.*, 15, 707, 2025, https://doi.org/10.1038/s41598-024-84648-x.
7. Bhat, M.A., Mishra, A.K., Jan, S., Bhat, M.A., Kamal, M.A., Rahman, S., Shah, A.A., Jan, A.T., Plant Growth Promoting Rhizobacteria in Plant Health: A Perspective Study of the Underground Interaction. *Plants*, 12, 3, 629, 2023, https://doi.org/10.3390/plants12030629.
8. Blanco Nouche, C., Paris, C., Dhalleine, T., Oger, P., Turpault, M., Uroz, S., The non-ribosomal peptide synthetase-independent siderophore (NIS) rhizobactin produced by *Caballeronia mineralivorans* PML1(12) confers the ability to weather minerals. *Appl. Environ. Microbiol.*, 89, e00453–23, 2023, https://doi.org/10.1128/aem.00453-23.
9. Briat, J., Dubos, C., Gaymard, F., Iron nutrition, biomass production, and plant product quality. *Trends Plant Sci.*, 20, 33–40, 2015.
10. Brick, M.B., Hussein, M.H., Mowafy, A.M., Hamouda, A.R., Ayyad, A.M., Refaay, D.A., Significance of siderophore-producing cyanobacteria on enhancing iron uptake potentiality of maize plants grown under iron-deficiency. *Microb. Cell Fact.*, 24, 3, 2025, https://doi.org/10.1186/s12934-024-02618-4.
11. Chamkhi, I., El Omari, N., Balahbib, A., El Menyiy, N., Benali, T., Ghoulam, C., Is the rhizosphere a source of applicable multi-beneficial microorganisms

for plant enhancement? *Saudi J. Biol. Sci.*, 29, 2, 1246–1259, 2022, https://doi.org/10.1016/j.sjbs.2021.09.032.

12. Cui, H., Cheng, Q., Jing, T., Chen, Y., Li, X., Zhang, M., Dengfeng, Q., Feng, J., Vafadar, F., Wei, Y., Kai, L., Zhao, Y., Zhou, D., Xie, J., *Trichoderma virens* XZ11-1 producing siderophores inhibits the infection of *Fusarium oxysporum* and promotes plant growth in banana plants. *Microb. Cell Fact.*, 24, 22, 2025, https://doi.org/10.1186/s12934-024-02626-4.
13. Cui, K., Xu, T., Chen, J., Yang, H., Liu, X., Zhuo, R., Peng, Y., Tang, W., Wang, R., Chen, L., Zhang, X., Zhang, Z., He, Z., Wang, X., Liu, C., Chen, Y., Zhu, Y., Siderophores, a Potential Phosphate Solubilizer from the Endophyte *Streptomyces* sp. CoT10, Improved Phosphorus Mobilization for Host Plant Growth and Rhizosphere Modulation. *J. Clean. Prod.*, 367, 133110, 2022, https://doi.org/10.1016/j.jclepro.2022.133110.
14. Cunha, I.C.M., Silva, A.V.R., Marcandalli-Boleta, E.H., Pellegrinetti, T.A., Guandalin-Zagatto, L.F., Zagatto, S.S., Chaves, M.G., Mendes, R., Maistro-Patreze, C., Tsai, S.M., Mendes, L.W., The interplay between the inoculation of plant growth-promoting rhizobacteria and the rhizosphere microbiome and their impact on plant phenotype. *Microbiol. Res.*, 283, 127706, 2024, https://doi.org/10.1016/j.micres.2024.127706.
15. de Andrade, L.A., Santos, C.H.B., Frezarin, E.T., Sales, L.R., Rigobelo, E.C., Rizobacterias promotoras del crecimiento de las plantas para la producción agrícola sostenible. *Microorganisms*, 11, 4, 1088, 2023.
16. Dimkpa, C.O., Merten, D., Svatos, A., Büchel, G., Kothe, E., Metal-induced oxidative stress impacting plant growth in contaminated soil is alleviated by microbial siderophores. *Soil Biol. Biochem.*, 41, 1, 154–162, 2009, https://doi.org/10.1016/j.soilbio.2008.10.010.
17. Ehinmitan, E., Losenge, T., Mamati, E., Ngumi, V., Juma, P., Siamalube, B., BioSolutions for Green Agriculture: Unveiling the Diverse Roles of Plant Growth-Promoting Rhizobacteria. *Int. J. Microbiol.*, 2024, 1, 6181491, 2024, https://doi.org/10.1155/2024/6181491.
18. Elsafy, M., Sir Elkhatim, K.A., Tia, N.A., Gobara Hamid, M., Rahmatov, M., Joergensen, R.G., Hassan, A.B., Abdelhalim, T.S., Arbuscular mycorrhiza fungi (AMF) mediated biofortification of micronutrients, protein characteristics, and γ-aminobutyric acid in sorghum grown on vertisols in the central clay plains of Sudan. *Arch. Agron. Soil Sci.*, 71, 1, 1–15, 2025.
19. Gomes, A.F.R., Almeida, M.C., Sousa, E., Resende, D.I.S.P., Siderophores and metallophores: Metal complexation weapons to fight environmental pollution. *Sci. Total Environ.*, 932, 173044, 2024, https://doi.org/10.1016/j.scitotenv.2024.173044.
20. Gu, S., Yang, T., Shao, Z., Wang, T., Cao, K., Jousset, A., Friman, V.P., Mallon, C., Mei, X., Wei, Z., Xu, Y., Shen, Q., Pommier, T., Siderophore-mediated interactions determine the disease suppressiveness of microbial consortia. *mSystems*, 5, 3, e010–1128, 2020, https://doi.org/10.1128/msystems.00811-19.

21. Gu, S., Shao, Z., Qu, Z., Zhu, S., Shao, Y., Zhang, D., Allen, R., He, R., Shao, J., Xiong, G., Jousset, A., Friman, V.P., Wei, Z., Kummerli, R., Li, Z., Siderophore synthetase–receptor gene coevolution reveals habitat- and pathogen-specific bacterial iron interaction networks. *Sci. Adv.*, 11, 3, eadq5038, 2025, https://doi.org/10.1126/sciadv.adq5038.
22. Hasan, A., Tabassum, B., Hashim, M., Khan, N., Role of Plant Growth Promoting Rhizobacteria (PGPR) as a Plant Growth Enhancer for Sustainable Agriculture: A Review. *Bacteria*, 3, 2, 59–75, 2024, https://doi.org/10.3390/bacteria3020005.
23. Hassani, M.A., Durán, P., Hacquard, S., Microbial interactions within the plant holobiont. *Microbiome*, 6, 58, 2018, https://doi.org/10.1186/s40168-018-0445-0.
24. Ruolin, H., Gu, S., Xu, J., Li, X., Chen, H., Shao, Z., Wang, F., Shao, J., Long-Qian, W.B.Y., Wei, Z., Li, Z., SIDERITE: Unveiling hidden siderophore diversity in the chemical space through digital exploration. *iMeta*, 3, e192, 2024, https://doi.org/10.1002/imt2.192.
25. Kamath, A. and Patel, D., *Bacillus amyloliquefaciens* D5 as a biocontrol agent against *Cercospora* Leaf Spot Disease in *Vigna radiata* L. *Physiol. Mol. Plant Pathol.*, 137, 102605, 2025, https://doi.org/10.1016/j.pmpp.2025.102605.
26. Kamath, A., Sharma, A., Shukla, A., Parmar, P., Patel, D., Fortifying plant fortresses: siderophores in defense against *Cercospora* leaf spot disease in *Vigna radiata* L. *Front. Microbiol.*, 15, 1492139, 2025, https://doi.org/10.3389/fmicb.2024.1492139.
27. Khasheii, B., Mahmoodi, P., Mohammadzadeh, A., Siderophores: Importance in bacterial pathogenesis and applications in medicine and industry. *Microbiol. Res.*, 250, 126790, 2021, https://doi.org/10.1016/j.micres.2021.126790.
28. Khanna, K., Ohri, P., Bhardwaj, R., Ahmad, P., Unsnarling Plausible Role of Plant Growth-Promoting Rhizobacteria for Mitigating Cd-Toxicity from Plants: An Environmental Safety Aspect. *J. Plant Growth Regul.*, 41, 2514–2542, 2022, https://doi.org/10.1007/s00344-021-10445-9.
29. Kobayashi, T., Nozoye, T., Nishizawa, N.K., Transporte de hierro y su regulación en plantas. *Free Radic. Biol. Med.*, 133, 11–20, 2019, https://doi.org/10.1016/j.freeradbiomed.2018.10.439.
30. Li, W.F., Zhu, H.Y., Lan, P., Research Progress of Iron Homeostasis Regulation in Strategy I Plants. *Soil*, 53, 1101–1106, 2021.
31. Lindsay, W.L., Iron-oxide solubilization by organic matter and its effect on iron availability. *Plant Soil*, 130, 27–34, 1991.
32. Long, L., Zhu, H., Yao, Q., Analysis of Bacterial Communities Associated with Spores of *Gigaspora margarita* and *Gigaspora rosea*. *Plant Soil*, 310, 1–9, 2008.
33. Lozano-González, J.M., Valverde, S., Montoya, M., Martín, M., Rivilla, R., Lucena, J.J., López-Rayo, S., Evaluation of siderophores generated by *Pseudomonas* bacteria and their possible application as Fe biofertilizers. *Plants*, 12, 23, 4054, 2023.

34. Mengwei, X., Xu, L., Tongyao, C., Yige, Z., Liya, M., Xiaoyu, S., Xiao, C., Yu, Y., Anderson, J.M., Dynamics of wheat rhizosphere microbiome and its impact on grain production across growth stages. *Sci. Total Environ.*, 964, 178524, 2025, https://doi.org/10.1016/j.scitotenv.2025.178524.
35. Murdoch, C.C. and Skaar, E.P., Inmunidad nutricional: la batalla por los metales nutritivos en la interfaz huésped-patógeno. *Nat. Rev. Microbiol.*, 20, 657–670, 2022, https://doi.org/10.1038/s41579-022-00745-6.
36. Myers, S.S., Zanobetti, A., Kloog, I., Huybers, P., Leakey, A.D.B., Bloom, A.J., Carlisle, E., Dietterich, L.H., Fitzgerald, G., Hasegawa, T., Holbrook, N.M., Nelson, R.L., Ottman, M.J., Raboy, V., Sakai, H., Sartor, K.A., Schwartz, J., Seneweera, S., Tausz, M., Usui, Y., Increasing CO_2 Threatens Human Nutrition. *Nature*, 510, 7503, 139–142, 2014, https://doi.org/10.1038/nature13179.
37. Nithyapriya, S., Sundaram, L., Eswaran, S., Perveen, K., Alshaikh, N.A., Sayyed, R.Z., Mastinu, A., Purification and Characterization of Desferrioxamine B of *Pseudomonas fluorescens* and Its Application to Improve Oil Content, Nutrient Uptake, and Plant Growth in Peanuts. *Microb. Ecol.*, 87, 60, 2024, https://doi.org/10.1007/s00248-024-02377-0.
38. Pantigoso, H.A., Newberger, D., Vivanco, J.M., The rhizosphere microbiome: Plant–microbial interactions for resource acquisition. *J. Appl. Microbiol.*, 133, 5, 2864–2876, 2022, https://doi.org/10.1111/jam.15686.
39. Parniske, M., Arbuscular mycorrhiza: the mother of plant root endosymbioses. *Nat. Rev. Microbiol.*, 6, 10, 763–775, 2008, https://doi.org/10.1038/nrmicro1987.
40. Pecoraro, L., Wang, X., Shah, D., Song, X., Kumar, V., Shakoor, A., Tripathi, K., Ramteke, P.W., Rani, R., Biosynthesis pathways, transport mechanisms and biotechnological applications of fungal siderophores. *J. Fungi*, 8, 1, 21, 2021, https://doi.org/10.3390/jof8010021.
41. Pérez-Miranda, S., Cabirol, N., George-Téllez, R., Zamudio-Rivera, L.S., Fernández, F.J., O-CAS, a fast and universal method for siderophore detection. *J. Microbiol. Methods*, 70, 127–131, 2007.
42. Rajapitamahuni, S., Kang, B.R., Lee, T.K., Exploring the roles of arbuscular mycorrhizal fungi in plant–iron homeostasis. *Agriculture*, 13, 10, 1918, 2023.
43. Rana, K.L., Kour, D., Kaur, T., Sheikh, I., Yadav, A.N., Kumar, V., Suman, A., Dhaliwal, H.S., Endophytic Microbes from Diverse Wheat Genotypes and Their Potential Biotechnological Applications in Plant Growth Promotion and Nutrient Uptake. *Proc. Natl. Acad. Sci., India, Sect. B Biol. Sci.*, 90, 4, 969–979, 2020, https://doi.org/10.1007/s40011-020-01168-0.
44. Ravet, K., Touraine, B., Boucherez, J., Briat, J.F., Gaymard, F., Cellier, F., Ferritins control interaction between iron homeostasis and oxidative stress in *Arabidopsis*. *Plant J.*, 57, 400–412, 2009.
45. Romero, F., Labouyrie, M., Orgiazzi, A., Ballabio, C., Panagos, P., Jones, A., Tedersoo, L., Bahram, M., Guerra, C.A., Eisenhauer, N., *et al.*, Soil Health Is

Associated with Higher Primary Productivity across Europe. *Nat. Ecol. Evol.*, 8, 1847–1855, 2024.
46. Roesti, D., Ineichen, K., Braissant, O., Redecker, D., Wiemken, A., Aragno, M., Bacteria associated with spores of the arbuscular mycorrhizal fungi *Glomus geosporum* and *Glomus constrictum*. *Appl. Environ. Microbiol.*, 71, 11, 6673–6679, 2005.
47. Roskova, Z., Skarohlid, R., McGachy, L., Siderophores: an alternative bioremediation strategy? *Sci. Total Environ.*, 819, 153144, 2022, https://doi.org/10.1016/j.scitotenv.2022.153144.
48. Saeed, Q., Xiukang, W., Haider, F.U., Kučerik, J., Mumtaz, M.Z., Holatko, J., Naseem, M., Kintl, A., Ejaz, M., Naveed, M., Brtnicky, M., Mustafa, A., Rhizosphere Bacteria in Plant Growth Promotion, Biocontrol, and Bioremediation of Contaminated Sites: A Comprehensive Review of Effects and Mechanisms. *Int. J. Mol. Sci.*, 22, 19, 10529, 2021, https://doi.org/10.3390/ijms221910529.
49. Sagar, S., Singh, A., Bala, J., Chauhan, R., Kumar, R., Badiyal, A., Walia, A., Plant growth-promoting bacteria from dung of indigenous and exotic cow breeds and their effect on the growth of pea plant in sustainable agriculture. *Biotechnol. Environ.*, 2, 3, 2025, https://doi.org/10.1186/s44314-025-00017-6.
50. Singh, P.K., Singh, M., Tripathi, B.N., Glomalin: an arbuscular mycorrhizal fungal soil protein. *Protoplasma*, 250, 663–669, 2013, https://doi.org/10.1007/s00709-012-0453-z.
51. Singh, V.S., Dubey, B.K., Rai, S., Singh, S.P., Tripathi, A.K., La ingeniería de la utilización de D-glucosa en *Azospirillum brasilense* Sp7 promueve la colonización de la raíz del arroz. *Appl. Microbiol. Biotechnol.*, 106, 7891–7903, 2022, https://doi.org/10.1007/s00253-022-12250-0.
52. Smith, S.E. and Read, D.J., *Mycorrhizal Symbiosis*, 3rd, Academic Press, Cambridge, UK, 2008.
53. Srivastava, N., Siderophore production in iron uptake and plant biofortification, in: *Plant Microbiome for Plant Productivity and Sustainable Agriculture. Microorganisms for Sustainability*, S. Chhabra, R. Prasad, N.R. Maddela, N. Tuteja (Eds.), vol. 37, pp. 313–329, Springer, Singapore, 2023, https://doi.org/10.1007/978-981-19-5029-2_13.
54. Trasmundi, F., Molina, M.H.J.B., Galieni, A., Campanelli, G., Leteo, F., Polilli, W., Platani, C., Chaves-Lopez, C., Stagnari, F., Impact of organic and conventional cropping systems on plant growth-promoting rhizobacteria (PGPR). *Biocatal. Agric. Biotechnol.*, 64, 103527, 2025, https://doi.org/10.1016/j.bcab.2025.103527.
55. Traxler, M.F., Seyedsayamdost, M.R., Clardy, J., Kolter, R., Interspecies modulation of bacterial development through iron competition and siderophore piracy. *Mol. Microbiol.*, 86, 3, 628–644, 2012.
56. Turrini, A., Avio, L., Giovannetti, M., Agnolucci, M., Functional complementarity of arbuscular mycorrhizal fungi and associated microbiota: The challenge of translational research. *Front. Plant Sci.*, 9, 1407, 2018, https://doi.org/10.3389/fpls.2018.01407.

57. Vining, L.C., Secondary Metabolites: Their Function and Evolution, in: *Ciba Foundation Symposium*, vol. 171, Wiley, pp. 184–198, 2007.
58. Wahab, A., Bibi, H., Batool, F., Muhammad, M., Ullah, S., Zaman, W., Abdi, G., Plant growth-promoting rhizobacteria biochemical pathways and their environmental impact: a review of sustainable farming practices. *Plant Growth Regul.*, 104, 637–662, 2024, https://doi.org/10.1007/s10725-024-01218-x.
59. Wogene, S., Tibor, J., Zoltán, M., Unveiling the significance of rhizosphere: Implications for plant growth, stress response, and sustainable agricultura. *Plant Physiol. Biochem.*, 206, 108290, 2024, https://doi.org/10.1016/j.plaphy.2023.108290.
60. Xiaolei, X., Lingyan, Z., Yamin, C., Ruiqiang, L., Liqi, G., Nan, L., Kai, A., Kaiyan, Z., Guiyao, Z., Xuhui, Z., Ectomycorrhizal fungi explain more variation in rhizosphere nutrient availability than root traits in temperate forests. *Appl. Soil Ecol.*, 207, 105923, 2025, https://doi.org/10.1016/j.apsoil.2025.105923.
61. Xie, B., Wei, X., Wan, C., Zhao, W., Song, R., Xin, S., Song, K., Exploring the biological pathways of siderophores and their multidisciplinary applications: A comprehensive review. *Molecules*, 29, 10, 2318, 2024.
62. Xinyue, Y., Yujie, S., Jiazhi, S., Litao, S., Kai, F., Hao, C., Kangwei, S., Zhaotang, D., Yu, W., The role of rhizosphere microbial community structure in the growth and development of different tea cultivars. *Appl. Soil Ecol.*, 206, 105817, 2025, https://doi.org/10.1016/j.apsoil.2024.105817.
63. Yan, A., Wang, Y., Tan, S.N., Mohd Yusof, M.L., Ghosh, S., Chen, Z., Fitorremediación: un enfoque prometedor para la revegetación de tierras contaminadas con metales pesados. *Front. Plant Sci.*, 11, 359, 2020, https://doi.org/10.3389/fpls.2020.00359.
64. Zhang, Y., Wang, E., Feng, B., Xu, L., Xue, Y., Chen, Y., Siderophore Production Capability of Nitrogen-Fixing Bacterium (NFB) GXGL-4A Regulates Cucumber Rhizosphere Soil Microecology. *Microorganisms*, 13, 2, 346, 2025.
65. Dlamini, S.P., Akanmu, A.O., Babalola, O.O., Rhizospheric microorganisms: The gateway to sustainable plant health. *Front. Sustain. Food Syst.*, 6, 925802, 2022, https://doi.org/10.3389/fsufs.2022.925802.
66. Mishra, S., Zhang, X., Yang, X., Plant communication with rhizosphere microbes revealed through microbial functional gene composition. *Microbiol. Res.*, 284, 127726, 2024, https://doi.org/10.1016/j.micres.2024.127726.
67. Oluwaseyi, S.O. and Olubukola, O.B., The rhizosphere microbial complex in plant health: A review of interaction dynamics. *J. Integr. Agric.*, 21, 8, 2168–2182, 2022, https://doi.org/10.1016/S2095-3119(21)63817-0.
68. Shibin, L., Fakun, H., Yakov, K., Huxuan, X., Duyen, T.T.H., Shengyan, P., Bahar, S.R., Nutrients in the rhizosphere: A meta-analysis of content, availability, and influencing factors. *Sci. Total Environ.*, 826, 153908, 2022, https://doi.org/10.1016/j.scitotenv.2022.153908.

69. Ahmed, E. and Holmstrom, S.J., Siderophores in environmental research: roles and applications. *Micro. Biotech*, 7, 196–208, 2014, https://doi.org/10.1111/1751-7915.12117.
70. Mushtaq, Z., Nazir, A., Asghar, H.N., Zahir, Z.A., Interactive Effect of Siderophore-Producing Bacteria and l-Tryptophan on Physiology, Tuber Characteristics, Yield, and Iron Concentration of Potato. *Potato Res.*, 65, 1015–1027, 2022, https://doi.org/10.1007/s11540-022-09565-w.
71. Pattnaik, S., Mohapatra, B., Gupta, A., Plant Growth-Promoting Microbe Mediated Uptake of Essential Nutrients (Fe, P, K) for Crop Stress Management: Microbe–Soil–Plant Continuum. *Front. Agron.*, 3, 689972, 2021, https://doi.org/10.3389/fagro.2021.689972.

4

Biosynthesis and Regulation of Siderophores in Microorganisms

Oyindamola John Samson[1,2*], Racheal Oluwayemisi Fashogbon[1], Stephen Olaosebikan Makanjuola[3], Moses Sunday Afariogun[4], Jamiu Oluwatosin Adeyemi[2] and Ismail Babatunde Onajobi[2]

[1]*Department of Microbiology and Biotechnology, Faculty of Natural Sciences, Ajayi Crowther University, Oyo, Nigeria*
[2]*Department of Microbiology, Faculty of Science, Olabisi Onabanjo University, Ago-Iwoye, Nigeria*
[3]*Department of Medical Laboratory Science, Faculty of Basic Medical Science, Olabisi Onabanjo University, Ago-Iwoye, Nigeria*
[4]*Department of Cell Biology and Genetics, University of Lagos, Akoka, Nigeria*

Abstract

Siderophores are low-molecular-weight, high-affinity iron-chelating compounds produced by microorganisms to acquire iron from their environment. This chapter aims to explore on the intricate processes of siderophore biosynthesis and regulation in various microorganisms. We begin by discussing the structural diversity of siderophores and their classification based on chemical composition. The biosynthetic pathways for major siderophore types, including catecholates, hydroxamates, and carboxylates, are examined in detail, highlighting key enzymes and intermediates. Special attention is given to the non-ribosomal peptide synthetase (NRPS) and NRPS-independent siderophore (NIS) pathways. It then explains more into the complex regulatory mechanisms controlling siderophore production, focusing on iron-dependent and iron-independent regulation systems. We explore the roles of transcriptional regulators, small RNAs, and post-transcriptional modifications in fine-tuning siderophore biosynthesis. The importance of siderophores in microbial ecology, pathogenesis, and potential biotechnological applications is also discussed. Finally, we address current challenges and future directions in siderophore research, including the development of novel antimicrobial strategies

**Corresponding author*: oyindamolasamson1997@gmail.com; ORCID: https://orcid.org/0000-0003-0404-7634

Debasis Mitra, Ayush Madan and Anju Rani (eds.) Microbial Siderophores for Agricultural Productivity: Perspectives of Use in Agriculture, (69–88) © 2026 Scrivener Publishing LLC

targeting siderophore systems and the exploitation of siderophores for environmental remediation and drug delivery.

***Keywords*:** Siderophores, iron acquisition, biosynthetic pathways, regulatory mechanisms, microbial ecology

4.1 Introduction

Siderophores are small, high-affinity iron-chelating compounds produced by microorganisms to sequester iron from their environment. Iron is an essential micronutrient for nearly all forms of life, playing a crucial role in various biochemical processes, including respiration, DNA synthesis, and metabolic pathways [41]. However, the bioavailability of iron in many environments is limited due to its tendency to form insoluble ferric hydroxides under aerobic conditions and neutral pH. Microorganisms, therefore, rely on siderophores as a sophisticated strategy to overcome iron limitation, enabling their survival and proliferation in diverse ecological niches. This capability underscores the importance of siderophores in microbial physiology, ecological dynamics, and interspecies interactions [11].

The biosynthesis of siderophores is a complex and highly regulated process influenced by environmental conditions and the metabolic state of the microorganism. These molecules are typically synthesized through nonribosomal peptide synthetases (NRPS) or polyketide synthases (PKS), although other pathways may also contribute [36]. The structure of siderophores is highly variable, ranging from hydroxamates and catecholates to carboxylates, reflecting the adaptive evolution of microorganisms to specific iron-deficient environments. The genetic regulation of siderophore biosynthesis involves intricate signaling networks and feedback mechanisms, ensuring efficient utilization of cellular resources while maintaining metabolic balance [44].

The production of siderophores is not solely a survival strategy but also a competitive advantage in microbial communities. Microorganisms capable of producing and efficiently utilizing siderophores can outcompete others in iron-limited environments, influencing community structure and dynamics [14]. In addition, siderophore-mediated iron acquisition often plays a pivotal role in pathogenicity for many bacteria and fungi, allowing them to establish infections in host organisms. The biosynthesis and regulation of siderophores is, therefore, essential for unraveling microbial ecology and the mechanisms underlying microbe–host interactions [16].

Beyond their biological significance, siderophores hold immense potential for biotechnological and agricultural applications. They have been exploited as biocontrol agents to suppress plant pathogens, biofertilizers to enhance crop productivity, and components in medical applications, such as drug delivery systems and diagnostic tools [31]. The growing interest in siderophores as sustainable and environmentally friendly solutions has spurred extensive research into their synthesis, regulation, and functional diversity [15]. These applications highlight the broader implications of siderophore research, transcending microbiology to address global challenges in agriculture, medicine, and environmental management. This chapter aims to delve into the intricate processes of siderophore biosynthesis and regulation, shedding light on their structural diversity, biosynthetic pathways, and regulatory mechanisms.

4.2 Structural Diversity and Classification of Siderophores

Siderophores are specialized molecules produced by microorganisms to sequester iron from the environment, a critical process for survival and growth under iron-limiting conditions. Their structural diversity reflects the intricate adaptations of microorganisms to varying ecological niches.

4.2.1 Chemical Composition and Structural Features

Siderophores are low-molecular-weight chelators with a high affinity for ferric iron (Fe^{3+}). Their structures are characterized by functional groups that coordinate iron ions, including catechols, hydroxamates, and carboxylates. These functional groups form strong complexes with Fe^{3+}, making siderophores highly effective in capturing and solubilizing iron from insoluble sources [7].

The backbone of siderophores typically consists of peptide or non-peptide components, contributing to their structural diversity. For instance, peptide-based siderophores are synthesized by non-ribosomal peptide synthetases, while non-peptide siderophores are often assembled through polyketide pathways. This chemical versatility enables microorganisms to produce siderophores tailored to specific environmental conditions, enhancing their competitiveness and ecological fitness [35].

4.2.2 Major Classes of Siderophores

Siderophores are broadly classified based on the nature of their iron-chelating groups. The primary classes include catecholates, hydroxamates, carboxylates, and mixed-type siderophores, each exhibiting unique structural and functional properties.

4.2.2.1 Catecholates

Catecholate siderophores, also known as phenolates, derive their iron-binding capability from catechol groups. These compounds are synthesized predominantly by bacteria, including species of *Escherichia* and *Pseudomonas*. Examples include enterobactin, a highly efficient siderophore produced by *E. coli* [21].

The catechol groups form strong bidentate chelates with Fe^{3+}, resulting in exceptionally stable iron complexes. This high affinity makes catecholates particularly effective in iron-limited environments, providing a competitive advantage to producing organisms [45].

4.2.2.2 Hydroxamates

Hydroxamate siderophores are characterized by hydroxamate functional groups, which coordinate Fe^{3+} through oxygen atoms. These siderophores are commonly produced by fungi and some bacteria, with examples including ferrichrome and desferrioxamine [17].

Hydroxamates exhibit moderate binding affinity compared to catecholates but are highly versatile due to their stability across a range of pH levels. This property enables their utilization in diverse ecological settings, from soil to aquatic environments [3].

4.2.2.3 Carboxylates

Carboxylate siderophores employ carboxyl functional groups for iron binding. These siderophores are predominantly produced by plant-associated microorganisms, such as rhizobacteria. An example is rhizobactin, a siderophore secreted by *Rhizobium* species [43].

Carboxylates have relatively lower binding affinities than catecholates and hydroxamates but are highly effective in alkaline environments. Their production is often associated with symbiotic interactions, where they facilitate nutrient exchange between plants and microbes [13].

4.2.2.4 Mixed-Type Siderophores

Mixed-type siderophores incorporate multiple functional groups, such as catechols, hydroxamates, and carboxylates, within a single molecule. These siderophores exhibit broad versatility in iron acquisition, as they can adapt to varying environmental conditions. An example is pyoverdine, a siderophore produced by *Pseudomonas* species [44].

The structural complexity of mixed-type siderophores allows them to outcompete simpler siderophores in diverse ecological niches, enhancing the survival and proliferation of producing organisms [11]. This classification highlights the structural diversity and ecological significance of siderophores (Table 4.1), underscoring their critical role in microbial survival and environmental adaptation.

4.3 Biosynthetic Pathways of Siderophores

Siderophore biosynthesis is a highly regulated and complex process, reflecting the critical role of these molecules in microbial survival under iron-limiting conditions. The pathways can be broadly categorized into non-ribosomal peptide synthetase (NRPS) and NRPS-independent siderophore (NIS) systems, each employing distinct mechanisms and enzymes for siderophore assembly.

4.3.1 Non-Ribosomal Peptide Synthetase Pathway

The NRPS pathway is a modular enzymatic system responsible for assembling peptide-based siderophores. It relies on non-ribosomal peptide synthetases, large multi-functional enzymes, to catalyze the stepwise addition of amino acids or amino acid derivatives. These enzymes operate independently of ribosomes, allowing for the incorporation of unusual amino acids and structural modifications [27].

NRPS pathways are particularly common in bacterial and fungal siderophores, such as ferrichrome and pyoverdine. The modularity of NRPS systems enables structural diversity, as each module is responsible for the

Table 4.1 Classification and features of siderophores.

Class	Functional groups	Producers	Examples	Features	References
Catecholates	Catechol	*E. coli*, *Pseudomonas*	Enterobactin	High affinity, stable iron complexes	Xu *et al.* [45]
Hydroxamates	Hydroxamate	Fungi, some bacteria	Ferrichrome, desferrioxamine	Versatile, stable across pH ranges	Kircheva and Dudev [17]
Carboxylates	Carboxyl	Rhizobacteria	Rhizobactin	Effective in alkaline environments	van der Helm and Winkelmann [43]
Mixed-type siderophores	Catechols, hydroxamates, carboxylates	*Pseudomonas* species	Pyoverdine	Broad versatility, competitive advantage	Gomes *et al.* [11]

addition of specific building blocks. This diversity is advantageous for microorganisms, allowing them to produce siderophores tailored to specific ecological niches [20].

4.3.2 NRPS-Independent Siderophore Pathway

The NIS pathway synthesizes siderophores without relying on peptide bonds. It instead utilizes enzymes that catalyze the condensation of organic acids, amines, and other small molecules. Key examples include enterobactin and desferrioxamine, both of which exhibit exceptional iron-chelating properties [40].

This pathway is characterized by high specificity and efficiency, as it often involves fewer enzymatic steps than the NRPS system. The simplicity of the NIS pathway makes it an attractive target for biotechnological applications, such as engineering microbial strains to produce siderophores for agricultural or medical use [42].

4.3.3 Important Enzymes Involved in Siderophore Biosynthesis

Numerous key enzymes drive the biosynthesis of siderophores, including synthetases, transferases, and ligases. In the NRPS pathway, adenylation domains activate amino acids, while thiolation domains transfer activated intermediates to carrier proteins. Cyclization domains introduce structural rings, enhancing the stability and functionality of the siderophore [23].

In the NIS pathway, enzymes like enterobactin synthetase catalyze the formation of catechol groups, while specific ligases link monomeric units into larger molecules. These enzymes' roles provide insights into the molecular basis of siderophore diversity and offer opportunities for synthetic biology applications [27].

4.3.4 Biosynthetic Pathways of Specific Siderophores

Enterobactin, a catecholate siderophore produced by *Escherichia coli*, is synthesized *via* the NIS pathway. The process begins with the activation of chorismate, followed by its conversion to 2,3-dihydroxybenzoate. These intermediates are subsequently linked to serine residues by enterobactin synthetase, forming the cyclic trimeric structure characteristic of enterobactin. The high affinity of enterobactin for Fe^{3+} enables *E. coli* to thrive in iron-depleted environments, making it a model system for studying siderophore biosynthesis [37].

Pyoverdine, a fluorescent siderophore produced by *Pseudomonas* species, is assembled through the NRPS pathway. Its biosynthesis involves multiple NRPS modules that sequentially incorporate amino acids and other precursors into the peptide backbone. Modifications, such as the addition of chromophore groups, enhance its iron-chelating capabilities. Pyoverdine's dual role as an iron carrier and signaling molecule highlights its ecological importance, particularly in competitive microbial communities [10].

Desferrioxamine, a hydroxamate siderophore produced by *Streptomyces* species, is synthesized *via* the NIS pathway. The biosynthesis begins with the activation of succinate and its condensation with hydroxylamine, catalyzed by specific ligases. Additional steps introduce the hydroxamate groups essential for iron binding. Desferrioxamine is widely used in clinical settings as a chelating agent for treating iron overload, underscoring the translational potential of microbial siderophores [33].

4.4 Regulation of Siderophore Production

The regulation of siderophore production is a complex process, tightly controlled to balance microbial iron acquisition with metabolic costs. This regulation occurs at multiple levels, including transcriptional, post-transcriptional, and environmental controls, ensuring that siderophores are synthesized only when needed.

4.4.1 Iron-Dependent Regulation

The Fur (ferric uptake regulator) and DtxR (diphtheria toxin regulator) families are key players in iron-dependent regulation. These transcriptional regulators sense intracellular iron levels and modulate gene expression accordingly. Under iron-replete conditions, Fur binds to conserved DNA sequences known as Fur boxes, repressing siderophore biosynthesis genes [24]. Similarly, DtxR family regulators control siderophore production in response to varying iron availability in species like *Corynebacterium diphtheriae*. The repression by Fur and DtxR prevents unnecessary siderophore production, conserving cellular resources. When iron levels are low, these regulators dissociate from DNA, derepressing genes involved in siderophore synthesis and iron transport, thus ensuring optimal iron uptake [38].

Beyond Fur and DtxR, other transcriptional regulators like IdeR and IscR also contribute to iron-responsive control. IdeR functions similarly to DtxR, while IscR is involved in regulating genes linked to iron–sulfur cluster assembly, indirectly influencing siderophore production. These regulators operate within complex networks, integrating iron availability with broader cellular processes [19].

4.4.2 Iron-Independent Regulation

Quorum sensing, a cell-density-dependent signaling mechanism, plays a pivotal role in iron-independent regulation of siderophore production. Many bacteria, including *Pseudomonas aeruginosa*, use quorum-sensing systems to coordinate siderophore synthesis with population density. Signal molecules like N-acyl homoserine lactones activate transcriptional regulators, inducing siderophore production as bacterial populations grow. This regulatory strategy ensures that siderophore production is synchronized within microbial communities, enhancing collective iron acquisition efficiency. Quorum sensing also integrates environmental cues, allowing bacteria to adapt to dynamic conditions [2].

Environmental factors such as pH, temperature, and nutrient availability significantly influence siderophore production. For instance, acidic conditions can enhance siderophore secretion by increasing the solubility of ferric iron. Similarly, nutrient limitation, particularly nitrogen or carbon scarcity, can upregulate siderophore biosynthesis to support survival in nutrient-depleted environments. These environmental cues operate in conjunction with iron-dependent regulation, creating a multifaceted control system that optimizes siderophore production [28].

4.4.3 Post-Transcriptional Regulation

Small RNAs (sRNAs) are emerging as critical regulators of siderophore production at the post-transcriptional level. These non-coding RNAs modulate mRNA stability and translation efficiency, fine-tuning the expression of siderophore biosynthesis genes. For example, in *Escherichia coli*, RyhB represses genes involved in iron storage and utilization, indirectly influencing siderophore production. When targeting multiple genes, sRNAs provide a rapid and efficient means of adapting siderophore production to fluctuating environmental conditions, complementing transcriptional regulation [26].

Post-translational modifications (PTMs) of biosynthetic enzymes add another layer of regulation. Phosphorylation, acetylation, and proteolytic cleavage can alter enzyme activity, stability, or interactions, affecting the efficiency of siderophore biosynthesis. These modifications allow rapid responses to environmental changes, ensuring that siderophore production aligns with cellular needs. For instance, phosphorylation of NRPS enzymes can enhance or inhibit their activity, providing a dynamic mechanism for controlling siderophore assembly under different conditions [5]. This comprehensive regulatory framework ensures efficient and context-specific siderophore production (Table 4.2), reflecting the adaptability of microbial systems to diverse environments.

Table 4.2 Regulation of siderophore production.

Study	**Regulation type**	**Mechanism**	**Examples**	**Impact on siderophore production**
Shankar and Akhter [38]	Iron-dependent A	Fur/DtxR-mediated transcriptional repression; other transcriptional regulators (IdeR, IscR)	Fur in *E. coli*, DtxR in *C. diphtheriae*, IdeR in *M. tuberculosis*	Prevents unnecessary production in iron-replete conditions; integrates iron availability with cellular processes
Patel *et al.* [28]	Iron-independent	Quorum sensing and environmental cues; environmental factors (pH, temperature, nutrients)	N-acyl homoserine lactones in *P. aeruginosa*; acidic pH enhances siderophore secretion	Coordinates production with population density; adapts production to environmental changes
Papenfort and Melamed [26]	Post-transcriptional	Small RNAs (sRNAs) modulating mRNA stability	RyhB in *E. coli*	Fine-tunes the expression of biosynthesis genes
Campbell and Philips [5]	Post-translational	Enzyme modifications (phosphorylation, acetylation)	NRPS phosphorylation in bacteria	Dynamically regulates enzymatic activity

4.5 Siderophores in Microbial Ecology and Pathogenesis

4.5.1 Role in Microbial Communities and Competition

Siderophores play a crucial role in microbial ecology by mediating competition and cooperation among microbial populations. In iron-limited environments, such as soil and aquatic systems, microbes secrete siderophores to chelate ferric iron, making it accessible for cellular processes. This capability often determines the survival and competitive advantage of microbial species. Siderophore production is energetically costly, but it enables microbes to colonize and persist in iron-scarce niches [18].

In microbial communities, siderophore-mediated iron acquisition often influences community structure and dynamics. For example, species producing high-affinity siderophores can outcompete others for limited iron resources, altering the balance of microbial populations. Conversely, some non-producing microbes can exploit siderophores secreted by others, benefiting from iron acquisition without incurring the metabolic cost of siderophore synthesis. This phenomenon highlights the dual role of siderophores in fostering both competition and mutualism within microbial ecosystems [8].

The ecological significance of siderophores extends to nutrient cycling and ecosystem functioning. The process of modulating microbial interactions and resource availability, siderophores contribute to the stability and resilience of microbial communities in various environments, from soil ecosystems to aquatic habitats [28].

4.5.2 Siderophores as Virulence Factors in Pathogens

In pathogenic microbes, siderophores are critical virulence factors that enhance the ability of pathogens to infect and proliferate within hosts. Iron is a limiting nutrient in host tissues, as it is sequestered by host proteins such as transferrin and ferritin. To overcome this barrier, pathogens produce siderophores to scavenge host-bound iron, thereby supporting their growth and virulence [25].

Pathogens such as *Escherichia coli*, *Klebsiella pneumoniae*, and *Pseudomonas aeruginosa* rely on siderophores like enterobactin and pyoverdine to establish infections. These siderophores exhibit high affinity for iron, enabling pathogens to outcompete host defenses and other microbes for this essential nutrient. Moreover, siderophore-mediated iron acquisition is often linked to the expression of other virulence factors, including toxins and adhesion molecules, further enhancing pathogenicity [16].

The role of siderophores in pathogenesis extends to immune evasion. During the process of scavenging iron from the host, siderophores can suppress host immune responses, creating an environment conducive to infection. This dual role in iron acquisition and immune modulation underscores the importance of siderophores in the survival and virulence of pathogenic microbes [12].

4.5.3 Siderophore Piracy and Cheating in Microbial Populations

Siderophore piracy and cheating represent intriguing phenomena in microbial populations, reflecting the complex interplay of cooperation and conflict. Siderophore piracy occurs when one microbial species uses siderophores produced by another, bypassing the metabolic cost of siderophore synthesis. This strategy is common in mixed microbial communities, where opportunistic microbes exploit public goods like siderophores to enhance their fitness [9].

Cheating, on the other hand, involves individuals within a species that do not produce siderophores but still benefit from those secreted by others. These cheats avoid the energetic expense of siderophore production while gaining access to iron. Although cheating can destabilize cooperative systems, it is often counterbalanced by regulatory mechanisms that maintain cooperation, such as spatial structuring and kin selection [39].

The dynamics of piracy and cheating have significant ecological and evolutionary implications. They influence the stability of microbial communities, the evolution of cooperative traits, and the outcomes of microbial interactions. Having understand these processes provides insights into microbial behavior and the factors driving community structure and function [9].

The roles of siderophores in microbial ecology and pathogenesis is shown in Figure 4.1. The integration of the microbial ecology and pathogenesis showcase how siderophores mediate interactions within and between microbial species, as well as their role in host-pathogen interactions.

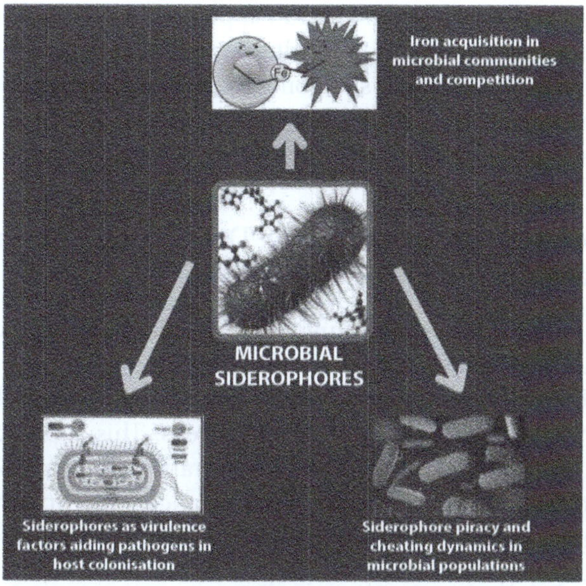

Figure 4.1 Siderophores in microbial ecology and pathogenesis.

4.6 Biotechnological Applications of Siderophores

4.6.1 Siderophores in Agriculture and Bioremediation

Siderophores play a pivotal role in agriculture by enhancing plant growth and mitigating soil nutrient deficiencies. Many plant-associated microbes, such as *Pseudomonas* and *Bacillus* species, produce siderophores that chelate iron, making it more bioavailable to plants. This promotes root development, enhances chlorophyll synthesis, and improves overall plant health. Additionally, siderophore-producing microbes can outcompete plant pathogens for iron, thereby reducing the incidence of diseases and supporting sustainable agricultural practices [30].

In bioremediation, siderophores contribute to the detoxification of heavy metals in contaminated environments. By forming stable complexes with metals like cadmium, lead, and arsenic, siderophores reduce their bioavailability and toxicity. This property makes them valuable tools for cleaning up polluted soils and water bodies. Furthermore, integrating siderophore-producing microbes into phytoremediation systems can enhance the efficiency of metal uptake by plants, creating a synergistic approach to environmental restoration [34].

The agricultural and bioremediation applications of siderophores highlight their versatility as eco-friendly alternatives to chemical fertilizers and remediation agents. Research is ongoing to develop bioinoculants and biostimulants based on siderophore-producing microbes, which could revolutionize sustainable farming and environmental management [30].

4.6.2 Medical Applications: Drug Delivery and Antimicrobial Strategies

In medicine, siderophores offer innovative solutions for drug delivery and combating antimicrobial resistance. Siderophore–drug conjugates, also known as "Trojan horse" systems, exploit microbial iron uptake pathways to deliver antibiotics directly into bacterial cells. This approach enhances the efficacy of antibiotics against drug-resistant pathogens by bypassing traditional resistance mechanisms. For instance, siderophore–antibiotic conjugates like cefiderocol have shown promise in treating multidrug-resistant Gram-negative infections [32].

Beyond drug delivery, siderophores exhibit intrinsic antimicrobial properties by depriving pathogens of essential iron. Siderophore-based therapies can be designed to disrupt microbial iron acquisition, weakening pathogen survival and virulence. Additionally, synthetic and modified siderophores are being explored for their potential to target specific pathogens without harming beneficial microbes, paving the way for precision antimicrobial therapies [22].

The medical applications of siderophores extend to diagnostic tools. The iron-binding properties of siderophores can be leveraged to develop biosensors for detecting infections and monitoring microbial activity. These advancements underscore the transformative potential of siderophores in addressing global health challenges, particularly in the fight against antibiotic resistance [4].

4.6.3 Industrial Applications of Siderophores

Siderophores have diverse applications in industrial processes, including bioleaching, fermentation, and material synthesis. In bioleaching, siderophore-producing microbes facilitate the extraction of metals from low-grade ores by solubilizing iron and other metals, improving the efficiency of mineral recovery. This environmentally friendly alternative to traditional mining techniques reduces chemical usage and waste generation [29].

In fermentation industries, siderophores are employed to optimize microbial growth and metabolite production. Iron is a critical cofactor for many enzymatic reactions, and siderophore-mediated iron acquisition can enhance the yield of bioactive compounds, including antibiotics, enzymes, and biofuels. Furthermore, siderophores are being explored for their role in stabilizing nanoparticles and catalyzing biochemical reactions in green chemistry applications [1].

The versatility of siderophores in industrial applications reflects their potential to drive innovation in sustainable manufacturing processes. Ongoing research focuses on engineering microbes for enhanced siderophore production and designing synthetic siderophores tailored to specific industrial needs, highlighting their future significance in biotechnology [6].

4.7 Conclusion

Siderophores play a pivotal role in the survival and adaptability of microorganisms, particularly under iron-limiting conditions. These specialized iron-chelating compounds are essential for microbial metabolism, virulence, and ecological interactions. The biosynthesis and regulation of siderophores are intricate processes governed by genetic, environmental, and metabolic cues. The regulation of siderophore production is tightly linked to iron availability, with microorganisms employing sophisticated systems such as the ferric uptake regulator (Fur) and other global regulators to optimize resource utilization. Advances in molecular biology and bioinformatics have further elucidated the pathways and genes involved in siderophore biosynthesis, offering a framework for genetic engineering and synthetic biology approaches to enhance or modulate siderophore production. In agriculture, siderophores are being explored as biofertilizers and biocontrol agents to improve crop productivity and protect plants from pathogens. In medicine, they hold promise in antimicrobial drug development and iron-overload therapies. Additionally, their potential for bioremediation, especially in heavy metal detoxification and recovery, underscores their environmental significance. Despite these advances, challenges remain, including the complexity of regulatory networks, strain-specific variability, and the interplay between siderophore production and other metabolic pathways. Continued exploration of these processes will not only deepen our understanding of microbial ecology but also pave the way for innovative solutions to address pressing challenges in health, agriculture, and environmental sustainability.

AI disclosure statement: The author acknowledges the use of AI-based tools for minor language editing or grammatical correction. However, all intellectual contributions, analyses, and conclusions presented in this chapter are entirely the author's own.

References

1. Abo-Zaid, G.A., Soliman, N.A.M., Abdullah, A.S., El-Sharouny, E.E., Matar, S.M., Sabry, S.A.F., Maximization of siderophores production from biocontrol agents, Pseudomonas aeruginosa f2 and Pseudomonas fluorescens JY3 using batch and exponential fed-batch fermentation. *Processes*, *8*, 4, 455, 2020, https://doi.org/10.3390/pr8040455.
2. Beasley, K.L., Cristy, S.A., Elmassry, M.M., Dzvova, N., Colmer-Hamood, J.A., Hamood, A.N., During bacteremia, Pseudomonas aeruginosa PAO1 adapts by altering the expression of numerous virulence genes including those involved in quorum sensing. *PLoS One*, *15*, 10, e0240351, 2020, https://doi.org/10.1371/journal.pone.0240351.
3. Bíró, L., Buglyó, P., Farkas, E., Factors Determining the Metal Ion Binding Ability and Selectivity of Hydroxamate-Based Compounds. *Curr. Med. Chem.*, *28*, 35, 7209–7237, 2021, https://doi.org/10.2174/0929867328666210405123952.
4. Bohac, T.J., Fang, L., Banas, V.S., Giblin, D.E., Wencewicz, T.A., Synthetic mimics of native siderophores disrupt iron trafficking in Acinetobacter baumannii. *ACS Infect. Dis.*, *7*, 8, 2138–2151, 2021, https://doi.org/10.1021/acsinfecdis.1c00119.
5. Campbell, S.L. and Philips, M.R., Post-translational modification of RAS proteins. *Curr. Opin. Struct. Biol.*, *71*, 180–192, 2021, https://doi.org/10.1016/j.sbi.2021.06.015.
6. Chabhadiya, S., Acharya, D.K., Mangrola, A., Shah, R., Pithawala, E.A., Unlocking the potential of biosurfactants: innovations in metabolic and genetic engineering for sustainable industrial and environmental solutions. *Biotechnol. Notes*, 2024, https://doi.org/10.1016/j.biotno.2024.07.001.
7. Drechsel, H. and Winkelmann, G., Iron chelation and siderophores, in: *Transition metals in microbial metabolism*, pp. 1–49, CRC Press, 2022.
8. Du, X., Liu, N., Yan, B., Li, Y., Liu, M., Huang, Y., Proximity-based defensive mutualism between Streptomyces and Mesorhizobium by sharing and sequestering iron. *ISME J.*, *18*, 1, wrad041, 2024, https://doi.org/10.1093/ismejo/wrad041.
9. Fritts, R.K., McCully, A.L., McKinlay, J.B., Extracellular metabolism sets the table for microbial cross-feeding. *Microbiol. Mol. Biol. Rev.*, *85*, 1, 10–1128, 2021, https://doi.org/10.1128/mmbr.00135-20.

10. Ghssein, G. and Ezzeddine, Z., A review of Pseudomonas aeruginosa metallophores: Pyoverdine, pyochelin and pseudopaline. *Biology*, *11*, 12, 1711, 2022, https://doi.org/10.3390/biology11121711.
11. Gomes, A.F., Almeida, M.C., Sousa, E., Resende, D.I., Siderophores and metallophores: Metal complexation weapons to fight environmental pollution. *Sci. Total Environ.*, 173044, 2024, https://doi.org/10.1016/j.scitotenv.2024.173044.
12. Haschka, D., Hoffmann, A., Weiss, G., Iron in immune cell function and host defense, in: *Seminars in Cell & Developmental Biology*, vol. 115, pp. 27–36, Academic Press, 2021, https://doi.org/10.1016/j.semcdb.2020.12.005.
13. Hider, R.C., Silva, A.M., Cilibrizzi, A., The iron (III) coordinating properties of citrate and α-hydroxycarboxylate containing siderophores. *BioMetals*, 1–13, 2024, https://doi.org/10.1007/s10534-024-00607-z.
14. Joshi, H. and Khan, A., Competition-driven phenotypic plasticity in Iron acquisition and aromatic utilization confers a fitness advantage to Pseudomonas putida in an Iron-limited rhizospheric environment. *World J. Microbiol. Biotechnol.*, *40*, 12, 386, 2024, https://doi.org/10.1007/s11274-024-04192-8.
15. Kapoor, D., Sharma, P., Sharma, M.M.M., Yadav, S., Husen, A., Exploring soil microbiota and their role in plant growth, stress tolerance, disease control and nutrient immobilizer. *Biocatal. Agric. Biotechnol.*, 103358, 2024, https://doi.org/10.1016/j.bcab.2024.103358.
16. Khasheii, B., Mahmoodi, P., Mohammadzadeh, A., Siderophores: Importance in bacterial pathogenesis and applications in medicine and industry. *Microbiol. Res.*, *250*, 126790, 2021, https://doi.org/10.1016/j.micres.2021.126790.
17. Kircheva, N. and Dudev, T., Gallium as an antibacterial agent: a DFT/SMD study of the Ga3+/Fe3+ competition for binding bacterial siderophores. *Inorg. Chem.*, *59*, 9, 6242–6254, 2020, https://doi.org/10.1021/acs.inorgchem.0c00367.
18. Kramer, J., Özkaya, O., Kümmerli, R., Bacterial siderophores in community and host interactions. *Nat. Rev. Microbiol.*, *18*, 3, 152–163, 2020, https://doi.org/10.1038/s41579-019-0284-4.
19. Krüger, A., Weber, U., Frunzke, J., Genome-wide analysis of DtxR and HrrA regulons reveals novel targets and a high level of interconnectivity between iron and heme regulatory networks in *Corynebacterium glutamicum*. bioRxiv, 2025-01, 2025, https://doi.org/10.1101/2025.01.02.631128.
20. Lorenzi, A.S. and Chia, M.A., Cyanobacteria's power trio: auxin, siderophores, and nitrogen fixation to foster thriving agriculture. *World J. Microbiol. Biotechnol.*, *40*, 12, 1–18, 2024, https://doi.org/10.1007/s11274-024-04191-9.
21. Luscher, A., Gasser, V., Bumann, D., Mislin, G.L., Schalk, I.J., Köhler, T., Plant-derived catechols are substrates of TonB-dependent transporters and sensitize Pseudomonas aeruginosa to siderophore-drug conjugates. *MBio*, *13*, 4, e01498–22, 2022, https://doi.org/10.1128/mbio.01498-22.

22. Mangal, S., Singh, V., Chhibber, S., Harjai, K., Headway in Synthetic Chemistry: A Gateway for Antimicrobial Trojan Horse Strategy for Drug Repurposing Against Multi-Drug Resistant Pseudomonas aeruginosa. *Indian J. Microbiol.*, 1–22, 2025, https://doi.org/10.1007/s12088-024-01439-8.
23. Martín, J.F. and Liras, P., Targeting of Specialized Metabolites Biosynthetic Enzymes to Membranes and Vesicles by Posttranslational Palmitoylation: A Mechanism of Non-Conventional Traffic and Secretion of Fungal Metabolites. *Int. J. Mol. Sci.*, *25*, 2, 1224, 2024, https://doi.org/10.3390/ijms25021224.
24. Martinez Pastor, M., Sakrikar, S., Hwang, S., Hackley, R.K., Soborowski, A.L., Maupin-Furlow, J.A., Schmid, A.K., TroR is the primary regulator of the iron homeostasis transcription network in the halophilic archaeon Haloferax volcanii. *Nucleic Acids Res.*, *52*, 1, 125–140, 2024, https://doi.org/10.1093/nar/gkad997.
25. Moreira, A.C., Mesquita, G., Gomes, M.S., Ferritin: an inflammatory player keeping iron at the core of pathogen-host interactions. *Microorganisms*, *8*, 4, 589, 2020, https://doi.org/10.3390/microorganisms8040589.
26. Papenfort, K. and Melamed, S., Small RNAs, large networks: posttranscriptional regulons in gram-negative bacteria. *Annu. Rev. Microbiol.*, *77*, 1, 23–43, 2023, https://doi.org/10.1146/annurev-micro-041320-025836.
27. Park, H., Jin, H., Kim, D., Lee, J., Cell-Free Systems: Ideal Platforms for Accelerating the Discovery and Production of Peptide-Based Antibiotics. *Int. J. Mol. Sci.*, *25*, 16, 9109, 2024, https://doi.org/10.3390/ijms25169109.
28. Patel, N.P., Shimpi, G.G., Haldar, S., Evaluation of heterotrophic bacteria associated with healthy and bleached corals of Gulf of Kutch, Gujarat, India for siderophore production and their response to climate change factors. *Ecol. Indic.*, *113*, 106219, 2020, https://doi.org/10.1016/j.ecolind.2020.106219.
29. Pathak, A., Kothari, R., Vinoba, M., Habibi, N., Tyagi, V.V., Fungal bioleaching of metals from refinery spent catalysts: A critical review of current research, challenges, and future directions. *J. Environ. Manage.*, *280*, 111789, 2021, https://doi.org/10.1016/j.jenvman.2020.111789.
30. Pattnaik, S., Mohapatra, B., Gupta, A., Plant growth-promoting microbe mediated uptake of essential nutrients (Fe, P, K) for crop stress management: Microbe–soil–plant continuum. *Front. Agron.*, *3*, 689972, 2021, https://doi.org/10.3389/fagro.2021.689972.
31. Prabhakar, P.K., Bacterial siderophores and their potential applications: a review. *Curr. Mol. Pharmacol.*, *13*, 4, 295–305, 2020, https://doi.org/10.2174/1874467213666200518094445.
32. Rayner, B., Verderosa, A.D., Ferro, V., Blaskovich, M.A., Siderophore conjugates to combat antibiotic-resistant bacteria. *RSC Med. Chem.*, *14*, 5, 800–822, 2023, https://doi.org/10.1039/D2MD00465H.
33. Rehan, M., Barakat, H., Almami, I.S., Qureshi, K.A., Alsohim, A.S., Production and potential genetic pathways of three different siderophore types in Streptomyces tricolor strain HM10. *Fermentation*, *8*, 8, 346, 2022, https://doi.org/10.3390/fermentation8080346.

34. Roskova, Z., Skarohlid, R., McGachy, L., Siderophores: an alternative bioremediation strategy? *Sci. Total Environ., 819*, 153144, 2022, https://doi.org/10.1016/j.scitotenv.2022.153144.
35. Salwan, R. and Sharma, V., Molecular and biotechnological aspects of secondary metabolites in actinobacteria. *Microbiol. Res., 231*, 126374, 2020, https://doi.org/10.1016/j.micres.2019.126374.
36. Schalk, I.J., Bacterial siderophores: diversity, uptake pathways and applications. *Nat. Rev. Microbiol.*, 1–17, 2024, https://doi.org/10.1038/s41579-024-01090-6.
37. Sestok, A.E., Lee, M.A., Smith, A.T., Prokaryotic Ferrous Iron Transport: Exploiting Pools of Reduced Iron Across Multiple Microbial Environments, in: *Microbial Metabolism of Metals and Metalloids*, pp. 299–357, Springer International Publishing, Cham, 2022, https://doi.org/10.1007/978-3-030-97185-4_12.
38. Shankar, G. and Akhter, Y., Stealing Survival: Iron Acquisition Strategies of Mycobacterium tuberculosis. *Biochimie*, 2024, https://doi.org/10.1016/j.biochi.2024.06.006.
39. Shao, J., Li, Y., Lu, J., Gu, S., Li, Z., Siderophore Piracy Promotes Dynamical Coexistence in Microbial Community. bioRxiv, 2023-11, 2023, https://doi.org/10.1101/2023.11.21.568182.
40. Su, L., Souaibou, Y., Hôtel, L., Paris, C., Weissman, K.J., Aigle, B., Biosynthesis of novel desferrioxamine derivatives requires unprecedented crosstalk between separate NRPS-independent siderophore pathways. *Appl. Environ. Microbiol., 90*, 3, e02115–23, 2024, https://doi.org/10.1128/aem.02115-23.
41. Sun, Y., Wu, Z., Lan, J., Liu, Y., Du, Y., Ye, H., Du, D., Effect of sulfate-reducing bacteria (SRB) and dissimilatory iron-reducing bacteria (DIRB) coexistence on the transport and transformation of arsenic in sediments. *Water Res., 270*, 122834, 2025, https://doi.org/10.1016/j.watres.2024.122834.
42. Terlouw, B.R., Biermann, F., Vromans, S.P., Zamani, E., Helfrich, E.J., Medema, M.H., RAIChU: automating the visualisation of natural product biosynthesis. *J. Cheminf., 16*, 1, 106, 2024, https://doi.org/10.1186/s13321-024-00898-x.
43. van der Helm, D. and Winkelmann, G., Hydroxamates and polycarboxylates as iron transport agents (siderophores) in fungi, in: *Metal ions in fungi*, pp. 39–98, CRC Press, 2020.
44. Xie, B., Wei, X., Wan, C., Zhao, W., Song, R., Xin, S., Song, K., Exploring the Biological Pathways of Siderophores and Their Multidisciplinary Applications: A Comprehensive Review. *Molecules, 29*, 10, 2318, 2024, https://doi.org/10.3390/molecules29102318.
45. Xu, Y.C., Li, N., Yan, X., Zou, H.X., DFT-based analysis of siderophore-metal ion interaction for efficient heavy metal remediation. *Environ. Sci. Pollut. Res., 30*, 40, 91780–91793, 2023, https://doi.org/10.1007/s11356-023-28854-6.

5

Iron Chelation and Transport Mechanisms in Plants Mediated by Siderophores

Manjoo Rani[1], Saty Dev[2*], Ayush Madan[3], Nand Kumar Singh[4] and Kartik Singh[5]

[1]*Department of Pharmacy, Shri Ram Murti Smarak College of Engineering and Technology Bareilly, UP, India*
[2]*Department of Mechanical Engineering, Shri Ram Murti Smarak College of Engineering and Technology Bareilly, UP, India*
[3]*Department of Biotechnology, School of Research & Technology (SORT), People's University, Bhanpur, Bhopal, Madhya Pradesh, India*
[4]*Department of Biotechnology, Motilal Nehru National Institute of Technology Allahabad, Prayagraj, India*
[5]*Department of Mechanical Engineering, Rajshree Institute of Management & Technology Bareilly, UP, India*

Abstract

Iron is an essential micronutrient for plant growth and development and is involved in many of the physiological and biochemical processes, including photosynthesis, respiration, and DNA synthesis. However, its availability in the soil is limited because it tends to form insoluble hydroxides and oxides in aerobic and alkaline conditions. Research has proven the value of microbial siderophores as sustainable solutions because they function as low molecular weight chelating compounds for iron acquisition in plants. The mechanisms of siderophores in iron chelation along with transport and uptake procedures in plants are examined throughout this chapter. This research examines molecular iron–siderophore transport pathways and gene networks for keeping nutrients through plant–microbe interactions. The chapter also examines siderophore applications in cultivation through the analysis of their productivity effects alongside their advantages for earth enrichment and pathogen disease management strategies. The chapter concludes by presenting future research paths along with existing hurdles and the role of microbial siderophores for fostering sustainable resilient agricultural systems.

**Corresponding author*: satyadevyadav08@gmail.com

Debasis Mitra, Ayush Madan and Anju Rani (eds.) Microbial Siderophores for Agricultural Productivity: Perspectives of Use in Agriculture, (89–120) © 2026 Scrivener Publishing LLC

Keywords: Iron chelation, siderophores, plant nutrient uptake, sustainable agriculture, soil fertility

Abbreviations

Fe	Iron
FDP	ferric ion-dissolving protein
SIA	siderophore-iron acquisition
SIDER	siderophore interactions and distribution in environmental resources
ROS	reactive oxygen species
ABC	ATP-binding cassette (transporter)
PCT	phytosiderophore chelate transporter
Fv/Fm	maximum quantum efficiency of photosystem II
NRR	nitrogen respiration regulation
R2R3-MYB	a family of transcription factors involved in plant stress response
IAA	indole-3-acetic acid (auxin)
APX	ascorbate peroxidase
PGR	plant growth regulators
SYN	synthesis
NADPH	nicotinamide adenine dinucleotide phosphate (reduced)
SNP	single-nucleotide polymorphisms
MIC	microbial inoculants for crops
BCA	biocontrol agents
PCA	phytochemical analysis
GM	genetically modified
PCR	polymerase chain reaction
ETC	electron transport chain
ABI	abscisic acid-induced
BSF	biological siderophore fertilizers
MST	metal siderophore transporter
NSF	National Siderophore Foundation

5.1 Introduction

The micronutrient iron (Fe) plays a fundamental role for optimal plant growth because it supports their development. It universally participates in critical physiological processes with essential biochemical components essential for plant survival. Plants depend on iron as a major element to produce chlorophyll which serves as a necessary factor for photosynthesis. The functional mechanism of the photosynthetic electron transport chain depends on ferredoxin and cytochromes along with other enzymes whose iron requirements the plant essentially needs for sustainability [1]. Plants need iron as a cofactor to run enzymes for respiration and nitrogen assimilation, thus making these processes important for energy creation and nitrogen processing [5]. Although iron constitutes most of the Earth's crust, it becomes difficult for plants to acquire because aerobic and alkaline soils transform iron into ferric hydroxides and other insoluble structures, thereby decreasing its bioavailability. Soils that lack bioavailable iron create deficiency problems, leading to chlorosis and growth restrictions and yield reduction in crops [4]. Plants thrive through adaptive strategies comprising organic acid and phytosiderophore secretion to unlock iron from the mineral soil substances. The sophisticated processes for iron acquisition illustrate why plants need iron for the preservation of their well-being and crop productive capacity. The necessity to improve soil iron accessibility stands vital for supporting enduring agricultural development, particularly within areas where iron limitations prove as major obstacles in farming [2]. Iron serves an essential role in basic plant metabolism while supplementing the defense responses of the plant. During pathogen attacks, reactive oxygen species (ROS) production through iron involvement leads plants to deploy an effective defense response. Furthermore, iron regulates the expression of genes involved in hormone biosynthesis, stress responses, and root development, making it a central element in plant growth and resilience [3]. Understanding and optimizing iron acquisition and utilization in plants is a cornerstone of improving agricultural productivity and combating global food insecurity.

5.1.1 Challenges of Iron Bioavailability in Soil

Iron bioavailability in soil poses a significant challenge for plant growth, even though iron (Fe) is the fourth most plentiful element in the Earth's crust. However, the major problem arises from the chemical nature of iron

(Fe) presence in soil, where it predominantly exists in its oxidized ferric (Fe^{3+}) form under aerobic and alkaline conditions. As per the previous research, ferric iron (Fe^{++}) is highly insoluble and forms stable hydroxides and oxides, making it inaccessible to plant roots [9]. Soil with high total iron but low bioavailable iron content causes iron deficiency problems for plants even when they have substantial amounts of iron. Many agricultural areas around the world have calcareous and alkaline soils where iron deficiency flourishes [7]. The issue is made worse by environmental factors. Iron solubility and uptake can be decreased by high pH, low soil moisture, and overuse of fertilizers based on phosphates. Additionally, competition with other metal ions, such as calcium, magnesium, and zinc, interferes with iron acquisition by plants [8]. Iron deficiency in crops demonstrates symptoms, such as interveinal chlorosis, that reduce photosynthetic efficiency and overall hinder crop yield growth. Developing iron-efficient crop varieties, enhancing soil management techniques, and using microbial solutions like siderophores to increase iron solubility and uptake in plants are some of the creative ways to address these issues [10]. Therefore, microbial siderophores play a significant role in agriculture, particularly in enhancing plant growth and soil health. Siderophores are small, high-affinity iron-chelating compounds produced by microorganisms like bacteria and fungi. These compounds are crucial for the efficient acquisition of iron, an essential nutrient for both plants and microbes. Below are key aspects of microbial siderophores in agriculture:

1.1 Iron acquisition for plants: The natural low solubility of iron makes it difficult to find in most soils. Microbial siderophores solve plant iron problems by binding iron so that plants can absorb it easier. The beneficial soil microbes *Pseudomonas*, *Bacillus*, and *Enterobacter* secrete siderophores that let these microbes help plants obtain more iron.

1.2 Plant growth promotion: Producers of siderophores typically demonstrate enhanced plant growth benefits as well. Along with iron delivery, they can generate plant hormones including auxins, cytokinins, and gibberellins which improve both plant growth and stress tolerance.

1.3 Suppression of pathogenic microorganisms: By taking iron away from soilborne pathogens, siderophores prevent the pathogens from multiplying. Beneficial microbe siderophore production restricts pathogenic microbes because these molecules lock iron during their formation which plant pathogens need to grow.

1.4 Stress tolerance: Approximately several siderophore-producing microbes protect plants from diverse types of stress by improving their ability to handle water shortages, salt damage, and harmful metal contamination. Siderophore-mediated enhancement of plant iron uptake along with other needed nutrients decreases their susceptibility to stress factors.

1.5 Biocontrol agents: Through biocontrol actions, microorganisms that produce siderophores offer protection for crops against diseases transmitted by soil pathogens. The beneficial microbes feed on harmful organisms to retrieve vital nutrients while releasing disease-fighting substances, enabling farmers to work beyond traditional chemical practices.

1.6 Improvement of soil fertility: Plants obtain nutrients through siderophore use which improves both soil aspects and promotes beneficial microbial growth. By making soil environments stronger, the use of chemical fertilizers can be decreased.

1.7 Sustainable agriculture: Microbial iron collection assists farmers through reducing their field dependence on synthetic fertilizers. Our sustainable agricultural methods combine support for organic farming practices with efforts to protect our environment. The use of functional microbial siderophores improves both plant nutrition levels and growth performance.

2 Iron chelation mechanisms: Physiologic analyses show how siderophores help plants absorb more iron during times of need in low iron solutions by binding to available iron sources.

2.1 Plant growth promotion: Plants receive beneficial developmental growth supports from microbes that produce siderophores. Bacteria facilitate plants to grow better despite stress through hormonal release of auxins, cytokinins, and gibberellins and promote plant development [18, 20].

2.2 Suppression of pathogenic microorganisms: Siderophores control soilborne pathogen growth as they grab available iron supplies which these bacteria and fungi need to survive.

Plants affected by fungi or bacterial pathogens function optimally with accessible iron supplies for pathogen development and growth. Beneficial microbes produce siderophores to regulate iron levels, thus suppressing pathogen growth [14, 23].

2.3 Stress tolerance: Research shows that microbial siderophores help plants better withstand dry soil conditions plus high salt levels and steel damage. Under stressful conditions, siderophores support cellular wellbeing in plants through increased nutrient access, especially iron [16, 21].

2.4 Biocontrol agents: Microorganisms that make siderophores protect the harvested crops from soil pathogens by controlling their growth. The microbes use their nutrients better than disease-causing microbes while making special substances that fight harmful germs to manage plant diseases sustainably [13, 15].

2.5 Improvement of soil fertility: By binding to iron toxicants in soil, the action of siderophores makes plants healthier while also helping soil microbes stay diverse and nutrients flow through the system naturally. Siderophores maintain a balanced soil environment and help farmers avoid using chemical fertilizers according to Cornelis and Dingemans [19].

2.6 Sustainable agriculture: Farmers using microbial siderophore methods can decrease their need for synthetic iron fertilizers and agrochemicals while adopting natural farming methods that protect our environment.

2.7 The microbial siderophore production system helps plants get essential nutrients and grow better by fighting disease-causing organisms. They help farms stay sustainable while making crops grow better and stay healthy. Low-molecular-weight organic compounds called siderophores that microbes produce allow these organisms and plants to capture and dissolve iron (Fe^{3+}) from their ecosystem. When plants or microorganisms face iron shortages, they create these specialized molecules to solve the problem of obtaining their necessary iron. Under iron shortages, the body uses siderophores to grab these iron atoms while providing efficient metal transport. They strongly take hold of Fe^{3+} molecules but only through specific fits before cells detect and take these bound materials.

5.2 Types of Siderophores

Siderophores can be classified based on their chemical structure, the type of metal they chelate, and the functional groups they contain. The primary types are shown in Figure 5.1.

5.2.1 Catecholate Siderophores

The siderophores in these bacteria have catechol groups that feature hydroxylated organic rings with a two-dimensional structure. Catecholate siderophores join to iron through two direct metal bonding connections, examples: enterobactin (*Escherichia coli* production) and pyoverdine (*Pseudomonas* species production) [39].

5.2.2 Hydroxamate Siderophores

The iron-binding properties of hydroxamate siderophores originate from their hydroxamate groups (-C=NOH) which act as derivatives of carboxylic acids and form solid coordination arrangements with iron ions.

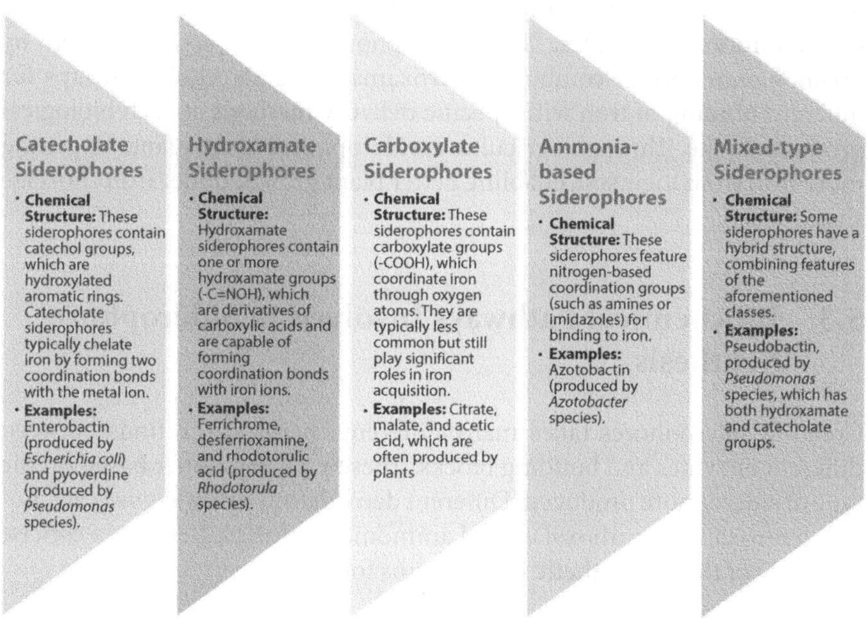

Figure 5.1 Types of siderophores.

Three examples of hydroxamate siderophores are fFerrichrome from fungi, desferrioxamine produced by *Streptomyces*, and rhodotorulic acid from *Rhodotorula* species.

5.2.3 Carboxylate Siderophores

Siderophores of these types use carboxylate groups (-COOH) to bind iron ions in oxygen-based coordination. They are typically less common but still play significant roles in iron acquisition [43]. Soil organisms and plants make citrate and other acids like malate during special environments.

5.2.4 Ammonia-Based Siderophores

The siderophores obtain iron using nitrogen-based connecting points like amines and imidazoles—example: azotobactin (produced by *Azotobacter species*).

5.2.5 Mixed-Type Siderophores

Several siderophores combine the important features of distinct iron-binding classes. These siderophores can be more versatile in their ability to bind iron and may be produced in response to diverse environmental conditions—examples: The siderophore pseudobactin released by *Pseudomonas* species combines hydroxamate and catecholate groups for different binding of iron with specific delivery methods at each biological production site. Through agriculture, siderophores assist plants in getting more iron from their soil, enabling better plant growth under iron shortage situations.

5.3 Biochemical Pathways Involved in Siderophore Synthesis

Creating siderophores takes many different chemical reactions between different enzymes and building blocks. These pathways differ based on the type of siderophore produced: Different deoxidation pathways use catecholate, hydroxamate, carboxylate and ammonia platforms. Below is a concise overview of the biosynthetic mechanisms for each type:

5.3.1 Catecholate Siderophores

Catecholate siderophores include enterobactin produced by *Escherichia coli* and pyoverdine produced by species of Pseudomonas. The synthesis of catecholate siderophores proceeds through pathways of aromatic biosynthesis; aromatic rings undergo hydroxylation to create the iron-chelating catecholate groups [25, 27]. Biosynthesis requires L-tyrosine as a precursor, and enzymes such as tyrosine aminotransferase (TyrB) and enterobactin synthetase (EntA–EntF) catalyze key steps [24, 29].

5.3.2 Hydroxamate Siderophores

Hydroxamate siderophores including ferrichrome (fungal producer) and desferrioxamine (Streptomyces producer) develop through enzyme pathways that adjust L-ornithine and L-glutamine derived amino acids. Through specialized enzymes hydroxamates appear and tightly hold iron ions according to Schalk *et al.* [28]. Siderophore synthetases serve as key enzymes by creating iron-chelating molecules according to Chimento *et al.* [26].

5.3.3 Carboxylate Siderophores

The production cycle of siderophores begins with central metabolic compounds including citrate and malate. The production of carboxylate siderophores by specialized enzymes (citrate synthase and aconitase) occurs as described by Moynie *et al.* [30]. Analysis by Perraud *et al.* [31] proved carboxylate groups in siderophores bond to iron through their oxygen atoms yet maintain weaker iron binding strength compared to hydroxamate and catecholate siderophores.

5.3.4 Ammonia-Based Siderophores

Azotobactin produced by Azotobacter microorganisms binds iron using nitrogen-based amines instead of traditional hydroxamate scaffoldings. Microbes transform nitrogen-based amino acids into siderophore molecules through aminotransferases along with synthetases according to Moynié *et al.* [29] and Perraud *et al.* [31]. Salient iron uptake mechanisms represent a microbial adaptation that enables cellular iron absorption needed for survival and population expansion.

5.3.5 Microbial Sources of Siderophores Relevant to Agriculture

Since microorganisms make most siderophores they steer iron absorption in soil when iron levels remain low. Through the production of siderophores microbes supply plants with better iron absorption which helps them grow better and improve soil quality. Bacterial fungi and actinomycetes microbial groups produce siderophores that help soil and plants gain more iron when there is not enough available iron in the soil [40]. Pseudomonas bacteria produce pyoverdine which helps them capture iron while protecting plants from diseases and is important for sustainable farming according to research published in 2000 and 2005 and updated in 2012 by Zhao and Poole, Hoegy *et al.* and Schalk *et al.* As part of plant promotion B. subtilis and other Bacillus species make bacillibactin to beat pathogens and help plant development [27, 31]. Under iron shortage Enterobacter spp. generate enterobactin that boosts crop health [26]. A. brasilense from the Azospirillum spp. enhances plant development by fixing nitrogen and producing azotobactin [25]. воздух rhizobium enhances how well plants take up iron and fixes nitrogen while creating beneficial root connections with legumes [28, 29]. Streptomyces spp. within the Actinomycetes group makes iron available to plants through desferrioxamine hydroxamate production which benefits soil fertility and plant growth while hindering harmful microorganisms. They can be used as biofertilizer and biocontrol solutions to decrease our need for chemical fertilizer and pesticide while helping crops grow better in healthy soil [30, 31].

5.4 Iron Chelation and Uptake Mechanisms in Plants

5.4.1 Mechanisms of Iron Chelation by Siderophores

Iron chelation by siderophores is one of the basic means by which microorganisms and plants take up the necessary nutrient, iron, from the environment. Siderophores are special molecules that bind with iron ions, in most cases in their ferric form, Fe^{3+}, producing stable complexes capable of transportation into microbial cells or plant systems. Various key mechanisms are involved in siderophore chelation of iron; these include Figure 5.2.

The mechanism of iron binding by siderophores involves many critical steps. Firstly, siderophores, containing functional groups such as

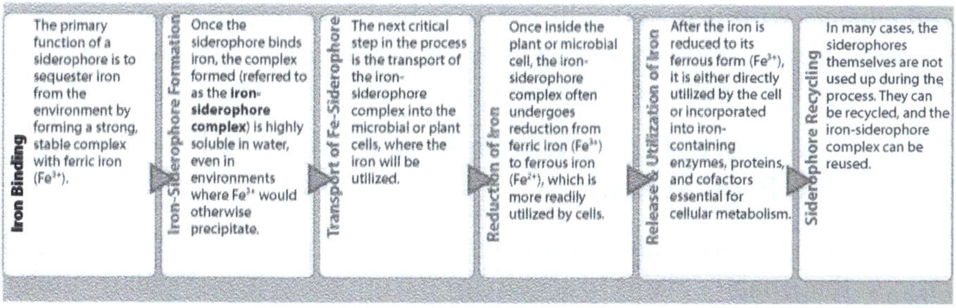

Figure 5.2 Key mechanisms are involved in siderophore chelation of iron.

hydroxamate (-C=NOH), catecholate (-C6H4O2), or hydroxypyridone (-C5H4NO), coordinate with ferric ions (Fe^{3+}) through coordinate bonds of lone pairs of electrons, forming stable iron-siderophore complexes. These siderophores have strong affinities for Fe^{3+}, particularly in environments where the available levels of iron are limited, as is found in alkaline or aerobic soils, thereby preventing iron from forming insoluble compounds with other elements, such as phosphate or carbonate. They are also selective for Fe^{3+} over other divalent and trivalent metals, ensuring thus efficient iron transport. Complexation of iron into an iron-siderophore complex increases iron solubility, enabling the passage of the metal through soils or microbial cell membranes. Typically, a siderophore forms a chelate with the ferric ion, in which one or more functional groups coordinate with Fe^{3+}, thus stabilizing it in solution. In microbes, the transport of the iron-siderophore complex involves a particular outer membrane receptor in Gram-negative bacteria. In this case, the receptors bind with the complex, allowing it to cross the outer membrane into the periplasmic space, where periplasmic binding proteins assist in translocation across the inner membrane *via* ATP-binding cassette (ABC) transporters or proton symporters. In plants, iron uptake follows a similar pathway, whereby specific transport proteins in the root cell membranes, such as iron-regulated transporters (IRT) or ferroportins, recognize and transport the iron-siderophore complex into plant cells. Ferric reductases, such as members of the FRO family of proteins, reduce Fe^{3+} to Fe^{2+} within the cells, facilitating cross-membrane transport into the cytoplasm, where it is used for several metabolic processes, including respiration, photosynthesis, and DNA synthesis. Microbes and plants use Fe^{2+} that is reduced for tasks like breathing, making proteins, and creating iron-based stuff like cytochromes and clusters with iron and sulfur. Also, a bunch of microbes keep reusing siderophores by always releasing them to snag iron from where they live.

5.4.2 Transport of Iron–Siderophore Complexes into Plant Systems

The transport of iron-siderophore complexes into plant systems is a crucial process that enables plants to acquire iron from the soil, particularly under conditions where iron is present in low bioavailability due to the soil's pH, composition, or iron's tendency to form insoluble compounds. This transport is mediated through specialized mechanisms and transporters, ensuring that plants can efficiently uptake the iron required for metabolic processes like chlorophyll synthesis, respiration, and enzyme activity [40]. The transport of iron-siderophore complexes into plant systems occurs in several key steps, as outlined in Figure 5.3.

Synthesis and secretion of siderophores by plant roots: Root exudation of siderophores: In some plant species, especially under iron-deficient conditions, roots excrete siderophores into the surrounding soil. These siderophores can chelate ferric iron (Fe^{3+}) from the soil and form soluble iron-siderophore complexes, which are more accessible for uptake by the plant. Example: Gramineous plants like grasses often excrete mugineic acid-type siderophores to solubilize iron from soil particles.

Endogenous siderophore production: Some plants, particularly those in the legume family, can also produce siderophores internally to facilitate iron uptake. These siderophores are synthesized within the plant and can transport iron through various plant tissues.

Recognition and iron–siderophore interception and binding to root cell membranes: Plasma membrane receptors: Having said that, the ground roots could present the iron-siderophore complexes because it has specific receptors located on the membranes of these cells. The latter are a part of the iron-specific transport system of plants, all bound to take up iron-siderophore complexes. For example, Iron-regulated transporters (IRT1) are the ones responsible for the incorporation of iron in plant elements, and the rootside plasma membranes are occupied by siderophore receptors that bind the iron-siderophore complex so they can get in.

Specificity in transport: In fact, its receptors are very distinctive, differentiating the different types of siderophores in the way they recognize and convey throughout the plant the iron complex with which they work.

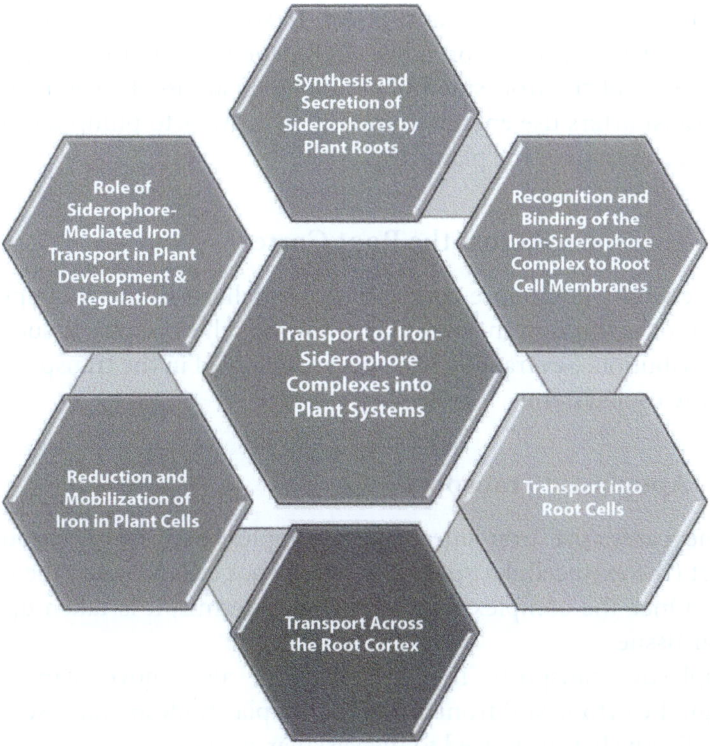

Figure 5.3 Transport of iron–siderophore into plant systems.

5.4.3 Transport into Root Cells

Once the iron-siderophore complex is recognized by the plasma membrane receptors, the next step is the translocation of the complex into the plant root cells. This occurs through several active transport mechanisms:

5.4.4 Proton-Coupled Symporters

The most common mechanism for transporting iron-siderophore complexes is through proton-coupled symporters. In this process, the transport of the iron-siderophore complex is coupled with the movement of protons (H^+) into the root cell. The proton gradient, created by proton pumps in the plasma membrane, provides the necessary energy for symporters to co-transport both the iron-siderophore complex and protons. Example: The Yellow Stripe-Like (YSL) family of transporters is known to facilitate the uptake of iron-siderophore complexes through proton-coupled symport. ATP-Dependent Transporters: In some cases, ATP-binding cassette (ABC) transporters or other ATP-dependent pumps assist in the translocation of the iron-siderophore complex across the cell membrane. These transporters use energy from ATP hydrolysis to pump the complex into the cytoplasm of the plant cell.

5.4.5 Transport Across the Root Cortex

After the iron-siderophore complex is internalized in the root epidermal cells, it moves through the root cortex towards the vascular tissue for further distribution. Several mechanisms are involved in the transport of iron across the root system:

5.4.6 Apoplastic Transport

In some cases, the iron-siderophore complex may travel through the apoplast (the extracellular spaces) between root cells to reach the vascular system. Once the complex reaches the endodermis, it is taken up by the vascular tissue.

Symplastic transport: The complex can also move symplastically (through the cytoplasm) from cell to cell *via* plasmodesmata, ensuring that iron is effectively transported to the vascular system.

5.4.7 Reduction and Mobilization of Iron in Plant Cells

Once the iron-siderophore complex reaches the vascular tissues and cells of the root or other plant tissues, the iron is often reduced from Fe^{3+} (ferric) to Fe^{2+} (ferrous), a more bioavailable form. This reduction is facilitated by ferric reductase enzymes located in the plasma membranes or endoplasmic reticulum.

Ferric reductases: Enzymes such as FRO (Ferric Reductase Oxidase) family members play a pivotal role in reducing Fe^{3+} to Fe^{2+}, making it accessible for transport and incorporation into the plant's metabolic processes.

Iron transport and distribution: After reduction, the iron is transported through the plant *via* the vascular system to various plant tissues, such as leaves and developing fruits. The iron is then utilized in processes such as chlorophyll synthesis, respiration, and the production of iron-dependent enzymes.

5.4.8 Role of Siderophore-Mediated Iron Transport in Plant Development

Enhancement of growth and productivity: Efficient iron acquisition through siderophore-mediated transport enhances overall plant growth and productivity, especially in iron-deficient soils. Iron is essential for chlorophyll production, photosynthesis, and general metabolic function.

Adaptation to iron-deficient conditions: Plants growing in iron-limited conditions have evolved sophisticated mechanisms, such as siderophore production and iron-siderophore complex uptake, to cope with iron scarcity and maintain essential biological functions.

Regulation of iron uptake and siderophore synthesis—Iron deficiency response: The plant regulates its production along with transport of iron-siderophore complexes through response mechanisms that determine iron deficiency strength. When plants detect insufficient iron levels they activate two strategies for iron acquisition: increasing siderophore biosynthesis while simultaneously boosting iron transporter efficiency. Example: The transcription factors bHLH and bZIP control iron uptake gene expression when iron levels fall low including the genes that manufacture siderophores and facilitate their transport.

The transport system depends on proton-coupled symporters together with ATP-dependent transporters to achieve active intake of the iron-siderophore complex. External iron-siderophore complexes get processed so that the system converts them into Fe^{2+} before distributing it through the plant vascular network for cell process requirements. The effective system developed by plants permits survival in low-iron environments thus leading to enhanced growth together with productivity improvement and ecological resistance.

5.4.9 Molecular and Genetic Insights into Plant Iron Uptake Systems

Plant iron acquisition depends on complex molecular and genetic systems that function specifically when plants face reduced iron levels. Plant organisms activate genes that reduce iron compounds and facilitate iron transport and siderophore production as a response to low amounts of iron in their environment. The ferric reductase FRO2 operates in Arabidopsis to convert ferric iron (Fe^{3+}) into the more bioavailable ferrous iron (Fe^{2+}). Under iron-limiting conditions transporters like IRT1 seriously uptakes materials because the ion reduction occurs under the control of FIT and bHLH family members that activate transcription. Through their regulatory functions these transcription factors manage the expression of genes important for iron absorption and mineral balance control [32, 33].

Plants use siderophores combined with direct iron transport as their second iron acquisition mechanism. Plants require the YSL family and ABC transporters to acquire iron-siderophore complexes through cellular import and intracellular movement respectively. Non-ribosomal peptide synthetases (NRPS) enzymes orchestrate siderophore formation while the bHLH and MYB transcription factors regulate the expression of these enzymes under iron deficient conditions. The biological molecules mugineic acid and other siderophores enhance iron solubility while improving uptake efficiency of the elements [34, 35]. Studying the complex molecular pathways dominates plant growth in low-nutrient conditions while enabling improved iron uptake strategies to counteract environmental stress [36].

Plants utilize different transport proteins to enable cellular and tissue movement of iron through their structure. These transporters are typically membrane proteins that enable the transfer of iron-siderophore complexes into the plant system.

Iron-regulated transporter (IRT) family: IRT family of transporters functions as fundamental iron uptake agents for rhizospheric sources. IRT1 serves as the most studied member of this family because it functions as a high-affinity transporter within root epidermal cells and draws Fe^{2+} into these cells based on iron availability. It is upregulated under iron-deficient conditions, enabling plants to acquire iron more efficiently. The expression of IRT1 is controlled by transcription factors like FIT, which are activated during iron deficiency.

YSL (yellow stripe-like) transporters: The YSL transporter family performs essential functions to shuttle iron-siderophore complexes inside plant cells. YSL transporters facilitate the uptake of Fe^{3+} bound to siderophores such as mugineic acid. YSL1 and YSL3 in *Arabidopsis* are involved in the transport of iron-siderophore complexes into root cells, particularly under iron-limiting conditions. These transporters are also involved in the movement of iron from roots to shoots in some plants, contributing to systemic iron distribution.

ABC (ATP-binding cassette) transporters: The ABC transporter family includes proteins that are involved in the active transport of various substrates, including metal ions such as iron. In plants, certain ABC transporters play a role in the movement of iron and iron-siderophore complexes.

Example: AtABCI1 is a putative ABC transporter that may participate in the transport of iron across cellular membranes in Arabidopsis [34, 35].

5.4.10 Iron Deficiency Signaling and Regulation in Plants

Plants have evolved sophisticated signaling networks to detect and respond to iron deficiency. The primary response to low iron availability involves the activation of specific genes involved in iron acquisition and transport.

Ferric iron reduction: The first step in iron uptake in many plants is the reduction of ferric iron (Fe^{3+}) to ferrous iron (Fe^{2+}), which is more readily absorbed by plant roots. This process is regulated by ferric reductases (e.g., FRO family), which are expressed in response to iron deficiency. FRO2 (Ferric Reductase Oxidase 2) in *Arabidopsis* plays a critical role in reducing Fe^{3+} to Fe^{2+} on the root surface [36]. Some transcription factors help in iron uptake regulation. The response to iron deficiency is tightly controlled by a range of transcription factors that regulate the expression of genes involved in iron uptake and siderophore biosynthesis.

bHLH (basic helix-loop-helix) transcription factors: In *Arabidopsis*, bHLH transcription factors, such as FIT (FER-LIKE IRON DEFICIENCY-INDUCED TRANSCRIPTION FACTOR), activate the expression of genes involved in iron uptake, such as those encoding FRO2 and IRT1 (Iron-Regulated Transporter 1). FIT is crucial for regulating the expression of iron uptake genes in response to iron stress and activating the expression of genes for root-based strategies like siderophore production.

MYB and bZIP transcription factors: The MYB and bZIP transcription factor families are also implicated in regulating iron homeostasis. These factors coordinate responses to both iron deficiency and oxidative stress in plants. Example: Under conditions of low iron levels AtbZIP19 and AtbZIP23 function as regulators of genes important for iron uptake [32, 33].

5.5 Regulation of Siderophore Production

The mechanisms that control the production of siderophores through genetics comprise a vital element of plant iron absorption strategies. In response to iron deficiency, plants upregulate the production of siderophores, which then chelate iron and facilitate its uptake into root systems.

Enzymes involved in siderophore biosynthesis: Non-ribosomal peptide synthetases (NRPS) functions within plants to produce the siderophores known as mugineic acid. The synthesis of these siderophores is regulated by transcription factors responsive to iron status.

Mugineic acid synthesis: The gramineous plant group including rice utilizes mugineic acid as its primary production type of siderophore. The biosynthesis of this siderophore is controlled by several enzymes, including MAM (mugineic acid synthesis) enzymes.

Regulation by transcription factors: The bHLH and MYB transcription factors control the gene expression of paths that lead to siderophore synthesis. At low iron concentration, these control factors activate genes responsible for siderophore manufacturing so that the body can produce more iron-chelating compounds [34, 35].

5.5.1 Root Hair and Rhizosphere Interaction for Iron Acquisition

Root hair development: Plants require root hairs to develop properly because they enlarge the area through which iron absorption can occur. A collection of genes controls root hair formation through iron-deficiency-responsive mechanisms.

IRON-DEFICIENCY-INDUCED ROOT HAIR (IDH) genes: Plants using effective intercropping also activate the IDH group of genes to develop longer root hairs and improve their ability to extract iron from soil solution. Root exudates have dual effects—they stimulate microbial siderophore-producing microorganisms which use their produced agents to enhance iron solubility while fostering mutually beneficial soil microbe–plant relationships.

Rhizobacterial interaction: Some plant-associated bacteria including *Pseudomonas* and *Bacillus* produce siderophores which dissolve iron in the soil while helping plants take up essential nutrients [36]. These beneficial microbes obtain nutrients simultaneously from plants in exchange for maintaining conducive rhizosphere conditions [36].

5.5.2 Iron Mobilization across Tissues

Iron mobilization within plants involves a coordinated system of transporters to ensure the distribution of iron from roots to various tissues, including leaves, stems, and seeds, where it supports essential functions like chlorophyll synthesis. The vascular system, particularly the xylem, plays a central role in this process. Key transporters, such as IRT1 and YSL proteins, facilitate long-distance iron transport, ensuring adequate supply to developing tissues. The development of genetic engineering has produced new methods that enable increased iron acquisition in plants. Plants with microbial siderophore biosynthesis genes or enhanced expression of iron-related transporters IRT1 FRO2 and YSL exhibit improved iron uptake and enriched stress tolerance in soils deficient in iron. Genetically modified crops developed from this research will survive nutrient-poor environments while improving both sustainability of agriculture and plant health outcomes. The convergence of these research findings with biotechnological innovations will produce enhanced agricultural productivity while solving worldwide farming dilemmas.

Genetic engineering for enhanced iron uptake: Plants injected with siderophore biosynthesis genes and transporters including IRT1, FRO2, and YSLs show improved iron acquisition capabilities and better iron deficiency adaptation. Plant iron uptake systems have become better understood through molecular and genetic research which showed the essential functions of specific transporters and enzymes and transcription factors. Studies in plant genetics have revealed methods to create new crop varieties which excel at surviving iron-deficient soil conditions and improve both farm sustainability and plant well-being. By integrating this knowledge with biotechnological approaches research holds potential to optimize plant performance and crop production in iron-deficient soils [36].

5.6 Applications in Agriculture

5.6.1 Enhancing Crop Yield and Soil Fertility through Siderophore-Mediated Iron Acquisition

As a fundamental micronutrient in plant development, both chlorophyll synthesis and respiration depend on iron. Due to its restricted bioavailability together with challenges from alkaline earth conditions and waterlogging of soils, plant productivity often suffers. Plants along with fungi and microbes produce siderophores which function as low-molecular-weight compounds that increase iron accessibility, leading to improved plant nutrition outcomes. Through siderophore chelation, iron in the soil becomes plant-root-absorbable forms [42]. Grasses deploy siderophores including mugineic acid during iron-deficient situations to extract iron from the soil. Crop yield enhancement is attained by improving iron uptake, as siderophores boost chlorophyll synthesis, enhancing photosynthesis and plant growth. They also support root development, enabling better absorption of macronutrients like nitrogen and phosphorus, leading to higher yields.

Microbial inoculants, such as siderophore-producing *Pseudomonas* or *Bacillus* strains, further enhance plant performance in nutrient-deficient soils. Figure 5.4 shows the major agricultural applications.

Soil fertility improvement: Siderophores support beneficial soil microbes that cycle nutrients, decompose organic matter, and enrich the rhizosphere. They also exhibit biocontrol properties by limiting pathogen growth through iron sequestration and can locally adjust the soil pH to enhance nutrient availability.

Sustainable agriculture: Siderophore-based biofertilizers offer an eco-friendly alternative to synthetic fertilizers, reducing environmental damage and input costs. These technologies also support biofortification, improving the nutritional content of crops in regions with widespread iron deficiency [40].

5.6.2 Control and Disease Resistance Mediated by Siderophores

Siderophores as antagonistic agents against plant pathogens: Plant pathogens which mainly originate from soil depend on iron for survival and

Figure 5.4 Major agricultural applications.

disease multiplication. The iron chelating activity of siderophores functions as natural competition against pathogens by restricting iron accessibility and helping decrease microbial population and thwart infections. The importance of plant interactions with beneficial microbes for supporting plant health involves the microbial ability to control pathogen growth according to Berg [37].

5.6.3 Iron Sequestration and Limitation

The iron requirement for growth and multiplication exists in three disease-causing pathogens of *Fusarium*, *Pythium*, and *Rhizoctonia*. Siderophore-producing microbial species including *Pseudomonas* and *Bacillus* find available iron by outcompeting pathogenic organisms when these microbes are placed in the rhizosphere. The ability of beneficial microbes to actively claim available iron resources over pathogens results in both diminished pathogen entry opportunities and reduced pathogen disease potential. *Pseudomonas fluorescens* functions as a well-studied plant-growth-promoting rhizobacterium (PGPR) capable of producing siderophores that suppress the growth of *Fusarium oxysporum* soil pathogen which causes wilt disease in numerous crops.

5.6.4 Direct Antimicrobial Activity

The direct antimicrobial properties combine with iron sequestration as possible activities of certain siderophore-producing microorganisms. Secondary metabolites produced by microbial siderophores exhibit both fungicidal as well as bactericidal effects which enhance pathogen combat capabilities.

5.6.5 Siderophore-Mediated Indirect Disease Resistance

Siderophore-producing microbes not only reduce pathogen populations through competition but also trigger indirect disease resistance mechanisms in plants.

5.6.6 Induced Systemic Resistance

The release of siderophores from particular bacteria triggers plant immunity through their innate defense systems in order to protect the plant against subsequent pathogens. The protective plant defense system becomes active to protect against further pathogen infections within a wide spectrum of possibilities.

Berg [37] conducted research that proved microorganisms possess induced systemic resistance-inducing capabilities, leading to improved disease protection of crops.

5.6.7 Quorum Sensing and Biofilm Formation

Microorganisms that produce siderophores have the ability to structure biofilms which serve as protective microbial communities. The rhizosphere biofilms enable microbial persistence along with removing barriers for harmful plant pathogens. Biofilms produced by ***Pseudomonas aeruginosa*** on plant roots increase bacterial growth and deny essential nutrients for pathogens while providing microbial defense mechanisms.

5.6.8 Practical Applications of Siderophore-Mediated Biocontrol

Siderophore-producing microorganisms represent a nature-friendly solution for managing plant diseases as biocontrol agents. Boari *et al.* [38]and Berg [37] recognized "microbigation" alongside controlled delivery mechanisms as important systems to deploy biological control agents for agricultural management.

5.6.9 Inoculants and Biofertilizers

Growing agricultural practices are now adopting inoculants that consist of bacteria-producing siderophores. Soil and seed applications of microbial inoculants function as disease preventions which both promote plant growth and enhance soil health conditions. The focus areas for sustainable agriculture include minimal chemical inputs according to Boari *et al.* [38].

5.6.10 Challenges and Future Directions

Research into microbial egress coupled with biocontrol efficiency measurements combined with delivery studies employing microbigation systems will improve siderophore usage in agriculture. The future of these biotechnological approaches depends on extensive research addressing their cost structure and scalability potential.

5.7 Future Perspectives and Challenges

While the potential of siderophore-mediated biocontrol is evident, several challenges remain.

Field effectiveness: Laboratory research demonstrates effective outcomes yet the availability of siderophore-based biocontrol in outdoor conditions remains inconsistent because environmental elements affecting soil variables control its implementation effectiveness.

Regulatory approvals and commercialization: Developing biocontrol products from siderophore-producing microbes necessitates expensive regulatory evaluations which take an extensive period of time.

Sustainability and cost: Scalability issues alongside production expenses serve as key barriers to the general implementation of siderophore-based biocontrol agents throughout agriculture. Scientific research must focus on production enhancement to develop cost-effective scale-up solutions for agricultural industry needs. Through competition for iron and activation of plant defense mechanisms and inhibition of plant pathogens, siderophores execute essential biocontrol functions. Biocontrols from natural sources serve as environment-friendly alternatives to chemicals for pest management, which build sustainable agricultural systems with high resilience.

While challenges remain in their widespread application, the future of siderophore-based biocontrol is promising, particularly with advancements in microbial biotechnology and better understanding of plant-microbe interactions. While siderophores offer promising solutions for enhancing plant growth, improving iron bioavailability, and supporting sustainable agriculture, several challenges and limitations currently hinder their widespread application. These limitations (Figure 5.5) encompass technical, ecological, and economic factors, which are critical to address for optimizing the practical use of siderophores in agriculture.

This chapter discussed the pivotal role of siderophores in iron chelation and transport mechanisms within plants, with a strong emphasis on their applications in agriculture. The key findings are as follows:

- Iron's importance in plant physiology: Iron is a crucial micronutrient in plants, essential for processes such as chlorophyll synthesis, respiration, and enzyme functions. However, its availability in soil is often limited, leading to iron deficiency in plants.

Figure 5.5 Key barriers to the general implementation of siderophore-based biocontrol agents throughout agriculture.

- Siderophores as key players in iron acquisition: Siderophores are small, high-affinity iron-chelating compounds produced by microbes and plants. They play a critical role in solubilizing and transporting iron from the soil to plant roots, thereby alleviating iron deficiency.
- Types and mechanisms of siderophore-mediated iron chelation: Different types of siderophores (e.g., hydroxamates, catecholates, and carboxylates) exist, each with unique structures and mechanisms of iron binding. These siderophores form complexes with ferric iron (Fe^{3+}), which are then taken up by plant root systems through specialized transporters.
- Microbial siderophores and their agricultural applications: Microbial siderophores, produced by bacteria and fungi, have significant potential in agricultural applications. These siderophores can not only enhance iron uptake in plants

but also promote soil fertility and act as biocontrol agents against plant pathogens.

- Genetic and molecular insights into iron uptake: Scientists have improved their knowledge about gene processes controlling siderophore production and plant absorption pathways through modern research techniques. Through direct applications of genetic modification technologies to microbes and plants, researchers have found opportunities to optimize iron delivery while simultaneously enabling better plant growth under iron-deficient conditions.
- Agricultural benefits of siderophores: When trees receive siderophores, they gain improved disease protective capabilities while agricultural productivity improves and soil richness extends across larger areas. Siderophores used in farming improve sustainability by reducing chemical inputs, which triggers better global access to nourishing food and security in food supplies.
- Challenges and future directions: Siderophore deployment faces production scalability restrictions together with affordability challenges and environmental factor complications. Future research must create perfect siderophore production systems and application methodologies while assessing how synthetic biology methods could maximize siderophore effectiveness.
- Sustainable agricultural solutions: The potential of siderophore-based systems appears promising for sustainable agricultural applications. Through these sustainable agricultural solutions, biofortification technology cooperates with biocontrol methods to minimize dependency on chemical fertilizers. Plant biology adopted through microbial research approaches needs to establish itself as a fundamental practice for increasing agricultural yields while preserving environmental stability.
- Innovations and opportunities for the future: Future advancements in siderophore research require blending synthetic biology methods and omics technology with gene editing techniques to enhance their application potential. By refining microbial inoculants and enhancing precision agriculture methods with plant microbiome regulatory techniques, siderophores will acquire iron more efficiently to achieve key crop production benefits.

5.8 Conclusions

The chapter reveals the substantial influence of siderophores in agricultural productivity. Iron is a major micronutrient that plants cannot survive without, but it is often at inadequate levels mostly in alkaline and calcareous soils. Siderophores are known to help solubilize other essential nutrients for crops, iron being perhaps the most significant source of concern. Thus, a more intelligent use of siderophores facilitates the solubilization and mobilization of iron, a very crucial nutrient, so it becomes more plant available. Improved bioavailability of iron through siderophores has a positive beneficial effect on plant growth and productivity, whereby the plants are protected against iron deficiency disorders including chlorosis. Siderophores play a role in enhancing iron acquisition and thus boosting the efficiency of photosynthesis, chlorophyll production, and different metabolic pathways. Thus, one finds themselves having healthier plants, with an increased rate of growth and increased resistance to environmental stresses, and consequently higher yields. Moreover, research demonstrates that plants inoculated with siderophore-producing microbes, compared to growth in iron-deficient soils, exhibit better biomass accumulation and overall productivity. Siderophores allow not only for the possible intake of iron but also for the solubility and mobility of other essential nutrients from the soil to provide improvement in soil overall fertility. Their ability to enhance the availability of not just iron but also other nutrients like phosphorus and nitrogen promotes a balanced nutrient supply, which is crucial for the optimal growth of plants Such biological fertilization would reduce the need for chemical fertilizers and also pave the way for sustainable agriculture, minimizing environmental pollution. Siderophores inhibit pathogenic microbes from securing an iron-rich soil and plant surfaces, hence lowering the risks of withering and disease, thus making the plant less vulnerable to chemical pesticides. This attribute abets the plant's susceptibility against illness and seriousness at the same time reduces dependency on chemical pesticides. Therefore, siderophores' biocontrol is proven as effective in combating root rot, blight, and other soil-borne diseases, giving plants more excellent health and healthier productivity. It also somewhat helps tolerance to the environmental stresses on plants, including drought, saline nature of the soil, and ambient temperature, by enhancing iron uptake and antioxidant defense systems. Meanwhile, what siderophores do next is help cope with inadequate environmental conditions—subcritical at sites so vis-a-vis recurring trends of climatic changes heightening the frequency and intensity of stressors. The capability to enrich crops' iron content could go a long way to realize the conceptual use of siderophores for biofortification of crop in the combat of human-micronutrient deficiency. Thus, it really does well contribute to combating iron deficiency anemia in HA, a disorder inherited by millions of people globally. Biofortification microbial inoculants put forward the environmentally friendly alternative to conventional fertilization practices from a sustainable perspective. The application of siderophores in agriculture correlates with the trend of switching from chemical fertilizers and pesticides to a lesser degree. We show, by harnessing context-relevant aspects from functioning microbiome outputs (e.g., siderophores), that agriculture can be made more robust, cheaper, and ecologically superior. It helps the soil in the long run and brushes off all those ill effects that synthetic inputs inflict on ecosystems. The use of siderophores is an exciting novel avenue that can increase agricultural productivity by facilitating the availability of nutrients, amplifying crop yield, providing tolerance to disease and promoting sustainable methods in agriculture. With their capacity to reduce iron deficiency, improve soil health, and promote the resilience of crops to abiotic stresses, they are key players in contemporary agriculture—tools with the potential to solve global food security and sustain farming for generations to come.

AI disclosure statement: The author acknowledges the use of AI-based tools for minor language editing or grammatical correction. However, all intellectual contributions, analyses, and conclusions presented in this chapter are entirely the author's own.

Bibliography

1. Adriano, D.C., *Trace elements in terrestrial environments: Biogeochemistry, bioavailability, and risks of metals*, Springer, 2001.
2. Buachoon, N., Heavy metal contamination in soil and plant samples by inductively coupled plasma-optical emission spectrometry. *Int. J. Chem. Environ. Biol. Sci.*, *2*, 197–200, 2014.
3. Zhao, K.L., Liu, X.M., Zhang, W.W., Xu, J.M., Wang, F., Spatial dependence and bioavailability of metal fractions in paddy fields on metal concentrations in rice grain at a regional scale. *J. Soils Sediments*, *11*, 1165–1177, 2011.

4. Benavides, M.P., Gallego, S.M., Tomaro, M.L., Cadmium toxicity in plants. *Braz. J. Plant Physiol.*, *17*, 21–34, 2005.
5. Radotic, K., Ducic, T., Mutavdric, D., Changes in peroxidase activity and isozymes in spruce needles after exposure to different concentrations of cadmium. *Environ. Exp. Bot.*, *44*, 105–113, 2000.
6. Bhushan, L., Ladha, J.K., Gupta, R.K., Singh, S., Tirol-Padre, A., Saharawat, Y.S., Gathala, M., Pathak, H., Saving of water and labour in a rice–wheat system with no tillage and direct seeding technologies. *Agron. J.*, *99*, 1288–1296, 2007.
7. Bingham, F.T., Page, A.L., McHier, R.J., Ganje, T.J., Cadmium availability to rice in sludge-amended soil under flood and nonflood culture. *Soil Sci. Soc. Am. J.*, *40*, 715–718, 1976.
8. Ramachandran, V. and Dsouza, T.J., Adsorption of cadmium by Indian soils. *Water Air Soil Pollut.*, *111*, 225–234, 1999.
9. Winkelmann, G., Microbial siderophore-mediated transport. *Biochem. Soc. Trans.*, *30*, 691–696, 2002.
10. Neilands, J.B., Siderophores: Structure and function of microbial iron transport compounds. *J. Biol. Chem.*, *270*, 8, 26723–26726, 1995.
11. Andersen, E. and Küpper, H., Cadmium toxicity in plants. *Met. Ions Life Sci.*, *11*, 395–413, 2013.
12. Sriprachote, A., Kanyawongha, P., Ochiai, K., Matoh, T., Current situation of cadmium-polluted paddy soil, rice, and soybean in the Mae Sot district, Tak province, Thailand. *Soil Sci. Plant Nutr.*, *58*, 349–359, 2012.
13. Diggle, S.P. and Whiteley, M., Microbe profile: *Pseudomonas aeruginosa*: Opportunistic pathogen and lab rat. *Microbiology*, *166*, 30–33, 2020, https://doi.org/10.1099/mic.0.000860.
14. Rahme, L.G., Stevens, E.J., Wolfort, S.F., Shao, J., Tompkins, R.G., Ausubel, F.M., Common virulence factors for bacterial pathogenicity in plants and animals. *Science*, *268*, 1899–1902, 1995, https://doi.org/10.1126/science.7604262.
15. Yahr, T.L. and Wolfgang, M.C., Transcriptional regulation of the *Pseudomonas aeruginosa* type III secretion system. *Mol. Microbiol.*, *62*, 631–640, 2006, https://doi.org/10.1111/j.1365-2958.2006.05412.x.
16. Bleves, S., Viarre, V., Salacha, R., Michel, G.P., Filloux, A., Voulhoux, R., Protein secretion systems in *Pseudomonas aeruginosa*: A wealth of pathogenic weapons. *Int. J. Med. Microbiol.*, *300*, 534–543, 2010, https://doi.org/10.1016/j.ijmm.2010.08.005.
17. Liu, P.V. and Shokrani, F., Biological activities of pyochelins: Iron-chelating agents of *Pseudomonas aeruginosa*. *Infect. Immun.*, *22*, 3, 878–890, 1978, https://doi.org/10.1128/iai.22.3.878-890.

18. Cornelis, P., Matthijs, S., Van Oeffelen, L., Iron uptake regulation in *Pseudomonas aeruginosa*. *Biometals*, *22*, 15–22, 2009, https://doi.org/10.1007/s10534-008-9193-0.
19. Cornelis, P. and Dingemans, J., *Pseudomonas aeruginosa* adapts its iron uptake strategies in function of the type of infections. *Front. Cell. Infect. Microbiol.*, 3, 75, 2013.
20. Nadal, J.P., Koch, G., Papaioannou, E., Wahjudi, M., Krzeslak, J., Coenye, T., Cool, R.H., Quax, W.J., Role of PvdQ in *Pseudomonas aeruginosa* virulence under iron-limiting conditions. *Microbiology*, *156*, 49–59, 2010, https://doi.org/10.1099/mic.0.030973-0.
21. Koebnik, R., TonB-dependent trans-envelope signaling: The exception or the rule? *Trends Microbiol.*, 13, 343–347, 2005, https://doi.org/10.1016/j.tim.2005.06.005.
22. Perraud, Q., Kuhn, L., Fritsch, S., Graulier, G., Gasser, V., Normant, V., Hammann, P., Schalk, I.J., Opportunistic use of catecholamine neurotransmitters as siderophores to access iron by *Pseudomonas aeruginosa*. *Environ. Microbiol.*, 24, 878–893, 2022, https://doi.org/10.1111/1462-2920.15372.
23. Schalk, I.J. and Guillon, L., Pyoverdine biosynthesis and secretion in *Pseudomonas aeruginosa*: Implications for metal homeostasis. *Environ. Microbiol.*, 15, 1661–1673, 2013, https://doi.org/10.1111/1462-2920.12013.
24. Zhao, Q. and Poole, K., A second tonB gene in *Pseudomonas aeruginosa* is linked to the exbB and exbD genes. *FEMS Microbiol. Lett.*, 184, 127–132, 2000, https://doi.org/10.1111/j.1574-6968.2000.tb09002.x.
25. Hoegy, F., Celia, H., Mislin, G.L., Vincent, M., Gallay, J., Schalk, I.J., Binding of iron-free siderophore, a common feature of siderophore outer membrane transporters of *Escherichia coli* and *Pseudomonas aeruginosa*. *J. Biol. Chem.*, 280, 20222–20230, 2005, https://doi.org/10.1074/jbc.M500776200.
26. Chimento, D.P., Kadner, R.J., Wiener, M.C., Comparative structural analysis of TonB-dependent outer membrane transporters: implications for the transport cycle. *Proteins*, 59, 240–251, 2005, https://doi.org/10.1002/prot.20416.
27. Shirley, M. and Lamont, I.L., Role of TonB1 in pyoverdine-mediated signaling in *Pseudomonas aeruginosa*. *J. Bacteriol.*, 191, 5634–5640, 2009, https://doi.org/10.1128/JB.00742-09.
28. Schalk, I.J., Mislin, G.L., Brillet, K., Structure, function and binding selectivity and stereoselectivity of siderophore-iron outer membrane transporters. *Curr. Top. Membr.*, 69, 37–66, 2012, https://doi.org/10.1016/B978-0-12-394390-3.00002-1.
29. Moynié, L., Luscher, A., Rolo, D., Pletzer, D., Tortajada, A., Weingart, H., Braun, Y., Page, M.G., Naismith, J.H., Köhler, T., Structure and function of the PiuA and PirA siderophore-drug receptors from *Pseudomonas*

aeruginosa and *Acinetobacter baumannii. Antimicrob. Agents Chemother.*, 61, e02531-16, 2017, https://doi.org/10.1128/AAC.02531-16.
30. Moynie, L., Milenkovic, S., Mislin, G.L.A., Gasser, V., Malloci, G., Baco, E., McCaughan, R.P., Page, M.G.P., Schalk, I.J., Ceccarelli, M., Naismith, J.H., The complex of ferric-enterobactin with its transporter from *Pseudomonas aeruginosa* suggests a two-site model. *Nat. Commun.*, 10, 3673, 2019, https://doi.org/10.1038/s41467-019-11508-y.
31. Perraud, Q., Cantero, P., Roche, B., Gasser, V., Normant, V.P., Kuhn, L., Hammann, P., Mislin, G.L.A., Ehret-Sabatier, L., Schalk, I.J., Phenotypic adaptation of *Pseudomonas aeruginosa* by hacking siderophores produced by other microorganisms. *Mol. Cell. Proteomics*, 19, 589–607, 2020, https://doi.org/10.1074/mcp.RA119.001829.
32. Vishnupradeep, R., Bruno, L.B., Taj, Z., *et al.*, Plant Growth Promoting Bacteria Improve Growth and Phytostabilization Potential of Zea Mays under Chromium and Drought Stress by Altering Photosynthetic and Antioxidant Responses. *Environ. Technol. Innov.*, 25, 102154, 2022.
33. Dimkpa, C.O., Merten, D., Svatoš, A., *et al.*, Siderophores Mediate Reduced and Increased Uptake of Cadmium by *Streptomyces Tendae* F4 and Sunflower (*Helianthus Annuus*), Respectively. *J. Appl. Microbiol.*, 107, 1687–1696, 2009.
34. Carrillo-Castañeda, G., Muñoz, J.J., Peralta-Videa, J.R., *et al.*, Modulation of Uptake and Translocation of Iron and Copper from Root to Shoot in Common Bean by Siderophore-Producing Microorganisms. *J. Plant Nutr.*, 28, 1853–1865, 2005.
35. Dimkpa, C., Svatoš, A., Merten, D., *et al.*, Hydroxamate Siderophores Produced by *Streptomyces Acidiscabies* E13 Bind Nickel and Promote Growth in Cowpea (*Vigna unguiculata* L.) under Nickel Stress. *Can. J. Microbiol.*, 54, 163–172, 2008.
36. Burd, G.I., Dixon, D.G., Glick, B.R., Plant Growth-Promoting Bacteria That Decrease Heavy Metal Toxicity in Plants. *Can. J. Microbiol.*, 46, 237–245, 963401, 2000.
37. Berg, G., Plant-microbe interactions promoting plant growth and health: Perspectives for controlled use of microorganisms in agriculture. *Appl. Microbiol. Biotechnol.*, *84*, 1, 11–18, 2009, https://doi.org/10.1007/s00253-009-2092-7.
38. Boari, A., Zuccari, D., Vurro, M., 'Microbigation': Delivery of biological control agents through drip irrigation systems. *Irrig. Sci.*, *26*, 1, 101–107, 2008, https://doi.org/10.1371/journal.pone.0081082.
39. Lozano-González, J.M., *et al.*, Evaluation of Siderophores Generated by Pseudomonas Bacteria and Their Possible Application as Fe Biofertilizers. *Plants*, 12, 23, 2023.
40. Luscher, A., Gasser, V., Bumann, D., Mislin, G.L., Schalk, I.J., Köhler, T., Plant-Derived Catechols Are Substrates of TonB-Dependent. *mBio*, 13, 4, 1–14, 2022, Available at: https://doi.org/10.3390/plants12234054.

41. Singh, A., *et al.*, Siderophore mediated attenuation of cadmium toxicity by paddy field cyanobacterium Anabaena oryzae. *Algal Res.*, 16, 63–68, 2016, Available at: https://doi.org/10.1016/j.algal.2016.02.030.
42. Singh, D.P., Singh, H.B., Prabha, R., *Microbial Inoculants in Sustainable Agricultural Productivity: Vol. 2: Functional Applications*, pp. 1–308, 2016, Available at: https://doi.org/10.1007/978-81-322-2644-4.
43. Winkelmann, G. and Drechsel, H., Microbial siderophores, in: *Biotechnology, Second, Completely Revised Edition, Volume 7: Products of Secondary Metabolism*, 2008, Available at: https://doi.org/10.1002/9783527620890.ch5.

6

Influence of Siderophores on Plant Stress Tolerance

Rahul Kumar[1], Debasis Mitra[1*], Yousef Rasmi[2], Ayush Madan[3,4], Rachan Karmakar[5] and Snežana Adjelković[6]

[1]Department of Microbiology, Graphic Era (Deemed to be University), Clement Town Dehradun, Uttarakhand, India
[2]Department of Biochemistry, Urmia University of Medical Sciences (UMSU), Urmia, Iran
[3]Department of Biotechnology, School of Research & Technology (SORT), People's University, Bhanpur, Bhopal, Madhya Pradesh, India
[4]Institute of Tropical Aquaculture and Fisheries (AKUATROP), Universiti Malaysia Terengganu, Kuala Nerus, Terengganu, Malaysia
[5]Department of Environmental Science, Graphic Era (Deemed to be University), Clement Town Dehradun, Uttarakhand, India
[6]Institute for Forage Crops Kruševac, Globoder, Kruševac, Serbia

Abstract

Siderophores are a group of iron-chelating molecules categorized into two different sorts of producers: microorganisms and plants. They play an important role in the acquisition as well as regulation of iron. The influence of siderophores has spread so widely in strengthening plant stress tolerance under various adverse conditions. The roles of siderophores include alleviating both abiotic and biotic stresses imposed by iron deficiency, heavy metal toxicity, drought, salinity, and pathogens. Siderophores improve plant growth and the tolerance of plants to adverse environments by promoting iron uptake and modulating plant–microbe interactions. This chapter describes the mechanisms through which siderophores induce stress tolerance, including solubilization of iron, suppression of pathogens, and induction of systemic resistance. It further shows the prospect of using siderophore-producing microbes in sustainable agriculture, for instance, in biofertilizers and biocontrol agents. Based on experimental evidence and case studies, this chapter

**Corresponding author*: madanayushmadan@gmail.com; debasismitra3@gmail.com
Rahul Kumar: ORCID: https://orcid.org/0009-0000-1323-2220

Debasis Mitra, Ayush Madan and Anju Rani (eds.) Microbial Siderophores for Agricultural Productivity: Perspectives of Use in Agriculture, (121–158) © 2026 Scrivener Publishing LLC

provides insight into how siderophores can be useful in enhancing crop productivity as well as soil health in a way toward sustainable farming practices.

Keywords: Siderophores, plant stress tolerance, abiotic, biotic stress, crop productivity

6.1 Introduction

Throughout their life cycle, plants face a few stresses that can prevent them from growing, developing, and producing [29]. Such stressors may be categorized as either biotic, which usually means stress initiated by living things, or abiotic, which describes the impact of environmental variables that are not living [69]. Because of their ability to chelate iron and promote higher uptake of plant nutrients, siderophores are low molecular weight compounds excreted by microorganisms play an essential role in promoting tolerance of both stresses in plants [100]. Biotic stress factors arise from interactions between plants and living things such as weeds, pests, and pathogens, which can directly or indirectly influence the physiological processes of plants [24]. The pathogens that cause diseases in plants include rusts, blights, and wilts, among others, caused by bacteria, fungus, viruses, and nematodes [73]. Herbivorous pests such as insects, mites, and other pests cause harm to plants through consumption of their tissues or spread of diseases. Weeds indirectly increase plant stress by competing with crops for vital resources such as light, water, and nutrients [46]. Effective control of weeds is important in reducing their negative effects on agricultural output. Abiotic stress factors include environmental factors such as temperature extremes, water availability, salinity, light intensity, and nutritional imbalances that negatively affect plant growth and development [56]. Because they are extremely sensitive to temperature changes, plants produce heat shock and antifreeze proteins to help them cope with these stresses. Light, salt, and water stress significantly impact plant growth [7]. In a nutshell, biotic and abiotic stressors are just a few of the many problems that plants face in their life cycle and may influence their development, growth, and yield. Siderophores that are secreted by microorganisms are low-molecular-weight chemicals and are important in increasing plant resistance to various stresses.

6.1.1 Role of Siderophores in Enhancing Stress Tolerance

Siderophores are highly specialized molecules produced by many microorganisms, such as fungi and bacteria, that chelate iron from the environment and make it bioavailable to plants [109]. Iron is essential for the growth, development, and survival of plants under stressful conditions. Thus, siderophores can prevent and curtail infections by outcompeting plant pathogens for iron. They increase the systemic resistance of plants by triggering defense responses. Siderophores increase the solubility of iron in nutrient-poor soils and, in that way, facilitate more efficient uptake by plant roots [53]. Increasing photosynthetic efficiency and promoting chlorophyll synthesis would be essential elements under abiotic stress situations, such as salinity or drought. Binding of heavy metals by siderophores will lower their toxicity and bioavailability to plants [25]. This is particularly useful in contaminated soils as heavy metals can cause oxidative stress and interference in plant metabolism, and microbes that produce siderophores improve antioxidant capacity, thus reducing oxidative stress and increasing resistance [42]. They indirectly help in drought tolerance by promoting root development and improving water uptake. Moreover, microbial communities associated with siderophore production improve the soil's structure, which helps in ensuring available and retained water [82]. The capability of siderophores to reduce plant stress has important ramifications on the area of sustainable farming.

The production of siderophores by microorganisms can reduce dependence on chemical pesticides and fertilizers by promoting advantageous plant-microbe interactions [13]. Addition of these microorganisms to biopesticides and biofertilizers can increase crop output, bringing environmental consequences down to zero. Genetic engineering techniques that enhance microbial colonization or introduce siderophore synthesis in plants can lead to long-term solutions for stress tolerance [38]. The research on the synergistic effects of siderophores with other features that promote plant growth may also be useful for developing robust crop varieties. All of these reactions depend on an essential component that is iron within plant physiology-nitrogen metabolism, hormone biosynthesis, heavy metal toxicity, nutrient deficiencies, and secondary production of metabolisms, respiration, and even photosynthesis [62]. Iron is the most important component in plant life, which plays an important role in photosynthesis and cellular respiration and in producing essential proteins, such as cytochromes and Fe–S (iron–sulfur) clusters. It also plays a role in chlorophyll synthesis so that leaves grow healthy and green and absorb the light energy appropriately [74]. Iron plays a crucial role in cellular respiration because it is part of the electron transport chain in mitochondria. Nitrate reductase and nitrite reductase, being iron-containing enzymes, reduce nitrate and nitrite into ammonia, which the plants use in synthesizing organic nitrogen molecules. In the production of amino acids and proteins, it is essential that nitrogen metabolism occur [101].

Iron further participates in the manufacturing of plant hormones, such as ethylene, which controls the process of senescence, stress reactions, and fruit ripening. The process of growth and development also hinges on the manufacture of gibberellin [89]. Iron aids in the acclimation of plants and reduction of effects of biotic and abiotic stressors; hence, it is an essential component of plant stress responses. It also regulates metal uptake and oxidative stress by increased antioxidant activity, reducing toxicity in soils contaminated with metals [27]. In agriculture, microorganisms such as bacteria and fungi produce siderophores that chelate iron from the environment and make its uptake easier at low levels of iron. Siderophores have a very high affinity for ferric iron (Fe^{3+}) and form soluble complexes that are absorbed by transport systems of either plants or microorganisms [4]. The mechanism has huge impacts on the resilience, development, and health of plants and is central to plant–microbe interactions. Applications of siderophores in agriculture include sustainable and eco-friendly means of boosting agricultural yields [106]. By incorporating them into biopesticides and biofertilizers, they can lessen reliance on artificial chemicals, thus promoting more robust and healthy agricultural systems. The technological advancements in biotechnology have the potential to produce plants with enhanced siderophore use capacities or capture bacteria that produce siderophores, thereby opening new avenues in sustainable agriculture (Figure 6.1).

6.1.2 Siderophores and Iron Homeostasis in Plants

For plants, iron is an essential micronutrient for many physiological processes, such as DNA synthesis, respiration, photosynthesis, and nitrogen metabolism [17]. However, because iron is poorly soluble in alkaline and aerobic soils, plants are often unable to access it. To circumvent this limitation, plants and the microbes that coexist with them have evolved ways to mobilize and acquire iron, maintaining iron homeostasis without

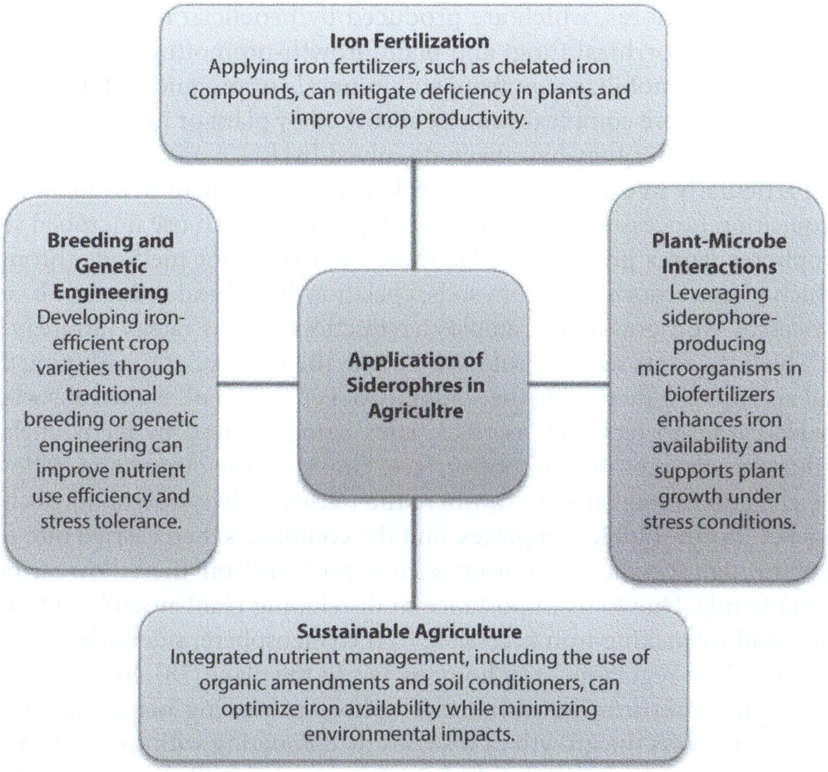

Figure 6.1 Application of siderophore in agriculture.

being toxic [48]. Microorganisms release certain iron-chelating molecules called siderophores that play an important role in enhancing iron availability and balancing it within plants. For binding and solubilizing ferric iron from the surroundings, bacteria, fungus, and certain cyanobacteria secrete low-molecular-weight compounds called siderophores [81]. Because of their extraordinary affinity for Fe^{2+}, these compounds can form stable complexes which render iron accessible for uptake by plants or microbes. This is the strategy of microbial survival through adaptation, and it usually starts siderophore synthesis in the presence of iron deficiency [9]. The four major categories of siderophores are hydroxamates, catecholates, carboxylates, and mixed kinds. These structural differences have a significant impact on the interactions of siderophores with plant systems and their iron-binding capabilities in the rhizosphere, where both microorganisms and plants coexist. Siderophores play a crucial role in enhancing plants' uptake of iron in the rhizosphere

[100]. Siderophores, which are produced by beneficial microorganisms including mycorrhizal fungi and plant growth-promoting rhizobacteria (PGPR), help mobilize iron from soil minerals or organic materials. The iron-siderophore complexes can be absorbed by plant or microbial transport systems after they have been dissolved [81].

Particularly in calcareous or iron-deficient soils where traditional iron acquisition processes are less successful, plants that can use microbial siderophores gain a great deal [91]. There are two major methods through which plants absorb iron from soils: chelation-based and reduction-based processes. Non-grass plants employ a reduction-based strategy (Strategy I), in which the roots secrete protons (H^+) into the rhizosphere, acidifying the soil and thereby increasing the solubility of Fe^{3+} [12]. Fe^{3+} is then absorbed by iron transporters such as IRT1, after being reduced by ferric-chelate reductase enzymes on the root surface. Grass, on the other hand, follows the chelation-based approach, where the released phytosiderophores chelate Fe^{3+}, form soluble complexes, and the complex is then carried into the root through specific transporters such as proteins from the Yellow Stripe1 (YS1) family. Due to its importance in developing plant-microbe interactions and by making iron accessible in the rhizosphere, siderophores play a great role in regulating iron homeostasis in the plant [78]. In addition to improving plant iron nutrition, siderophore-producing beneficial bacteria also suppress the growth of diseases by competing with them for iron. Siderophores are crucial for plant resilience and health because of their dual role in nutrient uptake and disease control. Siderophores have significant agricultural implications due to their role in iron homeostasis. In calcareous soils, shortage of iron leads to chlorosis, low crop production, and a decrease in nutrient quality [64]. A sustainable method for the enhancement of plant growth and increased availability of iron, while lowering the dependence on artificial fertilizers and iron chelates and promoting ecologically friendly agricultural practices, is using siderophore-producing microorganisms as biofertilizers. Moreover, crop output in siderophore-replete areas can be completely reoriented by breeding or genetically modifying plants to take better advantage of microbial siderophores.

6.1.3 Siderophores and Plant Stress Tolerance

There are many biotic and abiotic environmental stressors occurring continuously in plants, which can have a great influence on the development, growth, and productivity of plants [105]. Iron is an essential nutrient for plant health because it has direct implications on photosynthesis, respiration, enzyme activity, and cellular redox processes. However, due to low solubility in alkaline and aerobic soils, its bioavailability is often limited. Plants have complex processes to absorb and control iron, including their interactions with siderophores, which increase their ability to withstand stress [122]. In addition to increasing iron availability, these specific iron-chelating molecules, which are produced by rhizosphere microorganisms, greatly increase plant resistance to a variety of stressors. When there is an iron shortage, bacteria, fungus, and certain cyanobacteria release siderophores, which are low-molecular-weight substances [6]. They are highly affine to ferric iron (Fe^{3+}) and, through soluble complexes, will help in the uptake of iron by plants and microbes. Through increased availability of iron in the rhizosphere, a microbial survival strategy known as siderophore production is beneficial to plants, especially those operating in calcareous or nutrient-depleted soils. Siderophore-producing microbes and plants frequently interact in a mutualistic manner, with the microbes obtaining access to carbon-rich root exudates and the plants getting improved nutrient absorption, greater iron bioavailability, and increased stress tolerance [45].

Under stressful circumstances, when plant nutrient uptake and homeostasis are disturbed, this relationship is especially important. The capacity of siderophores to mitigate iron shortage is one of their most important contributions to plant stress tolerance [80]. Overall, common signs of iron deficiency in plants include interveinal chlorosis, stunted growth, and decreased photosynthetic efficiency, which ultimately lead to poorer agricultural yields. Soils and organic matter can contain iron that will be dissolved by siderophore-producing bacteria in the rhizosphere and can create Fe^{3+}-siderophore complexes that are absorbed by roots [16]. Thus, through these complexes, specific plants can utilize either *via* special transporters or through ferric reductases that convert Fe^{3+} to Fe^{2+}. Even in cases of nutrient limitation, this enhanced iron acquisition helps the plants to perform their critical metabolic activities. Apart from combating abiotic stress, siderophores are crucial in minimizing abiotic stressors, such as heavy metal toxicity, salt, and drought [55]. Drought stress enhances the susceptibility of iron deficiency and the impairment of physiological functions of plants by reducing moisture in the soil and solubility of nutrients. By chelating and making iron mobile, siderophores counterbalance such effects and ensure that the plant can use it even in dry soils.

While siderophores are significant for their capacity to protect against abiotic stressors, they also play a key role in defending plants from biotic stressors like pests and pathogens [77]. In the rhizosphere, beneficial microorganisms decrease the pathogen's potential to infect plants by withholding the iron available for its use by the secretion of siderophores. This mechanism, known as "nutritional immunity," allows a plant to develop increased resistance to diseases caused by bacteria, fungi, and other pathogenic microorganisms. The potential of siderophores to be helpful in improving food security and ensuring environmental sustainability is underscored through their dual role in nutrient acquisition and stress mitigation [14]. Advances in biotechnology and microbial ecology have unlocked new ways to apply siderophores in agriculture. The identification and characterization of siderophore producing microbial strains that work more efficiently under specific environmental conditions have made targeted applications in areas of high stress possible. Genetic engineering techniques have also been pursued for enhancing the use of microbial siderophores by plants or for enhancing the synthesis of siderophores in microorganisms [109].

6.2 Types of Siderophores Involved in Plant Stress Tolerance

The siderophores are classified by their chemical structure and iron-chelating capabilities and classify the plant stress tolerance by promoting plant-microbe interactions, reducing the toxicity of heavy metals, and increasing iron availability [109]. It includes the following types: hydroxamates, carboxylates, catecholates, and mixed type siderophores. They are fundamental to soil remediation techniques and sustainable farming methods because of their structural diversity, which enables them to operate well under a range of environmental circumstances [58]. Further research into the applications of siderophores could lead to innovative agricultural solutions. They are essential to soil remediation techniques and sustainable farming methods due to their structural variety (Table 6.1).

6.3 Mechanisms of Action of Siderophores in Stress Tolerance in Plants

Many problems, which include biotic stressors such as pathogen invasion, abiotic stressors like salt, drought, and heavy metals toxicity and nutrient

Table 6.1 Types of siderophore involved in plant stress tolerance.

Types	Producers	Chemical structure	Action of mechanisms	Role in plant stress tolerance
Phytosiderophores	Graminaceous plants (e.g., rice, wheat, maize)	Amino acid derivatives	Secreted by roots; from stable complexes with Fe3+	Essential for iron acquisition in grass species; enhance nutrient uptake under calcareous soil conditions; contribute to drought and salinity tolerance
Mixed types	Bacteria (*Pseudomonas fluorescens*)	Combination of groups	Exhibit diverse iron-binding properties; flexible in binding other metals	Versatile function in iron mobilization under varying soil conditions; mitigate multiple stresses, including salinity and drought; exhibit antimicrobial activity against soil-borne pathogens
Carboxylates	Bacteria (*Rhizobium, Sinorhizobium*)	Carboxylic acid groups	Weak iron chelation; work synergistically with other siderophores	Improve iron solubility in slightly acidic soils; reduce heavy metal toxicity by binding toxic ions; support nitrogen-fixing bacterial in symbiosis with legumes

(*Continued*)

Table 6.1 Types of siderophore involved in plant stress tolerance. (*Continued*)

Types	Producers	Chemical structure	Action of mechanisms	Role in plant stress tolerance
Catecholates	Bacteria (*Escherichia coli, Bradyrhizobium*)	Catechol groups	Bind iron through hydroxyl groups; participate in iron mobilization in competitive environments	Promote plant growth under iron-limited conditions aid in bioremediation by chelating heavy metals like cadmium and lead enhance stress resistance by modulating iron uptake and antioxidant defense
Hydroxamates	Bacteria (*Pseudomonas, Bacillus*), fungi (*Aspergillus*)	Contain hydroxamate groups	High-affinity binding to Fe3+; efficient solubilization of iron; transport to microbial of plants cells	High-affinity binding to Fe3+; efficient solubilization of iron transport to microbial or plant cells

deficit may encounter by plants in their natural habitats [72]. Plants frequently depend on their interactions with helpful bacteria in the rhizosphere for living and flourishing. Bacteria release specific iron-chelating molecules, known as siderophores, to scavenge iron from the environment and make it bioavailable. Their function can be described to improve plant stress tolerance, even though they mainly collect iron. Some of the modes of action include improvement of nutritional availability, reduction of oxidative stress, chelating of harmful metals, and the inhibition of infections [84]. Due to the all-round role that siderophores play, they are of extreme importance in contemporary farming techniques aimed at producing higher crop yields in adverse situations.

6.3.1 Iron Acquisition and Nutrient Mobilization

Iron is essential to plant metabolism mainly because it plays a very critical role in photosynthesis, respiration, nitrogen fixation, and enzyme catalysis [44]. Despite being high in most soils, iron is often present in forms that are insoluble and thus not available to plants, especially in aerobic or calcareous soils. Microorganisms such as *Bacillus species* and *Pseudomonas* produce siderophores with high affinity for ferric iron (Fe^{3+}), forming soluble siderophore-iron complexes [6]. The roots absorb these compounds either directly or through related microorganisms subsequently. In situations where nutrients are scarce, plants can use these complexes because they have receptors and reduction mechanisms. By improving iron uptake, metabolic processes are improved and iron deficiency signs like chlorosis and stunted growth are avoided. Apart from iron, siderophores have the capability of mobilizing other micronutrients necessary for plant development, such as copper, manganese, and zinc [35]. Siderophores enhance plant development and tolerance through the availability of these nutrients when available in nutrient-poor soils. Enhanced uptake of nitrogen renders plants more resistant to stresses which are known to adversely affect nutrient uptake and utilization, such as drought and salinity.

6.3.2 Oxidative Stress Mitigation

In plants, abiotic stressors such as drought, salt, and heavy metal pollution often lead to an overabundance of reactive oxygen species (ROS) [98]. Excessive ROS can impair plant development and survival because it can damage lipids, proteins, and DNA, among other cellular components. Siderophores regulate the availability of iron, which helps reduce oxidative stress. While being important, iron could cause oxidative damage through catalysing the Fenton reaction leading to the generation of ROS [111]. Through binding surplus iron and suppressing its catalytic activity in generating ROS, siderophores will have helped in ensuring iron homeostasis. Furthermore, increased iron availability by siderophores catalyses antioxidant enzyme activity such as peroxidase, catalase, and SOD. All these enzymes have been shown crucial in maintaining redox balance in a cell and eliminating harmful ROS. Siderophores help plants manage oxidative stress under harsh environmental conditions by regulating the availability of iron and strengthening antioxidant defense [22].

6.3.3 Detoxification of Heavy Metals

Heavy metal toxicity is one of the major abiotic stressors found in many agricultural soils, particularly in mining and industrial areas [5]. These metals cause impairment in plant metabolism by interfering with nutrient intake, producing ROS, and harming cellular structures such as cadmium (Cd), lead (Pb), and arsenic (As). Their high-affinity binding capabilities make siderophores crucial for the reduction of the toxicity of these heavy metals [118]. These molecules chelate toxic metal ions, thereby reducing their mobility and bioavailability in the soil. For instance, rhizosphere-associated bacteria produce siderophores that selectively bind heavy metals like lead and cadmium, making them immobile and thus not available for uptake by plant roots. This chelation stabilizes pollutants in the soil, thereby aiding in phytoremediation besides protecting plants from heavy metal toxicity. These siderophore-producing bacteria also stimulate the expression of plant genes involved in detoxification and heavy metal transport to further enhance stress tolerance [122].

6.3.4 Suppression of Plant Pathogens

The availability of iron in the rhizosphere frequently determines the outcome of plant-pathogen interactions since iron is a critical nutrient both for plants and pathogens [78]. The siderophores of beneficial microorganisms limit the growth of pathogens by creating an iron-limiting environment; this phenomenon is known as "nutritional immunity." Siderophores lower the virulence and potential of viruses to infect plants because they deprive them of the access to iron, which is a vital nutrient. Apart from competing for iron, some siderophores possess direct antibacterial properties [66]. For example, it has been established that the siderophore pyoverdine, which is produced by *Pseudomonas fluorescens*, represses the growth of fungi like *Phytophthora* and *Fusarium*. Moreover, by eliciting defense signaling pathways and enhancing the production of antimicrobial compounds, siderophore-producing microbes can provoke systemic resistance in the plant [94]. Their dual role in pathogen suppression and activation of plant defense mechanisms points to the significance of siderophores in biocontrol techniques.

6.3.5 Support for Drought and Salinity Tolerance

Among the numerous abiotic stresses limiting plant growth, salinity and drought stress are considered the most severe [65]. These factors inhibit nutrient uptake, disrupt ionic balance, and reduce water availability. In these conditions, siderophores enhance the tolerance of plants to these stresses by increasing the availability of nutrients and maintaining cell homeostasis. In drought-stressed soils, siderophore-producing microorganisms also solubilize iron and other nutrients to ensure that the plant continues to access them despite the reduced moisture in the soil [75]. This nutrient supply supports important physiological functions such as photosynthesis and water-use efficiency. Excess sodium ions cause osmotic stress and interfere with nutrition intake in saltwater environments. By enhancing iron absorption and promoting the activity of antioxidant enzymes that lessen oxidative damage brought on by salt, siderophores lessen these consequences [71]. Furthermore, certain microorganisms that produce siderophores help plants to produce more osmoprotectants, such as proline and glycine betaine, that help maintain cellular osmotic balance.

6.3.6 Modulation of Plant Hormone Levels

Microbes that produce siderophores often influence the levels of plant hormones, which are crucial for stress responses [100]. For instance, microbes produce auxins and gibberellins, which enhance root architecture and growth as well as water and nutrient uptake in stressful conditions. Similarly, siderophore-producing bacteria can regulate the level of ethylene in plants by producing the enzyme 1-aminocyclopropane-1-carboxylate (ACC) deaminase, which reduces the synthesis of ethylene [68]. Since ethylene is often associated with stress responses, plants may avoid premature senescence and grow further under adverse conditions by regulating it.

6.3.7 Contribution to Soil Health and Microbial Community Structure

Siderophores indirectly contribute to plant stress tolerance by enhancing a diverse and beneficial microbial community in rhizospheres as well as the improvement of soil health [113]. Siderophores favor nutrient-poor soil fertility by increasing the solubility of iron and other nutrients in the soil. The colonization of favorable microbial communities is also enhanced by siderophore-producing microorganisms, and these populations together enhance the toughness and growth of plants [51]. This expanded microbial diversity produces a strong rhizosphere ecosystem that can inhibit pathogens, increase nutrient cycling, and strengthen plant–microbe interactions.

6.3.8 Siderophores in Sustainable Agriculture

The emphasis of sustainable agriculture focuses on such methods to reduce chemical inputs into the soil while enhancing the fertility of the soil and minimizing ecological disturbance to harmonize environmental health with productivity [119]. These have led to the use of siderophores, one of the more promising biological techniques under investigation. Microbes and plants produce siderophores, which are low-molecular-weight, iron-chelating substances, to obtain iron under specific circumstances. In addition to iron mobilization, siderophores have been used for improvement of nutrient availability and promotion of plant development with minimal biotic and abiotic stressors along with enhanced soil health [80]. The given traits thus satisfy the agenda for sustainable agriculture because they suggest some environmentally benign approaches to build resilience in the soil and consequently produce crops with the minimization of reliance on synthetic pesticides and chemical fertilizers.

6.3.9 Enhancing Nutrient Availability

Iron, which is one of the most essential minerals for plants, is not often available in soils despite its abundance. Iron occurs mainly in the insoluble ferric (Fe^{3+}) form in aerobic or calcareous soils and is not readily available to plants [87]. Siderophores are produced by fungus, bacteria, and some plants that relate to the rhizosphere to solubilize ferric iron for uptake by plants. For instance, in Fe-deficient soil, phytosiderophores produced by graminaceous plants like rice and wheat are key to iron uptake. Similarly, bacterial siderophores, produced by *Bacillus* and *Pseudomonas* species, enhance plant growth and productivity as they increase Fe availability in the rhizosphere [80]. Besides iron, siderophores can complex other essential micronutrients, such as copper, manganese, and zinc, and thus enhance their bioavailability. This broad-spectrum nutrient mobilization is particularly beneficial in nutrient-depleted or degraded soils, where crop yields are often limited by shortages of various micronutrients. Siderophores contribute to sustainable farming methods that minimize resource depletion and environmental degradation by facilitating nutrient uptake and reducing the demand for synthetic fertilizers [20].

6.3.10 Promoting Plant Growth

By assisting in several physiological and metabolic processes, siderophores generally play a significantly positive role in promoting plant growth. Mobilized iron is necessary for respiration, photosynthesis, nitrogen fixation, and the synthesis of important cofactors and enzymes [52]. Therefore, enhanced availability of iron boosts the production of chlorophyll, which improves photosynthetic efficiency as well as plant health. In addition to acquiring nutrients, microorganisms that produce siderophores frequently have characteristics that promote plant growth, like the synthesis of phytohormones (such as auxins and gibberellins) and the regulation of ethylene levels [123]. For instance, bacteria that produce 1-aminocyclopropane-1-carboxylate (ACC) deaminase and siderophores aid in lowering ethylene-induced stress in plants, promoting improved root and shoot development. Better root development will help in more nutrient uptake and soil exploration, hence improving plant growth in difficult environments [61].

6.3.11 Mitigating Abiotic Stress

In contemporary agriculture, abiotic stressors like salt, drought, and heavy metal contamination pose significant problems. Through a few methods, siderophores help plants withstand harsh circumstances [37]. Microbial siderophores enhance nutrient absorption and iron solubility in drought-stressed soils, guaranteeing the availability of vital resources for plant survival. Additionally, they increase the activity of antioxidant enzymes, reducing oxidative damage brought on by a lack of water. Siderophores increase the acquisition of iron and stimulate the production of osmoprotectants, such as proline and glycine betaine, under saline conditions that cause an excessive accumulation of sodium ions, upsetting ionic equilibrium and nutritional absorption [110]. These chemicals protect plants against salt-induced damage and help in maintaining the osmotic balance of cells. The detoxification process from heavy metals also requires siderophores. Siderophores reduce the soil's bioavailability of toxic metals such as cadmium, lead, and arsenic by sequestering it, which effectively prevents plant roots from absorbing such metals. Chelation serves to protect the plants from toxic effects of heavy metals besides making soil rehabilitation through this chelation a sustainable agriculture practice [19]. The importance of siderophores in sustainable agriculture is underscored by their ability to address a range of abiotic stresses, particularly under conditions of deteriorating soil condition and climate changes.

6.3.12 Suppressing Biotic Stress

A grave threat to agricultural output, plant diseases encompass nematodes, fungus, and bacteria. Siderophores help create an iron-limited environment within the rhizosphere, thus assisting in the suppression of these pathogens; the process is referred to as nutritional immunity [103]. Siderophores inhibit growth and virulence by depriving them of the availability of iron, which is an essential nutrient for pathogens. Some of these siderophores have direct antimicrobial activity against soil-borne diseases, including pyoverdine, which is produced by *Pseudomonas fluorescens* [21]. Siderophores protect plants from diseases like root rot by preventing the growth of the dangerous fungus such as *Fusarium* and *Phytophthora*. Moreover, producing siderophores could lead to a systemic resistance in plants by causing activation of defense signaling pathways and elevated synthesis of antimicrobial chemicals through microorganisms [67]. Siderophores are useful tools in integrated pest management under sustainable agriculture systems because they play a double role in suppressing pathogens and activating plant immunity.

6.3.13 Supporting Soil Health and Microbial Diversity

It has been acknowledged that siderophores increase nutrient cycling and stimulate microbial diversity, both of which are essential to soil health [85]. The microorganisms that are capable of synthesizing siderophores increase mineral solubilization, hence making it easier for the soil to recycle vital nutrients. In addition, their presence incites beneficial bacteria species to colonize, creating a rhizosphere environment that is both balanced and diversified [97]. In improving nutrient availability and soil strength and by controlling the number of pathogens, diverse microbes develop the plants with resistance and grow. A vast majority of such microbes which form siderophores frequently create mutualistic or symbiotic relations with plants like rhizobia to legumes [41]. These also benefit the sustainability of soil as well as soil fertility besides aiding the uptake of more nutrients.

6.3.14 Reducing Reliance on Chemical Inputs

Conventional agriculture's overuse of chemical pesticides and fertilizers has led to severe environmental issues, including biodiversity loss, water pollution, and soil degradation [39]. A biological alternative that can reduce reliance on these synthetic inputs is siderophores. Siderophores decrease the need for chemical fertilizers by enhancing the availability of nutrients and promoting natural plant growth. Their role in biocontrol and disease suppression reduces the dependency on chemical pesticides, thus reducing their harmful impacts on the environment [108]. In addition, siderophores can be incorporated into formulations for biopesticides and biofertilizers, providing farmers with environmentally friendly alternatives for sustainable crop management. These bio-based products promote long-term soil and ecosystem health by adhering to the principles of regenerative agriculture and organic farming [40].

6.3.15 Applications and Future Perspectives

Despite the enormous potential of siderophores for sustainable agriculture, much research and development must be done before they can be applied extensively [109]. Methods for increasing the effectiveness of siderophore-producing bacteria include genetic engineering, strain selection, and the creation of microbial consortia suited to certain crops and environmental circumstances. The targeted and effective delivery of siderophores can also be facilitated by developments in biotechnology and nanotechnology [93]. Moreover, for siderophores to be more useful in various agricultural systems, it is essential to understand how they interact with plants and soil microbiomes. Therefore, researchers, legislators, and farmers can work together to incorporate siderophores into sustainable agriculture methods so that their benefits can be felt globally [88].

6.4 Case Studies and Experimental Evidence

The use of siderophores in sustainable agriculture has received much attention lately due to their potential to enhance nutrient availability, alleviate stress, and reduce dependency on chemical fertilizers and pesticides [31]. It is supported by the increasing amount of experimental data and case studies which indicate the beneficial impact of siderophores in enhancing plant development, stress tolerance, and soil health. These case studies provide relevant information on how siderophores could be exploited to solve many real-world agricultural problems [2]. This chapter has included case studies demonstrating the wide application of siderophores in diverse agricultural practices and, thereby, highlighting their diverse, useful contributions toward sustainable agriculture.

6.4.1 Siderophore-Producing Microbes in Rice Cultivation

Iron deficiency in soils frequently limits the productivity of rice, a staple crop farmed all over the world, particularly in areas with calcareous or alkaline soils [15]. *Pseudomonas fluorescens*, which produces siderophores, was used to treat rice fields with iron chlorosis in a study by Kumar *et al.*, [54]. The results demonstrated that applying this bacterium improved rice growth and yield in addition to reducing the symptoms of iron shortage. The siderophores produced by *P. fluorescens* dissolved iron from the soil, thereby enhancing its availability for uptake by the plant. This bacterium promoted plant growth through the formation of hormones that regulate growth, which include cytokinin's and auxins [76]. This study supports the idea of applying siderophore-producing microbes as part of the sustainable agricultural technique since these can enhance nutrient supply, enhance crop growth, and increase crop production in crops grown at iron-limited conditions.

6.4.2 Role of Siderophores in Biocontrol of Soil-Borne Pathogens

Other than nutrient uptake, siderophores are important in contributing to plant protection due to their inhibitory effect on plant diseases. A notable case study by Abo-Zaid *et al.* [1] showed the biocontrol capability of *Pseudomonas* strains that produce siderophores against soil-borne diseases, such as *Rhizoctonia solani* and *Fusarium oxysporum*, which cause root rot in several crops [1]. Instead, the scientists discovered that the *Pseudomonas* strains that produced siderophores dramatically suppressed the growth of these pathogens by outcompeting them for accessible iron in the rhizosphere [96]. This was by induction of iron restriction by siderophores, which shielded the plants against illness by decreasing pathogen development and pathogenicity. This study showed that siderophores play a dual role in stimulating plant growth and as organic biocontrol agents, which reduces the use of chemical pesticides and encourages more environmentally friendly approaches to pest control.

6.4.3 Siderophores in Heavy Metal Detoxification

Another challenge that is often experienced in agricultural soils is heavy metal contamination, especially near mining or industrial sectors. In this study, Syed *et al.* [107] investigated the role of siderophores in mitigating harmful effects of heavy metals, specifically cadmium (Cd), to crop plants [107]. Wheat plants grown in contaminated soils with Cd were treated with siderophore-producing bacteria isolated from contaminated soils by the researchers. According to the results, the bacteria chelated Cd effectively, thereby reducing its availability and uptake by the plants. The siderophore-producing bacteria also enhanced the growth of stressed wheat plants, possibly by minimizing the oxidative damage caused by the heavy metal [60]. This case study showed how siderophores can be extremely important in detoxifying heavy metals in soil, offering an environmentally benign way to remediate soil, and shielding crops from stress caused by metals.

6.4.4 Siderophore-Mediated Drought Tolerance in Crops

One of the most important abiotic factors affecting agricultural productivity worldwide is drought. Azeem *et al.* [8] explored the possibility of using siderophore-producing bacteria to make wheat plants more drought-tolerant [8]. The researchers stated that the application of *Bacillus subtilis*, a well-known siderophore producer, increased the drought tolerance of wheat plants by improving root development and water uptake. *B. subtilis* developed siderophores that allowed the mobilization of vital nutrients, such as iron, to be easier in the rhizosphere. The bacteria stimulated the synthesis of proline and other osmoprotectants, thus helping the plants maintain their osmotic balance. This research demonstrated how siderophores can enhance drought tolerance, hence becoming a very useful tool in solving agricultural water scarcity and enhancing sustainable crop production [28].

6.4.5 Siderophores in Salt Stress Mitigation

Another major abiotic stressor inhibiting crop production, particularly under arid to semi-arid conditions, is salinity. Gaya Karunasinghe *et al.* [34] employed *Pseudomonas* strains to produce siderophores to reduce the salt stress effects on tomato plantlets [34]. Utilization of this bacterium by the plants increases their tolerance enormously. The siderophores of bacteria chelated iron in the rhizosphere, thus enabling tomato plants to absorb more amounts even in the salinized environment. These increased amounts of iron availability had helped in minimizing oxidative damage induced by salt-stress by facilitating the activity of antioxidant enzymes, such as catalase and superoxide dismutase (SOD) [121]. The bacteria also produced plant growth hormones that stimulate the development of roots and shoots in attempts to help the plants alleviate the osmotic stress from salinity. The case study unveiled how siderophore-producing microorganisms increase salt tolerance in crops and create a long-term solution for agriculture in areas prone to salinity.

6.4.6 Siderophores in Improving Soil Fertility and Microbial Health

A critical component of sustainable agriculture is maintaining soil health, which includes enhancement in nutrient cycling and beneficial soil microbes. A study conducted by Chin-A- Gao *et al.* [33] reported that the increase in microbial diversity and fertility of soil resulted from the addition of *Pseudomonas* strains that were siderophore-producing to agricultural soils [33]. This included an increase in the amount of solubilized phosphorus and other nutrients involved in plant growth through the application of these bacteria. The siderophores from these bacteria enhanced the development of useful microbes responsible for soil fertility maintenance, mycorrhizal fungi and nitrogen-fixing bacteria [63]. Moreover, it showed that the introduction of siderophore-producing bacteria has lowered dangerous pathogens; thus, the soil's health has been enhanced. The above case study described how siderophores played an important role in a healthy microbiological environment of soil-a major element in conservation agriculture.

6.5 Various Types of Challenges and Limitations of Siderophores in Agricultural Applications

Although siderophores have the potential to significantly enhance plant growth, reduce stress, and improve soil health in sustainable agriculture, there are a few obstacles and restrictions that prevent their widespread use [102]. The problems include those with the stability and manufacturing of siderophores and the intricacies of their interactions in various soil types. Even though siderophores possess several benefits, resolving these drawbacks is critical to the optimal application in practical agricultural systems [38]. The principal challenges and constraints related to siderophore utilization in agriculture are discussed in the sections below.

6.5.1 Production Costs and Efficiency

The cost and efficiency of siderophore synthesis are major barriers to their commercial utilization. The main microorganisms, which are bacteria and fungus, require certain conditions to produce them optimally. The ranges in which the variables are found are temperature, pH, oxygen level, and availability of substrates [43]. Some bacteria produce enormous quantities of siderophores, but the production often requires fermentation in contained environments which can be very expensive and time-consuming. Besides microbial fermentation, purification and extraction of siderophores for application use in agricultural products are expensive and time-consuming [112]. Most of the industrial fermentation methods are still not optimized for large-scale, cost-effective siderophore synthesis, and scalability of siderophore production remains a significant limitation. The stability of siderophores during field application, storage, and transit poses another challenge. The activity of siderophores degrades over time due to their susceptibility to environmental conditions like light, humidity, and temperature [95]. For these reasons, manufacturing and marketing of siderophore-based products proves difficult at prices that would prove profitable for the farmer.

6.5.2 Soil Complexity and Environmental Variability

The effectiveness of siderophores within an agricultural application depends largely upon environmental factors and composition of soil and can be far more variable by place [120]. Though siderophores dissolve iron along with other nutrients, dissolution efficacy is not just a mere result of various variables, that is, it is not necessarily affected by several factors such as pH levels, the presence of organic matter, and others, competing for the same kind of microbes [116]. For instance, in alkaline soils, where iron is mainly insoluble, siderophores might have difficulties mobilizing enough iron for absorption into plants. Likewise, siderophores might compete with other nutrients for the available nutrient in soils rich in organic matter or other organic chelators, thereby reducing their efficiency. The rhizosphere, a dynamic ecosystem where the majority of siderophore-producing microorganisms interact with plants, is affected by plant species, root exudates, and microbial communities [11]. Due to the potential for these interactions to vary between crops and soil types, predicting the outcome of siderophore applications is challenging. Thus, the complexity of the soil environment might limit the performance of siderophore-based biotechnologies, requiring site-specific modification for every application. Other environmental stressors that include drought, salt, or heavy metal contamination also alter the behavior and availability of siderophores in the soil, which in turn makes their utilization much more challenging [36].

6.5.3 Microbial Competition and Ecological Imbalance

As the iron-sequestering siderophores bind this critical resource, which is often in limited availability in soil, competition by microbes is critically subject to this element [57]. This will potentially lead to some unanticipated ecological consequence, if siderophore-producing microorganisms are introduced into an existing soil microbiome. Non-native or altered soil microbes, including those that produce siderophores, can disrupt the balance of resident soil microbial communities [99]. This might result in some species overgrowing while suppressing others. Some plant-pathogenic microbes that also require iron to survive might thrive more readily when siderophore-producing bacteria are present. By depriving infections of iron, siderophores can starve their growth, but they can also inadvertently promote an environment that fosters the growth of other diseases that use different kinds of iron or that have become accustomed to competing with siderophores [32]. Further research is needed to fully understand this delicate balance of microbial interactions to assess the long-term ecological effects of using siderophores in agricultural settings. Some pathogens or bacteria could eventually become resistant to this iron sequestration mechanism mediated by siderophores, and this is also a concern [50]. As microorganisms and plants develop resistance to antibiotics, they may also develop ways to circumvent the iron-limiting activity of siderophores. These adaptive reactions may reduce the efficiency of siderophores as a tool for disease prevention and plant growth promotion.

6.5.4 Limited Understanding of Plant–Siderophore Interactions

Their role in iron acquisition is well established, but wider impacts of siderophores on plant physiology and stress reactions have been poorly elucidated [117]. Siderophores have already proven capable of enhancing the impacts of iron deficiency on plant growth and development, but their impact on other aspects of plant metabolism-including photosynthesis, nutrient circulation, and oxidative stress resistance-may be complex and dependent on several factors [115]. More research is, however required for understanding how these siderophores relate to other networks of plant signalling, such as hormone regulation or defensive responses. One of the main drawbacks associated with the efficacy of siderophores in crop varieties and in different types of soil is the unavailability of standard techniques for measurement [3]. Since these interactions between plant and siderophores are complex and context-dependent, some studies showed conflicting results whereas others showed some good results. Indeed, this effectiveness depends upon the type of environmental fluctuations affecting conditions, change in microbial communities that inhabit rhizosphere of a plant and shifts in plant genetic makeup across the species. Siderophores application is applied in the fields of agriculture by requiring higher order knowledge of siderophore actions as they operate upon the host plants and other living organisms as a medium [106].

6.5.5 Regulatory and Safety Concerns

Release of genetically modified or non-native microbes into the environment may be under strict regulations, even if it helps plants, for safety and regulatory issues in an agricultural system by using microbial inoculants [117]. The possible dangers of introducing those creatures into an agricultural ecosystem need to be addressed and decided by regulatory agencies. These risks include concerns over the impact of these organisms on biodiversity, spreading genetically engineered traits, and possible unintended effects on organisms not targeted. Moreover, in case the bacteria or their metabolites are pathogenic or result in the formation of toxic compounds, there are concerns about the safety of employing microbial formulations for food crops [26]. Although many of the siderophore-producing bacteria are considered non-pathogenic, their safety should be considered for various agricultural purposes in which they are used on a large scale. This requires checking whether these bacteria do not disseminate dangerous genes to other organisms or whether they carry virulence factors.

6.5.6 Scalability and Commercial Viability

Even though the laboratory and greenhouse studies were encouraging, the use of siderophores is still very challenging to scale up in agriculture [10]. A variety of factors such as the cost of production, consistency, and efficacy across different agricultural applications make commercializing siderophore-based products problematic. Field applications of siderophore-based formulations should have reliable and maintainable performance during time as well as against environmental conditions though some small-scale studies may demonstrate benefits in promoting growth of the crop and inhibition of diseases through use of siderophore-producing microorganisms [104]. It must be taken into consideration for making field usable siderophore products cost-effectiveness, compatibility with other agriculture inputs (such pesticides, fertilizers, etc.) along with convenient to use. A challenge in producing goods that are efficient enough but also reasonable for farmers' pockets and scalable to quite large units, especially in resource-poor regions, creates a hurdle in the use of siderophores in sustainable agriculture [70].

6.6 Future Directions and Research Opportunities in Siderophores for Agriculture

Microbial metabolites known as siderophores have shown significant potential in agriculture, particularly in enhancing plant growth, inducing stress tolerance, and controlling soil-borne diseases [18]. They are capable of chelating and transporting iron. Although much promise is recognized, there are still many untapped areas and further research is needed to overcome the challenges in applying them in agricultural systems. This section describes the main research avenues and future directions which could improve our knowledge and use of siderophores in sustainable agriculture.

6.6.1 Enhancing Siderophore Production for Commercial Viability

Improving the manufacturing efficiency of siderophores is one of the most significant areas of future research. Siderophores are expensive to produce industrially because controlled fermentation procedures are required and substrate prices are high [81]. Future studies should focus on creating more scalable and economical manufacturing techniques, including the application of biotechnological advancements like synthetic biology and metabolic engineering [79]. Metabolic engineering of microbes may enhance the yield, stability, and production cost-effectiveness of siderophores. The development of more sustainable industrial systems is another area of study. For example, siderophore production may be made less costly and less harmful to the environment by utilizing cheap carbon sources or agricultural wastes as substrates for microbial fermentation. Other potential avenues for improving the productivity and cost-effectiveness of these bio-based fertilizers and biocontrol agents will include research into new microbial strains or GMOs capable of producing siderophores at higher yields [59].

6.6.2 Understanding Siderophore-Mediated Plant–Microbe Interactions

Although there is significant evidence that siderophores play a role in iron acquisition, the complete spectrum of their functions in plant-microbe interactions remains an open question [23]. In addition to providing iron, siderophores can influence plant development and stress responses because of changes in hormone production, defense systems, and other metabolic changes. To apply siderophores successfully in agriculture, more studies on the signaling mechanisms underlying such effects are required [81]. Future studies should aim at understanding the molecular mechanisms of siderophore interaction with plant roots and plant physiology. It is crucial to look into how siderophores influence plant stress responses, including those caused by infections, salt, and drought. This kind of study could lead to new approaches for crop protection and growth promotion in relation to the ways plants perceive siderophores and then mount a defensive or growth-promoting reaction. Moreover, which plant species benefit the most from siderophore applications will help in tailoring siderophore use for specific crops and habitats.

6.6.3 Exploring the Role of Siderophores in Abiotic Stress Tolerance

Abiotic stressors, including salt, drought, heavy metal toxicity, and extreme temperatures, severely limit plant development and output globally [86]. While siderophores have been shown to alleviate some forms of abiotic stress, it remains unclear how they precisely mediate stress tolerance. Later, further investigations should focus on the mechanisms of how siderophores help in the management of various types of abiotic stresses in plants and whether their usage can increase the tolerance of many crops to the stress [49]. Some examples of that are research work needed to investigate how siderophores help support ion homeostasis, the antioxidant defense system, and the osmotic equilibrium in stressful environments. Examining how siderophores interact with other stress tolerance mechanisms, such as the plant's synthesis of stress-related hormones like abscisic acid, might help determine how best to use them in saline or drought-prone environments [47]. New farming methods that use siderophores as a means of boosting climate change resilience may be the outcome of research in this field.

6.6.4 Exploring Siderophore-Producing Microbes as Biocontrol Agents

Due to their ability to sequester iron, an essential resource which microbial pathogens must obtain to be able to progress, siderophores have shown a good deal of potential in biocontrol [114]. Future research into the use of siderophore-producing bacteria for IPM could seek to explore means by which their use can preclude soil-borne diseases while also reducing dependence on chemical pesticides. This is crucial because over-reliance on chemical pesticides has made several organisms resistant to them and badly affected the environment. The number of pathogens and the mechanisms of action involved with siderophore-mediated biocontrol should be further researched by scientists [117]. Research on how well siderophore-producing microorganisms work with other biocontrol agents, such as nitrogen-fixing bacteria or mycorrhizal fungi, may help in the development of more holistic and sustainable pest management strategies. Additionally, it should be explored whether siderophore-producing bacteria can suppress fungal diseases that are resistant to conventional chemical fungicides.

6.6.5 Microbial Ecology and Siderophore Dynamics in the Rhizosphere

Rhizosphere is described as the intricately complex soil-microbe-habitat which involves the contact between soil microorganisms and root plants. Additional research is thus needed to effectively learn the dynamics about siderophores' nature in the rhizosphere along with their reaction with the positive and negative forms of micro-organisms that include the soil biota [30]. Future studies should focus on understanding the ecological function of siderophores in the rhizosphere, their interaction with other PGPR, and effects on soil health. Research should concentrate on the effects of siderophore-producing bacteria on the organization of the rhizosphere's microbial community and if they impact plant-microbe symbioses and soil nutrient cycling [38]. The nature of competition between siderophore producers and other microbial communities, such as mycorrhizal fungi and N fixing bacteria, will also be known by investigation of how siderophores can be introduced into agroecosystems to improve the productivity of plants and the health of the soils. Enhanced efficiency of siderophore-based biocontrol and biofertilizer products could also arise from identifying species of plants or soil conditions that enable the growth of beneficial siderophore-producing microorganisms [90].

6.6.6 Field-Scale Trials and Commercialization of Siderophore-Based Products

Although several studies have demonstrated the efficacy of siderophores in controlled environments—for example, in laboratories and greenhouses—field-scale experiments are necessary to assess their performance under actual agricultural conditions [92]. Future studies should concentrate on extensive, multi-location trials to evaluate the effectiveness of siderophore-based products in different soil types, climates, and cropping systems. These studies will determine the best application techniques, doses, and timing for various crops and environmental circumstances. Research is also needed to develop and assess commercial formulations of biofertilizers, and biocontrol agents based on siderophores [83]. To bring these items to market, there needs to be an understanding of the practical issues surrounding product formulation, shelf-life, storage, and ease of use. To overcome these hurdles and ensure successful commercialization of siderophore-based technology, cooperation between researchers, agricultural businesses, and farmers will be vital.

6.6.7 Regulatory and Safety Assessment

Proper safety and legal considerations will be necessary, as siderophore-producing microorganisms grow in application for agriculture. Future research should focus on the potential risks that genetically modified or non-indigenous bacteria present to agricultural environments. Many of these studies should address long-term ecological implications, effects on non-target organisms, and horizontal gene transfer capability. Indeed, studies need to be conducted to ensure that siderophore-based products are safe for use in food crops, especially if they are sprayed directly to edible plants. The cooperation between regulatory bodies, researchers, and industry stakeholders will be needed to ensure the safety and effectiveness of siderophore-based products in agricultural systems.

AI disclosure statement: The author acknowledges the use of AI-based tools for minor language editing or grammatical correction. However, all intellectual contributions, analyses, and conclusions presented in this chapter are entirely the author's own.

References

1. Abo-Zaid, G.A., Abdullah, A.S., Soliman, N.A.M., El-Sharouny, E.E., Al-Askar, A.A., Su, Y., Sabry, S.A.F., Evaluation of Bio-Friendly Formulations from Siderophore-Producing Fluorescent Pseudomonas as Biocontrol Agents for the Management of Soil-Borne Fungi, Fusarium oxysporum and Rhizoctonia solani. *Agriculture*, 13, 7, 1418, 2023.
2. Adeniji, A., Fadiji, A.E., Li, S., Guo, R., From lab bench to farmers' fields: Co-creating microbial inoculants with farmers input. *Rhizosphere*, 100920, 2024.
3. Albelda-Berenguer, M., Monachon, M., Joseph, E., Siderophores: From natural roles to potential applications. *Adv. Appl. Microbiol.*, 106, 193–225, 2019.
4. Albrecht-Gary, A.M. and Crumbliss, A.L., Coordination chemistry of siderophores: thermodynamics and kinetics of iron chelation and release. *Met. Ions Biol. Syst.*, 35, 239–328, 1998.
5. Alengebawy, A., Abdelkhalek, S.T., Qureshi, S.R., Wang, M.Q., Heavy metals and pesticides toxicity in agricultural soil and plants: Ecological risks and human health implications. *Toxics*, 9, 3, 42, 2021.
6. Årstøl, E. and Hohmann-Marriott, M.F., Cyanobacterial siderophores—physiology, structure, biosynthesis, and applications. *Mar. Drugs*, 17, 5, 281, 2019.
7. Aslam, M., Fakher, B., Ashraf, M.A., Cheng, Y., Wang, B., Qin, Y., Plant low-temperature stress: Signaling and response. *Agronomy*, 12, 3, 702, 2022.
8. Azeem, M., Javed, S., Zahoor, A.F., Bacillus species as potential plant growth promoting rhizobacteria for drought stress resilience. *Russ. J. Plant Physiol.*, 70, 4, 59, 2023.
9. Aznar, A., Chen, N.W., Rigault, M., Riache, N., Joseph, D., Desmaële, D., Dellagi, A., Scavenging iron: a novel mechanism of plant immunity activation by microbial siderophores. *Plant Physiol.*, 164, 4, 2167–2183, 2014.
10. Backer, R., Rokem, J.S., Ilangumaran, G., Lamont, J., Praslickova, D., Ricci, E., Smith, D.L., Plant growth-promoting rhizobacteria: context, mechanisms

of action, and roadmap to commercialization of biostimulants for sustainable agriculture. *Front. Plant Sci.*, 9, 1473, 2018.
11. Barra Caracciolo, A. and Terenzi, V., Rhizosphere microbial communities and heavy metals. *Microorganisms*, 9, 7, 1462, 2021.
12. Berger, N., Marin, A.J., Stassen, M.J., Lourenço, T., Li, M., Watanabe, S., Marino, D., Molecular Regulation of Iron Homeostasis, in: *Progress in Botany*, vol. 85, p. 75, 2024.
13. Bhatt, K., Suyal, D.C., Kumar, S., Singh, K., Goswami, P., New insights into engineered plant-microbe interactions for pesticide removal. *Chemosphere*, 309, 136635, 2022.
14. Bhupenchandra, I., Chongtham, S.K., Devi, A.G., Dutta, P., Sahoo, M.R., Mohanty, S., Swapnil, P., Unlocking the potential of arbuscular mycorrhizal fungi: exploring role in plant growth promotion, nutrient uptake mechanisms, biotic stress alleviation, and sustaining agricultural production systems. *J. Plant Growth Regul.*, 1–39, 2024.
15. Bolan, N., Srivastava, P., Rao, C.S., Satyanaraya, P.V., Anderson, G.C., Bolan, S., Kirkham, M.B., Distribution, characteristics and management of calcareous soils. *Adv. Agron.*, 182, 81–130, 2023.
16. Brick, M.B., Hussein, M.H., Mowafy, A.M., Hamouda, R.A., Ayyad, A.M., Refaay, D.A., Significance of siderophore-producing cyanobacteria on enhancing iron uptake potentiality of maize plants grown under iron-deficiency. *Microb. Cell Fact.*, 24, 1, 3, 2025.
17. Cakmak, I., Brown, P., Colmenero-Flores, J.M., Husted, S., Kutman, B.Y., Nikolic, M., Zhao, F.J., Micronutrients, in: *Marschner's mineral nutrition of plants*, pp. 283–385, Academic Press, London, United Kingdom, 2023.
18. Chen, Q., Song, Y., An, Y., Lu, Y., Zhong, G., Mechanisms and Impact of Rhizosphere Microbial Metabolites on Crop Health, Traits, Functional Components: A Comprehensive Review. *Molecules*, 29, 24, 5922, 2024.
19. Chengatt, A.P., Sarath, N.G., Sebastian, D.P., Mohanan, N.S., Sindhu, E.S., George, S., Puthur, J.T., Chelate assisted phytoextraction for effective rehabilitation of heavy metal (loid) s contaminated lands. *Int. J. Phytorem.*, 25, 8, 981–996, 2023.
20. de Andrade, L.A., Santos, C.H.B., Frezarin, E.T., Sales, L.R., Rigobelo, E.C., Plant growth-promoting rhizobacteria for sustainable agricultural production. *Microorganisms*, 11, 4, 1088, 2023.
21. Deb, C.R. and Tatung, M., Siderophore producing bacteria as biocontrol agent against phytopathogens for a better environment: A review. *S. Afr. J. Bot.*, 165, 153–162, 2024.
22. Desoky, E.S.M., Saad, A.M., El-Saadony, M.T., Merwad, A.R.M., Rady, M.M., Plant growth-promoting rhizobacteria: Potential improvement in antioxidant defense system and suppression of oxidative stress for alleviating salinity stress in Triticum aestivum (L.) plants. *Biocatal. Agric. Biotechnol.*, 30, 101878, 2020.

23. Dhankhar, R., Gupta, S., Gulati, P., Insights on plant–microbe interactions in soil in relation to iron dynamics. *Vegetos*, 36, 3, 750–767, 2023.
24. Dixit, S., Sivalingam, P.N., Baskaran, R.M., Senthil-Kumar, M., Ghosh, P.K., Plant responses to concurrent abiotic and biotic stress: unravelling physiological and morphological mechanisms. *Plant Physiol. Rep.*, 29, 1, 6–17, 2024.
25. El-Meihy, R.M., Abou-Aly, H.E., Youssef, A.M., Tewfike, T.A., El-Alkshar, E.A., Efficiency of heavy metals-tolerant plant growth promoting bacteria for alleviating heavy metals toxicity on sorghum. *Environ. Exp. Bot.*, 162, 295–301, 2019.
26. Elnahal, A.S., El-Saadony, M.T., Saad, A.M., Desoky, E.S.M., El-Tahan, A.M., Rady, M.M., El-Tarabily, K.A., The use of microbial inoculants for biological control, plant growth promotion, and sustainable agriculture: A review. *Eur. J. Plant Pathol.*, 162, 4, 759–792, 2022.
27. Emamverdian, A., Ding, Y., Barker, J., Mokhberdoran, F., Ramakrishnan, M., Liu, G., Li, Y., Nitric oxide ameliorates plant metal toxicity by increasing antioxidant capacity and reducing Pb and Cd translocation. *Antioxidants*, 10, 12, 1981, 2021.
28. Fadiji, A.E., Santoyo, G., Yadav, A.N., Babalola, O.O., Efforts towards overcoming drought stress in crops: Revisiting the mechanisms employed by plant growth-promoting bacteria. *Front. Microbiol.*, 13, 962427, 2022.
29. Fahad, S., Bajwa, A.A., Nazir, U., Anjum, S.A., Farooq, A., Zohaib, A., Huang, J., Crop production under drought and heat stress: plant responses and management options. *Front. Plant Sci.*, 8, 1147, 2017.
30. Fazeli-Nasab, B., Piri, R., Rahmani, A.F., Assessment of the role of rhizosphere in soil and its relationship with microorganisms and element absorption, in: *Plant Protection: From Chemicals to Biologicals*, p. 225, 2022.
31. Ferreira, C.M., Soares, H.M., Soares, E.V., Promising bacterial genera for agricultural practices: An insight on plant growth-promoting properties and microbial safety aspects. *Sci. Total Environ.*, 682, 779–799, 2019.
32. Galdino, A.C.M., Vaillancourt, M., Celedonio, D., Huse, K., Doi, Y., Lee, J.S., Jorth, P., Siderophores promote cooperative interspecies and intraspecies cross-protection against antibiotics in vitro. *Nat. Microbiol.*, 9, 3, 631–646, 2024.
33. Gao, B., Chai, X., Huang, Y., Wang, X., Han, Z., Xu, X., Wang, Y., Siderophore production in pseudomonas SP. strain SP3 enhances iron acquisition in apple rootstock. *J. Appl. Microbiol.*, 133, 2, 720–732, 2022.
34. Gaya Karunasinghe, T., Hashil Al-Mahmooli, I., Al-Sadi, A.M., Velazhahan, R., The effect of salt-tolerant antagonistic bacteria from tomato rhizosphere on plant growth promotion and damping-off disease suppression under salt-stress conditions. *Acta Agric. Scand. Sect B*, 70, 1, 69–75, 2020.
35. Gorain, B., Paul, S., Parihar, M., Role of soil microbes in micronutrient solubilization, in: *New and Future Developments in Microbial Biotechnology and Bioengineering*, pp. 131–150, Elsevier, Amsterdam, Netherlands, 2022.

36. Guo, J., Muhammad, H., Lv, X., Wei, T., Ren, X., Jia, H., Hua, L., Prospects and applications of plant growth promoting rhizobacteria to mitigate soil metal contamination: A review. *Chemosphere*, 246, 125823, 2020.
37. Gupta, A., Mishra, R., Rai, S., Bano, A., Pathak, N., Fujita, M., Hasanuzzaman, M., Mechanistic insights of plant growth promoting bacteria mediated drought and salt stress tolerance in plants for sustainable agriculture. *Int. J. Mol. Sci.*, 23, 7, 3741, 2022.
38. Hakim, S., Naqqash, T., Nawaz, M.S., Laraib, I., Siddique, M.J., Zia, R., Imran, A., Rhizosphere engineering with plant growth-promoting microorganisms for agriculture and ecological sustainability. *Front. Sustainable Food Syst.*, 5, 617157, 2021.
39. Hossain, M.E., Shahrukh, S., Hossain, S.A., Chemical fertilizers and pesticides: impacts on soil degradation, groundwater, and human health in Bangladesh, in: *Environmental Degradation: Challenges and Strategies for Mitigation*, pp. 63–92, Springer International Publishing, Cham, 2022.
40. Jain, N., Rose, H.M., Kumar, A., Mathiyarasi, K.V., Bhatia, A., Chakrabarti, B., Kumar, S., Organic Farming and Greenhouse Gas Emission, in: *Agricultural Greenhouse Gas Emissions: Problems and Solutions*, pp. 33–57, Springer Nature Singapore, Singapore, 2024.
41. Jaiswal, S.K., Mohammed, M., Ibny, F.Y., Dakora, F.D., Rhizobia as a source of plant growth-promoting molecules: potential applications and possible operational mechanisms. *Front. Sustainable Food Syst.*, 4, 619676, 2021.
42. Jan, R., Khan, M.A., Asaf, S., Lubna, Lee, I.J., Kim, K.M., Metal resistant endophytic bacteria reduces cadmium, nickel toxicity, and enhances expression of metal stress related genes with improved growth of Oryza sativa, via regulating its antioxidant machinery and endogenous hormones. *Plants*, 8, 10, 363, 2019.
43. Juneja, A., Ceballos, R.M., Murthy, G.S., Effects of environmental factors and nutrient availability on the biochemical composition of algae for biofuels production: a review. *Energies*, 6, 9, 4607–4638, 2013.
44. Kafeel, U., Jahan, U., Khan, F.A., Role of mineral nutrients in biological nitrogen fixation, in: *Sustainable Plant Nutrition*, pp. 87–106, Academic Press, Cambridge, Massachusetts, USA, 2023.
45. Kapoor, D., Sharma, P., Sharma, M.M.M., Yadav, S., Husen, A., Exploring soil microbiota and their role in plant growth, stress tolerance, disease control and nutrient immobilizer. *Biocatal. Agric. Biotechnol.*, 103358, 2024.
46. Kaur, R., Kumar, S., Ali, S.A., Kumar, S., Ezing, U.M., Bana, R., Singh, T., Impacts of climate change on crop-weed dynamics: Challenges and strategies for weed management in a changing climate. *Open J. Environ. Biol.*, 9, 015–021, 2024.
47. Kaya, C., Uğurlar, F., Adamakis, I.D.S., Epigenetic and Hormonal Modulation in Plant–Plant Growth-Promoting Microorganism Symbiosis for Drought-Resilient Agriculture. *Int. J. Mol. Sci.*, 24, 22, 16064, 2023.

48. Kermeur, N., Pédrot, M., Cabello-Hurtado, F., Iron availability and homeostasis in plants: a review of responses, adaptive mechanisms, and signaling, in: *Plant Abiotic Stress Signaling*, pp. 49–81, 2023.
49. Khan, N., Ali, S., Shahid, M.A., Mustafa, A., Sayyed, R.Z., Curá, J.A., Insights into the interactions among roots, rhizosphere, and rhizobacteria for improving plant growth and tolerance to abiotic stresses: a review. *Cells*, 10, 6, 1551, 2021.
50. Khasheii, B., Mahmoodi, P., Mohammadzadeh, A., Siderophores: Importance in bacterial pathogenesis and applications in medicine and industry. *Microbiol. Res.*, 250, 126790, 2021.
51. Kramer, J., Özkaya, O., Kümmerli, R., Bacterial siderophores in community and host interactions. *Nat. Rev. Microbiol.*, 18, 3, 152–163, 2020.
52. Kroh, G.E. and Pilon, M., Regulation of iron homeostasis and use in chloroplasts. *Int. J. Mol. Sci.*, 21, 9, 3395, 2020.
53. Ku, Y.S., Rehman, H.M., Lam, H.M., Possible roles of rhizospheric and endophytic microbes to provide a safe and affordable means of crop biofortification. *Agronomy*, 9, 11, 764, 2019.
54. Kumar, A., Verma, H., Singh, V.K., Singh, P.P., Singh, S.K., Ansari, W.A., Pandey, K.D., Role of Pseudomonas sp. in sustainable agriculture and disease management, in: *Agriculturally Important Microbes for Sustainable Agriculture: Volume 2: Applications in Crop Production and Protection*, pp. 195–215, 2017.
55. Kumar, R., Swapnil, P., Meena, M., Selpair, S., Yadav, B.G., Plant growth-promoting rhizobacteria (PGPR): Approaches to alleviate abiotic stresses for enhancement of growth and development of medicinal plants. *Sustainability*, 14, 23, 15514, 2022.
56. Kumari, V.V., Banerjee, P., Verma, V.C., Sukumaran, S., Chandran, M.A.S., Gopinath, K.A., Awasthi, N.K., Plant nutrition: An effective way to alleviate abiotic stress in agricultural crops. *Int. J. Mol. Sci.*, 23, 15, 8519, 2022.
57. Lahlali, R., Ezrari, S., Radouane, N., Kenfaoui, J., Esmaeel, Q., El Hamss, H., Barka, E.A., Biological control of plant pathogens: A global perspective. *Microorganisms*, 10, 3, 596, 2022.
58. Lal, R., Restoring soil quality to mitigate soil degradation. *Sustainability*, 7, 5, 5875–5895, 2015.
59. Lastochkina, O., Aliniaeifard, S., SeifiKalhor, M., Bosacchi, M., Maslennikova, D., Lubyanova, A., Novel approaches for sustainable horticultural crop production: Advances and prospects. *Horticulturae*, 8, 10, 910, 2022.
60. Lastochkina, O., Aliniaeifard, S., Seifikalhor, M., Yuldashev, R., Pusenkova, L., Garipova, S., Plant growth-promoting bacteria: biotic strategy to cope with abiotic stresses in wheat, in: *Wheat Production in Changing Environments: Responses, Adaptation and Tolerance*, pp. 579–614, 2019.
61. Li, X., Zeng, R., Liao, H., Improving crop nutrient efficiency through root architecture modifications. *J. Integr. Plant Biol.*, 58, 3, 193–202, 2016.

62. Licina, V., Krogstad, T., Simić, A., Akšić, M.F., Meland, M., Nutrition and fertilizer application to apple trees-a review. NIBIO Rapport, 2021.
63. Liu-Xu, L., González-Hernández, A.I., Camañes, G., Vicedo, B., Scalschi, L., Llorens, E., Harnessing Green Helpers: Nitrogen-Fixing Bacteria and Other Beneficial Microorganisms in Plant–Microbe Interactions for Sustainable Agriculture. *Horticulturae*, 10, 6, 621, 2024.
64. López-Pérez, M.C., Espinoza-Acosta, J.L., Pérez-Labrada, F., Iron nutrition management in calcisol soils as a tool to mitigate chlorosis and promote crop quality–An overview. *J. Appl. Biol. Biotechnol.*, 12, 1, 17–29, 2024.
65. Ma, Y., Dias, M.C., Freitas, H., Drought and salinity stress responses and microbe-induced tolerance in plants. *Front. Plant Sci.*, 11, 591911, 2020.
66. Marchetti, M., De Bei, O., Bettati, S., Campanini, B., Kovachka, S., Gianquinto, E., Ronda, L., Iron metabolism at the interface between host and pathogen: from nutritional immunity to antibacterial development. *Int. J. Mol. Sci.*, 21, 6, 2145, 2020.
67. Meena, M., Swapnil, P., Divyanshu, K., Kumar, S., Tripathi, Y.N., Upadhyay, R.S., PGPR-mediated induction of systemic resistance and physiochemical alterations in plants against the pathogens: Current perspectives. *J. Basic Microbiol.*, 60, 10, 828–861, 2020.
68. Moon, Y.S. and Ali, S., Possible mechanisms for the equilibrium of ACC and role of ACC deaminase-producing bacteria. *Appl. Microbiol. Biotechnol.*, 106, 3, 877–887, 2022.
69. Mosa, K.A., Ismail, A., Helmy, M., Mosa, K.A., Ismail, A., Helmy, M., Introduction to plant stresses, in: *Plant Stress Tolerance: An Integrated Omics Approach*, pp. 1–19, 2017.
70. Nair, K.P. and Nair, K.P., Technological advancements in coconut, arecanut and cocoa research: A century of service to the global farming community by the Central Plantation Crops Research Institute, Kasaragod, Kerala State, India, in: *Tree Crops: Harvesting Cash from the World's Important Cash Crops*, pp. 377–536, 2021.
71. Naveed, M., Ramzan, N., Mustafa, A., Samad, A., Niamat, B., Yaseen, M., Xu, M., Alleviation of salinity induced oxidative stress in Chenopodium quinoa by Fe biofortification and biochar—endophyte interaction. *Agronomy*, 10, 2, 168, 2020.
72. Nawaz, M., Sun, J., Shabbir, S., Khattak, W.A., Ren, G., Nie, X., Sonne, C., A review of plants strategies to resist biotic and abiotic environmental stressors. *Sci. Total Environ.*, 165832, 2023.
73. Nazarov, P.A., Baleev, D.N., Ivanova, M.I., Sokolova, L.M., Karakozova, M.V., Infectious plant diseases: etiology, current status, problems and prospects in plant protection. *Acta Nat.*, 12, 3, 46, 2020.
74. Nishio, J.N., Why are higher plants green? Evolution of the higher plant photosynthetic pigment complement. *Plant Cell Environ.*, 23, 6, 539–548, 2000.

75. Ojuederie, O.B., Olanrewaju, O.S., Babalola, O.O., Plant growth promoting rhizobacterial mitigation of drought stress in crop plants: implications for sustainable agriculture. *Agronomy*, 9, 11, 712, 2019.
76. Orozco-Mosqueda, M.D.C., Santoyo, G., Glick, B.R., Recent advances in the bacterial phytohormone modulation of plant growth. *Plants*, 12, 3, 606, 2023.
77. Oukala, N., Aissat, K., Pastor, V., Bacterial endophytes: The hidden actor in plant immune responses against biotic stress. *Plants*, 10, 5, 1012, 2021.
78. Pandey, S.S., The Role of Iron in Phytopathogenic Microbe–Plant Interactions: Insights into Virulence and Host Immune Response. *Plants*, 12, 17, 3173, 2023.
79. Patra, P., Das, M., Kundu, P., Ghosh, A., Recent advances in systems and synthetic biology approaches for developing novel cell-factories in non-conventional yeasts. *Biotechnol. Adv.*, 47, 107695, 2021.
80. Pattnaik, S., Mohapatra, B., Gupta, A., Plant growth-promoting microbe mediated uptake of essential nutrients (Fe, P, K) for crop stress management: Microbe–soil–plant continuum. *Front. Agron.*, 3, 689972, 2021.
81. Pecoraro, L., Wang, X., Shah, D., Song, X., Kumar, V., Shakoor, A., Rani, R., Biosynthesis pathways, transport mechanisms and biotechnological applications of fungal siderophores. *J. Fungi*, 8, 1, 21, 2021.
82. Philippot, L., Chenu, C., Kappler, A., Rillig, M.C., Fierer, N., The interplay between microbial communities and soil properties. *Nat. Rev. Microbiol.*, 22, 4, 226–239, 2024.
83. Pirttilä, A.M., Mohammad Parast Tabas, H., Baruah, N., Koskimäki, J.J., Biofertilizers and biocontrol agents for agriculture: How to identify and develop new potent microbial strains and traits. *Microorganisms*, 9, 4, 817, 2021.
84. Pisoschi, A.M., Pop, A., Iordache, F., Stanca, L., Predoi, G., Serban, A.I., Oxidative stress mitigation by antioxidants-an overview on their chemistry and influences on health status. *Eur. J. Med. Chem.*, 209, 112891, 2021.
85. Prasad, S., Malav, L.C., Choudhary, J., Kannojiya, S., Kundu, M., Kumar, S., Yadav, A.N., Soil microbiomes for healthy nutrient recycling, in: *Current Trends in Microbial Biotechnology for Sustainable Agriculture*, pp. 1–21, 2021.
86. Rahim, H.U., Ali, W., Uddin, M., Ahmad, S., Khan, K., Bibi, H., Alatalo, J.M., Abiotic stresses in soils, their effects on plants, and mitigation strategies: a literature review. *Chem. Ecol.*, 1–34, 2024.
87. Rajamuthuramalingam, T., da Silva, W., Zuverza-Mena, N., Dimkpa, C., White, J.C., Nano-sized metal oxide fertilizers for sustainable agriculture: balancing benefits, risks, and risk management strategies. *Nanoscale*, 2024.
88. Ram, R.M., Debnath, A., Negi, S., Singh, H.B., Use of microbial consortia for broad spectrum protection of plant pathogens: Regulatory hurdles, present status and future prospects, in: *Biopesticides*, pp. 319–335, 2022.
89. Raza, A., Salehi, H., Rahman, M.A., Zahid, Z., Madadkar Haghjou, M., Najafi-Kakavand, S., Zhuang, W., Plant hormones and neurotransmitter

interactions mediate antioxidant defenses under induced oxidative stress in plants. *Front. Plant Sci.*, 13, 961872, 2022.

90. Richa, Metabolite-based Bioformulation: Next Generation of Biofertilizers, in: *Metabolomics, Proteomics and Gene Editing Approaches in Biofertilizer Industry: Volume II*, pp. 53–81, Springer Nature Singapore, Singapore, 2024.
91. Roriz, M., Pereira, S.I., Castro, P.M., Carvalho, S.M., Vasconcelos, M.W., Iron metabolism in soybean grown in calcareous soil is influenced by plant growth-promoting rhizobacteria–a functional analysis. *Rhizosphere*, 17, 100274, 2021.
92. Saeed, Q., Xiukang, W., Haider, F.U., Kučerik, J., Mumtaz, M.Z., Holatko, J., Mustafa, A., Rhizosphere bacteria in plant growth promotion, biocontrol, and bioremediation of contaminated sites: a comprehensive review of effects and mechanisms. *Int. J. Mol. Sci.*, 22, 19, 10529, 2021.
93. Sarvepalli, M., Velidandi, A., Ramachandravarapu, A.K., Korrapati, N., Marine Actinomycetes Siderophores: Types, High Throughput Characterization Techniques, Applications, and Their Association with Nanotechnology: A Comprehensive Review. *Nanoworld J.*, 10, 1, 1–21, 2024.
94. Seth, K., Vyas, P., Deora, S., Gupta, A.K., Meena, M., Swapnil, P., Understanding plant-plant growth-promoting rhizobacteria (PGPR) interactions for inducing plant defense, in: *Plant-Microbe Interaction-Recent Advances in Molecular and Biochemical Approaches*, pp. 201–226, Academic Press, London, United Kingdom, 2023.
95. Shahid, M., Khan, M.S., Singh, U.B., Pesticide-tolerant microbial consortia: Potential candidates for remediation/clean-up of pesticide-contaminated agricultural soil. *Environ. Res.*, 116724, 2023.
96. Shao, Z., Gu, S., Zhang, X., Xue, J., Yan, T., Guo, S., Wei, Z., Siderophore interactions drive the ability of Pseudomonas spp. consortia to protect tomato against Ralstonia solanacearum. *Hortic. Res.*, 11, 9, uhae186, 2024.
97. Sharma, P., Sharma, M.M.M., Malik, A., Vashisth, M., Singh, D., Kumar, R., Pandey, V., Rhizosphere, rhizosphere biology, and rhizospheric engineering, in: *Plant Growth-Promoting Microbes for Sustainable Biotic and Abiotic Stress Management*, pp. 577–624, Springer International Publishing, Cham, 2021.
98. Shehzad, J. and Mustafa, G., Mechanism of reactive oxygen species regulation in plants, in: *Reactive Oxygen Species: Prospects in Plant Metabolism*, pp. 17–41, Springer Nature Singapore, Singapore, 2023.
99. Singh Rawat, V., Kaur, J., Bhagwat, S., Arora Pandit, M., Dogra Rawat, C., Deploying microbes as drivers and indicators in ecological restoration. *Restor. Ecol.*, 31, 1, e13688, 2023.
100. Singh, P., Chauhan, P.K., Upadhyay, S.K., Singh, R.K., Dwivedi, P., Wang, J., Jiang, M., Mechanistic insights and potential use of siderophores producing microbes in rhizosphere for mitigation of stress in plants grown in degraded land. *Front. Microbiol.*, 13, 898979, 2022.

101. Stitt, M., Müller, C., Matt, P., Gibon, Y., Carillo, P., Morcuende, R., Krapp, A., Steps towards an integrated view of nitrogen metabolism. *J. Exp. Bot.*, 53, 370, 959–970, 2002.
102. Suman, J., Rakshit, A., Ogireddy, S.D., Singh, S., Gupta, C., Chandrakala, J., Microbiome as a key player in sustainable agriculture and human health. *Front. Soil Sci.*, 2, 821589, 2022.
103. Sun, J., Xiao, S., Xue, C., The tug-of-war on iron between plant and pathogen. *Phytopathol. Res.*, 5, 1, 61, 2023.
104. Sun, Y., Wu, J., Shang, X., Xue, L., Ji, G., Chang, S., Emaneghemi, B., Screening of siderophore-producing bacteria and their effects on promoting the growth of plants. *Curr. Microbiol.*, 79, 5, 150, 2022.
105. Suzuki, N., Rivero, R.M., Shulaev, V., Blumwald, E., Mittler, R., Abiotic and biotic stress combinations. *New Phytol.*, 203, 1, 32–43, 2014.
106. Swarnalatha, G.V., Goudar, V., Reddy, E. C. R. G. S., Al Tawaha, A.R.M., Sayyed, R.Z., Siderophores and their applications in sustainable management of plant diseases, in: *Secondary Metabolites and Volatiles of PGPR in plant-growth promotion*, pp. 289–302, 2022.
107. Syed, A., Elgorban, A.M., Bahkali, A.H., Eswaramoorthy, R., Iqbal, R.K., Danish, S., Metal-tolerant and siderophore producing Pseudomonas fluorescence and Trichoderma spp. improved the growth, biochemical features and yield attributes of chickpea by lowering Cd uptake. *Sci. Rep.*, 13, 1, 4471, 2023.
108. Tariq, M., Khan, A., Asif, M., Khan, F., Ansari, T., Shariq, M., Siddiqui, M.A., Biological control: a sustainable and practical approach for plant disease management. *Acta Agric. Scand. Sect B*, 70, 6, 507–524, 2020.
109. Timofeeva, A.M., Galyamova, M.R., Sedykh, S.E., Bacterial siderophores: Classification, biosynthesis, perspectives of use in agriculture. *Plants*, 11, 22, 3065, 2022.
110. Tiwari, S., Sharma, B., Bisht, N., Tewari, L., Role of beneficial microbial gene pool in mitigating salt/nutrient stress of plants in saline soils through underground phytostimulating signalling molecules. *Pedosphere*, 33, 1, 153–171, 2023.
111. Touati, D., Iron and oxidative stress in bacteria. *Arch. Biochem. Biophys.*, 373, 1, 1–6, 2000.
112. Tripathi, A., Pandey, P., Tripathi, S.N., Kalra, A., Perspectives and potential applications of endophytic microorganisms in cultivation of medicinal and aromatic plants. *Front. Plant Sci.*, 13, 985429, 2022.
113. Verma, K.K., Song, X.P., Li, D.M., Singh, M., Wu, J.M., Singh, R.K., Li, Y.R., Silicon and soil microorganisms improve rhizospheric soil health with bacterial community, plant growth, performance and yield. *Plant Signaling Behav.*, 17, 1, 2104004, 2022.
114. Wang, H., Liu, R., You, M.P., Barbetti, M.J., Chen, Y., Pathogen biocontrol using plant growth-promoting bacteria (PGPR): Role of bacterial diversity. *Microorganisms*, 9, 9, 1988, 2021.

115. Wang, Y., Jia, X., An, S., Yin, W., Huang, J., Jiang, X., Nanozyme-Based Regulation of Cellular Metabolism and Their Applications. *Adv. Mater.*, 36, 10, 2301810, 2024.
116. Wolters, V., Invertebrate control of soil organic matter stability. *Biol. Fertil. Soils*, 31, 1–19, 2000.
117. Xie, B., Wei, X., Wan, C., Zhao, W., Song, R., Xin, S., Song, K., Exploring the Biological Pathways of Siderophores and Their Multidisciplinary Applications: A Comprehensive Review. *Molecules*, 29, 10, 2318, 2024.
118. Xu, Y.C., Li, N., Yan, X., Zou, H.X., DFT-based analysis of siderophore-metal ion interaction for efficient heavy metal remediation. *Environ. Sci. Pollut. Res.*, 30, 40, 91780–91793, 2023.
119. Yang, T., Siddique, K.H., Liu, K., Cropping systems in agriculture and their impact on soil health-A review. *Global Ecol. Conserv.*, 23, e01118, 2020.
120. Zhang, S., Deng, Z., Borham, A., Ma, Y., Wang, Y., Hu, J., Bohu, T., Significance of soil siderophore-producing bacteria in evaluation and elevation of crop yield. *Horticulturae*, 9, 3, 370, 2023.
121. Zia-ur-Rehman, M., Anayatullah, S., Irfan, E., Hussain, S.M., Rizwan, M., Sohail, M.I., Alharby, H.F., Nanoparticles assisted regulation of oxidative stress and antioxidant enzyme system in plants under salt stress: A review. *Chemosphere*, 314, 137649, 2023.
122. Singh, P., Chauhan, P.K., Upadhyay, S.K., Singh, R.K., Dwivedi, P., Wang, J., ... & Jiang, M., Mechanistic insights and potential use of siderophores producing microbes in rhizosphere for mitigation of stress in plants grown in degraded land. *Front. Microbiol.*, 13, 898979, 2022.
123. Qin, H., Wang, Z., Sha, W., Song, S., Qin, F., Zhang, W., Role of plant-growth-promoting rhizobacteria in plant machinery for soil heavy metal detoxification. *Microorganisms*, 12, 4, 700, 2024.

7

Siderophores as Biofertilizers: Enhancing Nutrient Availability

Sanya Sharma[1], Rachan Karmakar[2], Ayush Madan[3], Dilfuza Jabborova[4,5], Rahul Kumar[1]* and Debasis Mitra[1]

[1]*Department of Microbiology, Graphic Era (Deemed to be University), Uttarakhand, India*

[2]*Department of Environmental Science, Graphic Era (Deemed to be University), Uttarakhand, India*

[3]*Department of Biotechnology, School of Research & Technology, People's University, Bhanpur, Bhopal, Madhya Pradesh, India*

[4]*Institute of Genetics and Plant Experimental Biology, Uzbekistan Academy of Sciences, Kibray, Uzbekistan*

[5]*Faculty of Biology, National University of Uzbekistan, Tashkent, Uzbekistan*

Abstract

Siderophores are Fe-chelating compounds secreted by microorganisms that render nutrients available in the soil and thus play important roles as useful biofertilizers in sustainable agriculture. This chapter goes slightly beyond to discuss how siderophores assist in solubilization and nutrient uptake and hypothesize to emphasize their efficiency in enhancing plant availability of major elements like Fe, Zn, Cu, and Mn. We will outline the dominant microbial genera that synthesize siderophores and their symbiotic and nonsymbiotic associations with plant root systems. In addition, we will outline the different application strategies of siderophore-based biofertilizers such as soil, seed, and foliar and critically analyze recent field trials and case studies that demonstrate these advantages in the form of yield improvements as well as enhanced fertility. We will also focus on the advantages of siderophores as biofertilizers—advantages that obviate the need for chemical fertilizers and for the upkeep of good health of soils—and also explain challenges such as instability and effectiveness under diverse environmental conditions. Moreover, directions for future research that can drive innovative

**Corresponding author*: rahulkumar.phdmb@gmail.com;
ORCID: https://orcid.org/0009-0000-1323-2220

Debasis Mitra, Ayush Madan and Anju Rani (eds.) Microbial Siderophores for Agricultural Productivity: Perspectives of Use in Agriculture, (159–182) © 2026 Scrivener Publishing LLC

formulations of siderophores in precision agriculture will be provided. This chapter meets the opportunity of siderophores as biofertilizers for sustainable crop and nutrient production.

Keywords: Siderophores, biofertilizers, plant–microbe interactions, sustainable agriculture

7.1 Introduction

Biofertilizers are inexpensive materials with living organisms or microbes that enhance the growth of plants and trees by providing space for necessary nutrients [27]. Mycorrhizal fungi, blue-green algae, and bacteria are included in these biofertilizers, which enhance the growth of plants and trees by fixing nitrogen from the atmosphere, making phosphorus soluble, and synthesizing growth substances [48]. Inoculation with *Pseudarthrobacter* and *Vibrio neocaledonicus* has enhanced *Salicornia ramosissima* growth in polluted soils, implying their application in phytoremediation and stress tolerance [5]. Biofertilizers are an important component of sustainable agriculture due to their potential to decrease reliance on chemical fertilizers and limit pollution [36]. New technologies in the formulation of liquid biofertilizers have met improved shelf life, increased crop quality, and reduced application levels, all amenable to today's agricultural applications [60]. The increasing population of the world emphasizes the possible role of biofertilizers in sustainable food systems for food security, soil health, and environmental health [71]. Siderophores are low-molecular-weight organic compounds that are specific in structure and are produced by microorganisms and plants to capture iron from their environment [19]. Current research has pointed out the varied roles played by siderophores in nutrient cycling and environmental sustainability, assisting in iron acquisition. Microbes with siderophore production capacity have been demonstrated to be capable of alleviating stress in soils that are degraded, where nutrient imbalance and heavy metal pollution are significant barriers to plant growth [25]. They also make the soil fertile and plants healthy by enhancing other important elements' availability and improving iron acquisition. The application of siderophore-producing bacteria has been explored as a green approach to enhance plant productivity and soil fertility, offering a sustainable alternative to chemical fertilizers and helping combat climate change [53]. These biological approaches promote more environmentally friendly agricultural practices by enhancing nutrient acquisition efficiency and reducing the need for synthetic inputs. Siderophores, iron-chelating compounds excreted by microbes, are important in agriculture [70]. They enhance soil iron bioavailability and enhance plant growth while alleviating stress. A study emphasized the significance of these bacteria in alleviating iron deficiency in plants cultivated on degraded soils, emphasizing their ability to restore soil fertility and plant health [26]. The symbiotic relationship between plant roots and siderophore-producing bacteria enhances nutrient uptake and environmental tolerance [57]. The use of siderophore-producing bacteria can increase plant growth parameters by as much as 30%, illustrating their dramatic influence on crop yield and quality [17]. Siderophores are unique molecules secreted by bacteria and fungi that bind iron from the surrounding environment. They possess a high affinity for ferric ions (Fe^{3+}), leading to stable siderophore–iron complexes that microbes can readily take up [24]. Intracellular iron regulates siderophore synthesis and excretion tightly, and siderophore manufacture and transport genes are expressed in iron-deprived conditions. Siderophores effectively solubilize iron by chelating surrounding Fe^{3+} during excretion. Some cell surface receptors of microorganisms detect the resulting ferric–siderophore complexes and trigger their uptake [33]. The diversity of siderophore structure and uptake mechanisms illustrates how bacteria have developed to cope with diverse ecological niches and iron conditions. New discoveries have expanded our knowledge of such mechanisms, which has applications in medicine and agriculture, including designing strategies to prevent pathogenic bacteria by targeting their iron absorption mechanisms.

7.1.1 Role of Siderophores for Mobilizing

Siderophores, iron-specific molecules secreted by microbes and plants in iron-deficient environments, play an important role in mobilizing other micronutrients such as zinc, copper, and manganese, impacting their bioavailability in various habitats [46]. Research has indicated that siderophores are implicated in the mobility of Mn from minerals such as manganite and γ-MnOOH, pointing to their significance in Mn cycling [20]. However, this mobilization can have a contrary effect on iron solubility, which suggests a complex interaction between siderophore-mediated Fe and Mn biogeochemical cycles. Environmental *Pseudomonas* species were studied in an investigation that focused on the influence of siderophores on zinc and manganese mobilization and discovered that Mn^{2+} significantly enhanced

Table 7.1 Role of siderophore in mobilizing micronutrients.

Micronutrient	Siderophore involved	Mobilization mechanisms	Source/ organisms	Impact on other metals
Manganese	Pyoverdine, phytosiderophores	Mobilization from minerals (manganite, γ- MnOOH)	Environmental *Pseudomonas* spp., plants	Enhances pyoverdine biosynthesis but affects Fe solubility
Copper	Pyochelin, phytosiderophores	Sequestration and facilitated absorption	*Pseudomonas aeruginosa*, plants	Competitive uptake with Fe and Zn
Zinc	Pyoverdine, phytosiderophores	Chelation and uptake in Zn-deficient soils	*Pseudomonas* spp., Graminaceous plant roots	Mobilization linked with Fe and Mn uptake
Iron	Pyoverdine, enterobactin, phytosiderophores	Chelation and solubilization in Fe-deficient conditions	Graminaceous plants, bacteria (*Pseudomonas*, *E. coli*, etc.)	Competitive inhibition by high levels of Zn, Mn, Cu

pyoverdine biosynthesis, which suggests that siderophores may be involved in the sequestration of Mn, facilitating its absorption in limited-availability environments [54]. The dual role of siderophores in mobilizing iron and other micronutrients has significant implications for plant nutrition and soil health [45]. Under iron deficiency in graminaceous plants, their roots exude phytosiderophores, which chelate not only Fe^{3+} but also Zn^{2+}, Cu^{2+}, and Mn^{2+}. The mobilization of iron could be suppressed by the high levels of other metals, indicating a competitive process in siderophore-fueled micronutrient acquisition (Table 7.1).

7.1.2 Influence of Soil Properties on the Effectiveness of Siderophores

The chemical composition of soils, including their pH, contents of organic material, and competing ions, has a considerable effect on the functioning and performance of siderophores in agricultural production and health. Soil pH also determines iron solubility and availability, and it is mainly precipitated into insoluble species in calcareous soils. For the chelation and mobilization of iron, plants and microbes release additional siderophores in such alkaline environments, enhancing iron uptake and promoting plant growth. Organic matter in the soil is also essential for the working of siderophores [1]. Organic-matter-rich soils harbor diverse microbial populations, including siderophore-producing bacteria (SPB). Research has indicated that the number of SPB in soils is associated with higher plant growth parameters, e.g., plant weight, plant height, and germination rate. This enhances the uptake of nutrients by plants and improves crop yields sustainably. Environmental stresses and co-ions can also affect siderophore function. Environmental stresses like salinity, heavy metal contamination, and nutritional deficiencies can inhibit plant growth in soils with poor quality [18]. Certain rhizobacteria secrete siderophores that could chelate other heavy metals, reducing their toxicity and enhancing plant resistance. Through the modification of the rhizosphere environment, these microbes can enhance iron availability and solubility even in difficult conditions, ensuring healthier plant growth.

7.2 Siderophore-Producing Microorganisms in Agriculture

Siderophores, which are specific molecules synthesized by microorganisms such as *Pseudomonas*, *Bacillus*, and *Rhizobium*, are very important in sustainable agriculture [70]. The microbes chelate iron from the environment and enhance its availability for plants, which is necessary for physiological processes such as chlorophyll synthesis and enzymatic activities. Current studies have demonstrated the benefits of siderophore-producing bacteria in farming, for instance, *Pseudomonas monteilii* strain MN759447's antagonistic properties against fungal infections, implying a possible biocontrol strategy for tree death in forestry use [6]. The knowledge on siderophores' secretion processes and biosynthesis mechanisms can aid in the production of biofertilizers that enhance crop production in a sustainable manner. Also, siderophore-producing bacteria and intercropping systems complement each other, as in a study where faba bean cultivation with maize increased the diversity of siderophore-producing bacteria in the rhizosphere [67], inhibiting *Fusarium oxysporum*, lowering disease incidence, and promoting healthier plant growth.

7.3 Symbiotic and Non-Symbiotic Relationships with the Plants

Symbiotic relationships between microbes and plants have the potential to enhance iron uptake in iron-deficient soils [34]. Rhizobacteria and legumes form these associations by solubilizing and sequestering iron from the soil matrix through siderophore secretion, high-affinity iron-chelating compounds. Siderophore complexes augment iron nutrition and induce plant growth, maintaining the microbial population and enhancing plant growth [45]. In non-symbiotic associations, siderophore-producing bacteria exist independently in the rhizosphere, enhancing the bioavailability of iron and other micronutrients for the plant and the bacteria. This approach is beneficial in alkaline or calcareous soils where iron solubility is low and plants have difficulty obtaining sufficient iron [9]. Siderophores are also involved in avoiding soil-borne infections by competitive exclusion, blocking the access of harmful organisms to this important nutrient, inhibiting growth, and reducing crop disease incidence. Siderophore-producing microorganisms are integral parts of integrated pest and nutrient management practices since they minimize the use of chemical pesticides and fertilizers [56]. They increase microbial diversity and soil structure, enhancing nutrient uptake and promoting plant growth and soil health. Siderophores are critical parts of integrated pest and nutrient management practices since they act both in suppressing pathogens and in acquiring nutrients. Efficient siderophore strains have been found in various soils such as heavy-metal-polluted soil, and they have been proven to have potential in phytoremediation programs to stimulate plant growth and decontaminate polluted soils [51].

7.4 Improvement of Soil Fertility and Plant Health Through Microbial Inoculation

Siderophores are unique molecules secreted by bacteria and fungi that chelate iron, converting iron into soluble complexes that are plant-absorbable [70]. The process encourages a more robust and nutrient-dense soil environment, mitigating iron deficiency in plants. Incorporation of siderophore-producing bacteria (SPB) into soils has the potential to enhance plant growth by 30% since they increase the bioavailability of iron, a crucial element required for plant physiological processes such as enzyme activity and chlorophyll synthesis [21]. SPB also enhance soil health by facilitating the decomposition of organic matter and nutrient cycling. Siderophores also possess antagonistic properties against soil-borne pathogens, enhancing the health of plants. Under intercropping conditions, elevated bacterial diversity and the growth of bacteria that produce siderophores are capable of suppressing diseases such as *Fusarium oxysporum*, sequestering iron and promoting better crop stands [25]. The use of siderophores in agricultural practices is in harmony with sustainable agriculture principles, presenting an eco-friendly alternative to chemical fertilizers and pesticides and enhancing soil quality and plant growth.

7.5 Mechanisms of Action of Siderophores as Biofertilizers

Iron chelation is a process by which microorganisms, such as bacteria and fungi, release high-affinity iron-chelating compounds known as siderophores to chelate iron from the surroundings [13]. Siderophores are critical biofertilizers in sustainable agriculture, as they enhance the availability of iron in the rhizosphere, the zone around plant roots. Iron is crucial for respiration and chlorophyll synthesis but challenging for plants to take up in most soils, particularly calcareous or alkaline soils. Rhizospheric bacteria address this issue by secreting siderophores, which chelate iron and facilitate its uptake by plants [41]. The secretion of these compounds into the rhizosphere chelates ferric iron (Fe^{3+}) to yield soluble siderophore–iron complexes, thereby enhancing the bioavailability of iron in the soil. These complexes are taken up by plants through transporters present on their root cell membranes. Upon entering the plant, the siderophore complex liberates the iron, which is utilized for different metabolic functions [54]. This microbial iron acquisition system is particularly helpful in iron-limited soils. Siderophilic bacteria isolated from *Paris polyphylla* var. *yunnanensis* rhizosphere may contribute to plant growth under conditions of nutrient deficiency [49] and enhance soil fertility and plant health by facilitating iron uptake and exhibiting phosphate-solubilizing activity. Adding siderophilic bacteria to sustainable agriculture practices can boost crop yields and decrease reliance on synthetic fertilizers.

7.5.1 Interaction of Siderophores with Plant Root Systems

Microorganisms, especially plant-growth-promoting rhizobacteria (PGPR), secrete siderophores that chelate iron from the environment, which is not easily accessible to plants in calcareous soils with low iron solubility or high pH [16]. Siderophore production is important for the stimulation of iron uptake in plants and for their growth and development. Siderophores form soluble iron–siderophore complexes by chelating iron from the soil medium, which are recognized and transported by receptors on the membranes of plant roots. Iron is released from the complex inside the plant, utilized for physiological processes such as electron transport and chlorophyll synthesis [28]. This mechanism is beneficial in iron-deficient soils since direct iron uptake is minimal. Siderophores influence plant immunological responses as well, triggering plant defense mechanisms *via* recognition pathways or altering the availability of iron. Siderophores have been associated with systemic resistance in plants, enhancing their resistance to pathogenic infections [39]. Siderophores can serve as biofertilizers in sustainable agriculture as they possess two functions: immunity and nutrient acquisition.

7.5.2 Role in Increased Root Development

Siderophores are compounds that are iron-attracting, and they are very important nutrients found in plants since they are generally limited by forming insoluble hydroxides under aerobic conditions [59]. Siderophores facilitate plants to uptake iron and aid them in carrying out essential physiological activities such as photosynthesis and respiration. Iron-feeding to plants can greatly be improved through siderophore-producing microorganisms in iron-limiting soils. They also support root growth, enhancing branching and elongation, expanding the root surface area, and enhancing nutrient and water uptake [14]. This robust root system enhances plant stability and resistance to environmental stresses. Certain plant-growth-promoting rhizobacteria (PGPR) that are siderophore producers also release phytohormones, which induce root growth. Siderophores also inhibit soil-borne pathogens, enhancing plant productivity and health. They inhibit iron availability to pathogenic microorganisms, inhibiting their growth and reducing disease occurrence [64]. This biocontrol trait is paramount in sustainable agriculture and presents a greener option as compared to chemical pesticides. Research indicates that siderophore-producing bacteria inoculation can result in increased crop yields.

7.6 Siderophores Variants and Efficiency of Biofertilizers

7.6.1 Catecholates

Catecholate siderophores, with high specificity toward ferric ions (Fe^{3+}), play a critical role in iron solubilization and uptake, particularly under low iron concentrations [47]. Enterobactin from *Escherichia coli* is one such compound with well-documented iron-chelating activity. Siderophores play a fundamental role in promoting plant growth and facilitating host-associated microbes in their access to Fe. Siderophilic rhizospheric bacteria, for instance, are an example. *Yunnanensis* produce catecholate-type siderophores, which strongly enhance iron availability in the soil [37]. This enhances the uptake of nutrients by the plants, enhancing plant resistance and growth. Utilizing these bacteria as biofertilizers can minimize the reliance on chemical fertilizers and enhance eco-friendly farming practices. Catecholate siderophores influence soil health and microbial interactions and act as a natural biocontrol by competing with pathogenic agents for nutrients [12]. Microbial siderophores are eco-friendly alternatives to man-made chelating agents as they are biodegradable. Microbial and plant genetic manipulation has enhanced the benefits of siderophore production in agroecosystems.

7.6.2 Hydroxamates

Hydroxamate siderophores, which are synthesized by soil microorganisms, play a crucial role in mobilizing nutrients in iron-deficient soils. They enhance the solubility and availability of Fe(III) in plants by creating stable complexes [61]. Hydroxamate siderophores have been found to be promising biofertilizers in sustainable agriculture because they are effective in different soils. Hydroxamate siderophores, such as ferrichrome and ferricrocin, occur in podzolic forest soils and possess an acidic composition and low nutrient availability [46]. They have a critical function in mobilizing iron in nutrient-deficient soils, inducing plant growth and soil fertility. Their efficiency in mobilizing nutrients depends on the pH, organic matter content, and microbial population structure of the soil [10]. Plant-growth-promoting bacteria (PGPB) can produce hydroxamate-type siderophores, which induce plant growth and well-being by mobilizing iron and other minerals.

7.6.3 Carboxylates

Microorganisms also synthesize siderophores, specific molecules that bind iron from the environment and important for plant nutrition and soil health. Siderophores can enhance plant iron uptake, promoting growth and resilience [45]. *Pseudomonas fluorescens* C7 synthesizes Fe–pyoverdine complexes absorbed by *Arabidopsis thaliana*, enhancing iron content and growth. Siderophores also interact with other micronutrients, modulating their availability and uptake. Carboxylates, which are organic acids that chelate metal ions and enhance nutrient uptake, are employed to mobilize rare earth elements (REEs) in plants [73]. The method enhances plant nutrition and growth by promoting the uptake of essential micronutrients and aiding in phosphorus uptake. Carboxylates also contribute to the uptake of micronutrients, including iron, zinc, manganese, and other metal ions. This chelation is especially useful in high-pH or calcium carbonate soils where micronutrient availability is limited. Incorporating microorganisms that synthesize carboxylate into biofertilizer products can enhance the crop uptake of micronutrients [35]. For sustainable agriculture, siderophore-producing microorganisms and carboxylate-releasing microorganisms in biofertilizers can minimize the adverse environmental impacts of traditional farming practices, enhance nutrient use efficiency, and reduce dependence on synthetic fertilizers.

7.7 Field Application of Siderophore-Based Biofertilizers

Siderophore-producing microbial inoculums are biofertilizers applied in agriculture to enhance plant growth and resistance, particularly in low-fertilizer soils [30]. The biofertilizers minimize dependence on synthetic inputs and environmental degradation, promoting sustainable agriculture. Certain *Pseudomonas* strains produce siderophores that chelate iron atoms, enhancing plant growth. The inoculants, however, pose special challenges, including their survival and efficacy upon field application, storage, and transportation [62]. The development of formulation technology, including encapsulation techniques, has enhanced their efficacy and survival rates. Culture medium optimization and growth conditions have also been found to affect siderophore production. Multi-omics approaches have enhanced the knowledge on the metabolic processes and regulatory mechanisms governing siderophore production, facilitating genetically modified strains for crop requirements and soil types [65]. The translation of these findings into economically sustainable products is dependent on the cooperation between industry and academia. Using stringent quality control measures at all points of manufacturing and use, siderophore-based biofertilizers are assured to be consistent and reliable, promoting increased farmer uptake and facilitating sustainable agricultural practices [2].

7.8 Soil, Seed, and Foliar Application

Siderophore-derived biofertilizers have been reported to enhance plant growth and soil health using foliar spraying, seed treatment, and soil application. These biofertilizers enhance the availability of iron in the rhizosphere, which enhances the vigor of the plants by employing siderophores, iron-chelating compounds synthesized by bacteria [8]. Through soil microbiome enrichment, siderophore-producing bacteria could enhance soil fertility and support nutrient cycling. They can also enhance the resistance of plants against environmental stresses and inhibit soil-borne pathogens. One study revealed that *Bacillus subtilis* siderophore-producing bacteria can inhibit *Cephalosporium maydis* in maize, leading to healthier plants with increased yields [40]. When used on seeds, microbial inoculants are applied before sowing so that beneficial microbes infect the root system early, enhancing seed germination and seedling health. This plant–bacteria symbiotic relationship increases nutrient uptake and aids in healthy plant growth. Foliar spraying, in which biofertilizers are applied to the leaves of plants, has been effective in treating micronutrient deficiencies [66]. Seaweed and other natural source liquid biofertilizer formulations have provided improved crop quality and stress resistance when used as foliar spray. Siderophore-based biofertilizers integrated into sustainable agricultural practices can increase soil health and crop yields.

7.9 Environmental Factors Affecting Siderophore-Based Biofertilizers

Siderophores, iron-chelating compounds of high affinity that are secreted by microorganisms, are being employed as an eco-friendly alternative to synthetic iron chelates in agriculture. These biofertilizers enhance the pool of iron accessible to plants, stimulating plant growth and alleviating iron chlorosis [77]. *Pseudomonas* isolates have exhibited potential as iron biofertilizers by the secretion of pyoverdine, a high-affinity siderophore. These strains also exhibit plant-growth-promoting traits like the production of glutamic acid and indole-3-acetic acid and are thus suitable for use in agricultural fields. Environmental factors like soil pH, temperature, and soil moisture impact the efficacy of siderophore-based biofertilizers heavily [50]. Alkaline conditions might inhibit iron solubility and thus require the use of efficient siderophores to enable iron uptake by the plant. Certain *Pseudomonas* are adapted for calcareous soils because they can create stable iron complexes. Temperature and soil aeration also influence siderophore yield and microbial activity [63]. Complementary microorganisms present in the rhizosphere are another key environmental consideration since introduced strains that produce siderophores can be impacted by the indigenous microbial community. Co-inoculation tactics using compatible microbial consortia have been examined to enhance synergistic interactions and general plant health [74].

7.10 Case Studies: Applications of Siderophore-Based Biofertilizers

Siderophores, which are secreted by bacteria such as *Pseudomonas fluorescens* C7, can enhance iron uptake in plants considerably, enhancing their growth and yield [43]. This symbiotic relationship not only enhances the nutritional quality of cereals but also aids in sustainable agriculture. For example, oat plants utilize siderophores in iron-depleted conditions, promoting iron uptake and growth. This symbiotic relationship minimizes the use of artificial iron chelators, promoting sustainable agricultural practices. Siderophores of *Rhizobium* species in legume crops enhance iron availability within the rhizosphere, enhancing nitrogen and Fe nutrition of the legumes [23]. Inoculation of alfalfa with *Rhizobium lupini* also enhanced iron uptake and nitrogen fixation. Such synergies highlight the significance of using the appropriate microbial strains to maximize biofertilization for legume development. Siderophore-producing pathogens such as *Pseudomonas aeruginosa* inhibit metal toxicity in tomato plants under metal stress by chelating surplus metals and facilitating iron uptake [38]. The protective mechanism induces improved growth and yield. *Serratia marcescens* treatment of lead- and cadmium-exposed spinach plants resulted in enhanced growth parameters, which was due to the ability of the bacterium to chelate metals and produce siderophores. In general, siderophores can increase the yield and health of cereal, legume, and vegetable crops.

7.11 Relative Efficiencies of Siderophores in Different Agricultural Systems

Siderophore-producing microbes (SPMs) have revealed great promise in degraded lands, especially those under heavy metal pollution [15]. They enhance plant growth through iron chelation and a decrease in toxicity of heavy metals. *Pseudomonas fluorescens* forms pyoverdine siderophores, raising the mobility of iron and reducing toxicity in contaminated environments. Such microbes are ideal for rehabilitating degraded lands and promoting sustainable agriculture. The performance of siderophores differs in accordance with the type of microbe and agro-environment [61]. Siderophores produced by bacteria are widely superior to those of fungal or plant origin in terms of iron-binding affinity. Improvements in microbial inoculant preparations seek to enhance the field stability and efficacy of siderophore-producing microorganisms. These advancements offer environmentally friendly alternatives to chemical pesticides and fertilizers and thus are pertinent to sustainable farming [52].

7.12 Recent Field and Greenhouse Experiments

It has been demonstrated in research that *Pseudomonas fluorescens-putida* and other plant-growth-promoting rhizobacteria (PGPR) can secrete siderophores that efficiently complex ambient iron, enhancing plant growth and a healthier rhizosphere [32]. Field trials have indicated substantial yield enhancements in potato, sugar beet, and radish crops, with some studies indicating as high as 144% increases. Greenhouse tests have also indicated the mechanisms through which siderophore-producing bacteria enhance plant growth. *Paris polyphylla var. yunnanensis* rhizosphere was employed to screen siderophilic bacteria, and strains that could produce hydroxamate-type siderophores were identified [72]. These compounds promoted seed germination and seedling growth in crops such as maize and pepper. Evidence of the capability of these bacteria to act as biofertilizers was provided by siderophore synthesis and other bioactive metabolites, including ammonia, to enhance nutrient uptake and overall plant vigor, yet the advantages of siderophore-producing biofertilizers are hard to convert into field application because of environmental factors such as temperature, microbial richness, and soil chemistry [7]. Recent studies have emphasized the importance of rigorous standards in agronomic field trials to close the gap between laboratory findings and real-world applications.

7.13 Applications of Siderophores As Biofertilizers

7.13.1 Chemical Fertilizer Use Reduction

Siderophores, high-affinity iron-chelating compounds, are essential to improve plant growth and reduce the need for chemical fertilizers [68]. They increase iron availability to plants and make it easier for microorganisms to obtain iron, promoting development and reducing the need for synthetic fertilizers. Siderophores can be especially useful in calcareous soils where iron bioavailability is low and can naturally enhance iron uptake, minimizing the use of synthetic fertilizers and encouraging sustainable agriculture practices [4]. Siderophores chelate heavy metals, lowering phytotoxicity and enhancing a stronger rhizosphere. They also lower the bioavailability and toxicity of metals such as lead and cadmium by stabilizing complexes with them. This process of detoxification improves soil quality, making the environment conducive for plants to grow healthier. Pathogenic organisms can be inhibited by siderophore-producing bacteria for essential nutrients, decreasing the rate of soil-borne diseases [29]. Biofertilizers from siderophores can be integrated into the agricultural sector for sustainable and environmentally friendly agriculture practices. They can minimize the detrimental environmental impact of chemical fertilizers and increase the yields and the quality of nutrients in crops. However, additional research and field trials are required to exploit siderophore-based biofertilizers to their full potential in sustainable agriculture. In general, additional multidisciplinary research and field trials are required to exploit siderophore-based solutions to their full potential, enhancing food security and environmental health [76].

7.13.2 Soil Fertility and Microbial Diversity

Soil microbial communities are differentially important to soil fertility, influencing plant well-being and yield. Bacteria, fungi, and archaea play crucial roles in decomposing organic matter, cycling nutrients, and suppressing soil-borne pathogens [3]. A sub-humid tropical grassland found in Northeastern Brazil was reported to have soils with greater fertility and that possessed more diverse and functionally dense microbial communities, resulting in greater soil structure and nitrogen cycling. The maintenance or enhancement of soil fertility may enhance positive microbial communities that are useful in maintaining sustainable pasture systems [42]. Multi-omics has added insight into how the bacteria in soils relate to fertility, making models better predict the outcomes of plant growth. Innovative treatments of the soil have been explored to enhance soil health and microbial richness. For instance, thermophile-fermented compost favorably alters the composition of bacteria in soil, enhancing the quality and yield of carrot production. These changes are linked to improved soil quality and nutrient cycling, which show the capacity of some organic amendments to change microbial diversity and soil fertility [11]. Adding cattle to farming systems enhances biodiversity and the sequestration of carbon in soils. Adding livestock manure to arable crops and livestock boosts soil carbon storage by approximately one-third over arable-only farms, promoting more diverse plant populations and higher ecosystem resilience.

7.13.3 Crop Production and Quality Enhancement

Artificial intelligence (AI) is transforming agriculture by enhancing yields and the use of resources, especially for small farmers in Kenya. AI-based tools such as Plant Village and Virtual Agronomist enable farmers to detect pests and diseases, resulting in cost efficiency and better yields [58]. This innovation responds to the decline of conventional agricultural extension services. Scientists are also developing climate-resilient crop varieties in crop genetics, including climate-resilient rice seeds and high-quality potato varieties. Firms like Ohalo Genetics and Solynta are producing potatoes with enhanced yield and health qualities to meet consumer health issues and farm productivity [55]. Controlled environment agriculture is gaining momentum to mitigate climate change impacts, including increased heat and irregular rainfall in Kashmir. Agricultural producers are testing indoor farming solutions to enhance produce quality and safeguard crops against extreme weather [22]. Policy developments, like the US Department of Agriculture's acceptance of genetically engineered HB4 wheat cultivation, have also significantly contributed to modern-day agriculture.

Such embracement of cutting-edge farm techniques is instrumental for sustainable food supply and global warming mitigation.

7.14 Challenges and Limiting Factors

Biofertilizer formulation and commercialization are challenging, such as the choice of suitable microbial strains, carrier material optimization, and maintaining shelf life under different environmental conditions [75]. The intricacy of microbial interactions and maintaining viability upon storage and application adds to the complexity of formulations. Scaling up production from laboratory to industrial levels without loss of efficacy is also a challenge. Interdisciplinary research and collaboration are essential to develop robust, effective, and scalable biofertilizer solutions [69]. Regulatory systems and market acceptance also impede the extensive application of biofertilizers. Different definitions and regulations in most countries complicate standardization and product quality assurance. Detailed regulations for the production and marketing of biofertilizer production and marketing are not available in the United States, even though other countries such as the European Union, Brazil, India, and China have made some progress [31]. Regulatory uncertainty may result in variable product quality, undermine consumer confidence, and stifle market growth. Soil types, climate, and native soil microbiomes can also influence the effectiveness of biofertilizers. Site-specific formulations and the development of robust microbial strains are thus important to realize biofertilizer potential and enhance farmer and stakeholder uptake.

7.15 Future Research Directions and Prospects

Research is directed toward the creation of multifunctional biofertilizers that harbor synergistic associations of beneficial microbes to promote nutrient acquisition, resistance to diseases, and stress endurance in plants. Co-inoculation of phosphorus-solubilizing fungi with nitrogen-fixing bacteria can enhance agricultural productivity and soil fertility, curbing the requirement for chemical fertilizers and encouraging a more resilient agro-ecosystem [44]. Biotechnology advancements have made genetic modifications of microbial strains employed in biofertilizers possible to enhance efficiency in nutrient fixation and stress tolerance. Genetically modified strains of *Rhizobium* have demonstrated enhanced nitrogen-fixing activity, enhancing legume growth even under stressful environmental conditions. These developments hold promise for the customization of biofertilizers according to crop requirements and local soil conditions, maximizing their benefits. The international market for biofertilizers is anticipated to grow at a CAGR of 12.23% and reach USD 5.16 billion by 2032. Nevertheless, there are still challenges with large-scale applications, such as regulatory hurdles, environmental conditions, and farmer ignorance. Future studies need to focus on elucidating the intricate interactions among biofertilizers, plants, and soil microbiota to create more predictable and consistent products.

7.16 Conclusion

Siderophores, iron-chelating compounds, are important in sustainable agriculture. They enhance the iron bioavailability of the rhizosphere, supporting improved plant growth and development. Siderophore-producing bacteria, such as certain *Pseudomonas* species, possess antagonistic properties against soil-borne pathogens, encouraging plant health and resilience. The bacteria are of interest for biofertilizer production, lowering the use of artificial pesticides and fertilizers. Knowledge of siderophore mechanisms in agricultural scenarios enables their optimized application. Field applications of biofertilizers based on siderophores have been found to be effective in enhancing crop productivity and soil quality. Siderophore-mediated biofertilizers enhance the stronger and auto-sustained agricultural ecosystem with enhanced soil microbiota and necessary nutrients. Biofertilizers also minimize the adverse impacts of chemical fertilizers like soil erosion and water pollution. New biofertilizer formulations become possible due to continuous research efforts that continue to elucidate siderophore-mediated interactions in the soil–plant system. Siderophore-based solutions represent a natural and pragmatic means to enhance crop yield while preserving ecological balance against the backdrop of increasing food requirements and environmental conservation.

AI disclosure statement: The author acknowledges the use of AI-based tools for minor language editing or grammatical correction. However, all intellectual contributions, analyses, and conclusions presented in this chapter are entirely the author's own.

References

1. Albelda-Berenguer, M., Monachon, M., Joseph, E., Siderophores: From natural roles to potential applications. *Adv. Appl. Microbiol.*, *106*, 193–225, 2019.
2. Alzate Zuluaga, M.Y., Fattorini, R., Cesco, S., Pii, Y., Plant-microbe interactions in the rhizosphere for smarter and more sustainable crop fertilization: the case of PGPR-based biofertilizers. *Front. Microbiol.*, *15*, 1440978, 2024.
3. Bertola, M., Ferrarini, A., Visioli, G., Improvement of soil microbial diversity through sustainable agricultural practices and its evaluation by-omics approaches: A perspective for the environment, food quality and human safety. *Microorganisms*, *9*, 7, 1400, 2021.
4. Bolan, N., Srivastava, P., Rao, C.S., Satyanaraya, P.V., Anderson, G.C., Bolan, S., Kirkham, M.B., Distribution, characteristics and management of calcareous soils. *Adv. Agron.*, *182*, 81–130, 2023.
5. Bonaventure, P., Guentas, L., Burtet-Sarramegna, V., Amir, H., Potential of Halophytes-Associated Microbes for the Phytoremediation of Metal-Polluted Saline Soils. *Appl. Sci.*, *13*, 7, 4228, 2023.
6. Chaithra, M., Nishmitha, K., Patel, P.S., Mitra, D., Gangaraj, R., Sayyed, R.Z., Fungal Bio-stimulants: Cutting-Edge Bioinoculants for Sustainable Agriculture, in: *Plant Microbiome and Biological Control: Emerging trends and applications*, pp. 289–307, Springer Nature Switzerland, Cham, 2025.
7. Chaudhary, P., Singh, S., Chaudhary, A., Sharma, A., Kumar, G., Overview of biofertilizers in crop production and stress management for sustainable agriculture. *Front. Plant Sci.*, *13*, 930340, 2022.
8. Chen, Q., Song, Y., An, Y., Lu, Y., Zhong, G., Mechanisms and Impact of Rhizosphere Microbial Metabolites on Crop Health, Traits, Functional Components: A Comprehensive Review. *Molecules*, *29*, 24, 5922, 2024.
9. Colombo, C., Palumbo, G., He, J.Z., Pinton, R., Cesco, S., Review on iron availability in soil: interaction of Fe minerals, plants, and microbes. *J. Soils Sediments*, *14*, 538–548, 2014.
10. Coonan, E.C., Kirkby, C.A., Kirkegaard, J.A., Amidy, M.R., Strong, C.L., Richardson, A.E., Microorganisms and nutrient stoichiometry as mediators of soil organic matter dynamics. *Nutr. Cycling Agroecosyst.*, *117*, 3, 273–298, 2020.
11. Dai, Z., Xiong, X., Zhu, H., Xu, H., Leng, P., Li, J., Xu, J., Association of biochar properties with changes in soil bacterial, fungal and fauna communities and nutrient cycling processes. *Biochar*, *3*, 239–254, 2021.
12. Deb, C.R. and Tatung, M., Siderophore producing bacteria as biocontrol agent against phytopathogens for a better environment: A review. *S. Afr. J. Bot.*, *165*, 153–162, 2024.
13. Dhuldhaj, U. and Pandya, U., Diversity, function, and application of fungal iron chelators (Siderophores) for integrated disease management, in: *Role of Microbial Communities for Sustainability*, pp. 259–288, 2021.

14. Duque, L.O. and Villordon, A., Root branching and nutrient efficiency: status and way forward in root and tuber crops. *Front. Plant Sci.*, *10*, 237, 2019.
15. Dzięcioł, J., Szlachetka, O., Rodrigues Tavares, J.M., From Volcanic Popcorn to the Material of the Future: A Critical Review of Expanded Perlite Applications and Environmental Impacts. *Sustainability*, *17*, 4, 1454, 2025.
16. Etesami, H. and Adl, S.M., Plant growth-promoting rhizobacteria (PGPR) and their action mechanisms in availability of nutrients to plants, in: *Phyto-Microbiome in Stress Regulation*, pp. 147–203, 2020.
17. Feng, Z., Sun, H., Qin, Y., Zhou, Y., Zhu, H., Yao, Q., A synthetic community of siderophore-producing bacteria increases soil selenium bioavailability and plant uptake through regulation of the soil microbiome. *Sci. Total Environ.*, *871*, 162076, 2023.
18. Gavrilescu, M., Water, soil, and plants interactions in a threatened environment. *Water*, *13*, 19, 2746, 2021.
19. Gomes, A.F., Almeida, M.C., Sousa, E., Resende, D., II, Siderophores and metallophores: Metal complexation weapons to fight environmental pollution. *Sci. Total Environ.*, 173044, 2024.
20. Gorny, J., Billon, G., Noiriel, C., Dumoulin, D., Lesven, L., Madé, B., Chromium behavior in aquatic environments: a review. *Environ. Rev.*, *24*, 4, 503–516, 2016.
21. Guo, J., Muhammad, H., Lv, X., Wei, T., Ren, X., Jia, H., Hua, L., Prospects and applications of plant growth promoting rhizobacteria to mitigate soil metal contamination: A review. *Chemosphere*, *246*, 125823, 2020.
22. Hati, A.J. and Singh, R.R., Smart indoor farms: Leveraging technological advancements to power a sustainable agricultural revolution. *AgriEngineering*, 3, 4, 728–767, 2021.
23. Jaiswal, S.K., Mohammed, M., Ibny, F.Y., Dakora, F.D., Rhizobia as a source of plant growth-promoting molecules: potential applications and possible operational mechanisms. *Front. Sustainable Food Syst.*, *4*, 619676, 2021.
24. Kaipanchery, V., Sharma, A., Albericio, F., de la Torre, B.G., Insights into the chemistry of the amphibactin–metal (M3+) interaction and its role in antibiotic resistance. *Sci. Rep.*, *10*, 1, 21049, 2020.
25. Kapoor, D., Sharma, P., Sharma, M.M.M., Yadav, S., Husen, A., Exploring soil microbiota and their role in plant growth, stress tolerance, disease control and nutrient immobilizer. *Biocatal. Agric. Biotechnol.*, 103358, 2024.
26. Khan, N., Humm, E.A., Jayakarunakaran, A., Hirsch, A.M., Reviewing and renewing the use of beneficial root and soil bacteria for plant growth and sustainability in nutrient-poor, arid soils. *Front. Plant Sci.*, *14*, 1147535, 2023.
27. Kour, D., Rana, K.L., Yadav, A.N., Yadav, N., Kumar, M., Kumar, V., Saxena, A.K., Microbial biofertilizers: Bioresources and eco-friendly technologies for agricultural and environmental sustainability. *Biocatal. Agric. Biotechnol.*, *23*, 101487, 2020.
28. Kroh, G.E. and Pilon, M., Regulation of iron homeostasis and use in chloroplasts. *Int. J. Mol. Sci.*, *21*, 9, 3395, 2020.

29. Kumar, R., Singh, A., Shukla, E., Singh, P., Khan, A., Singh, N.K., Srivastava, A., Siderophore of plant growth promoting rhizobacterium origin reduces reactive oxygen species mediated injury in Solanum spp. caused by fungal pathogens. *J. Appl. Microbiol.*, *135*, 2, lxae036, 2024.
30. Kumar, S., Sindhu, S.S., Kumar, R., Biofertilizers: An ecofriendly technology for nutrient recycling and environmental sustainability. *Curr. Res. Microb. Sci.*, *3*, 100094, 2022.
31. Kumawat, K.C., Nagpal, S., Sharma, P., Present scenario of bio-fertilizer production and marketing around the globe, in: *Biofertilizers*, pp. 389–413, Woodhead Publishing, Cambridge, Cambridgeshire, United Kingdom, 2021.
32. Leila, B. and El-Hafid, N., Biofertilizers and biopesticides: Microbes for sustainable agriculture, in: *Advances in Plant Microbiome and Sustainable Agriculture: Diversity and Biotechnological Applications*, pp. 257–279, 2020.
33. Liu, L., Wang, W., Wu, S., Gao, H., Recent advances in the siderophore biology of Shewanella. *Front. Microbiol.*, *13*, 823758, 2022.
34. Liu, Y., Xiong, Z., Wu, W., Ling, H.Q., Kong, D., Iron in the Symbiosis of Plants and Microorganisms. *Plants*, *12*, 10, 1958, 2023.
35. Mącik, M., Gryta, A., Frąc, M., Biofertilizers in agriculture: An overview on concepts, strategies and effects on soil microorganisms, in: *Advances in Agronomy*, vol. *162*, pp. 31–87, 2020.
36. Mahapatra, D.M., Satapathy, K.C., Panda, B., Biofertilizers and nanofertilizers for sustainable agriculture: Phycoprospects and challenges. *Sci. Total Environ.*, *803*, 149990, 2022.
37. Majewska, M., Słomka, A., Hanaka, A., Siderophore-producing bacteria from Spitsbergen soils—novel agents assisted in bioremediation of the metal-polluted soils. *Environ. Sci. Pollut. Res.*, 1–11, 2024.
38. Majumdar, S., Muruganantham, L., Karthik, R., Wanniang, S.K., Prakash, N.B., Zinc and iron solubilizing microbial biofertilizer: a potential tool for sustainable horticultural crop production, in: *Bio-Inoculants in Horticultural Crops*, pp. 197–234, Woodhead Publishing, Cambridge, Cambridgeshire, United Kingdom, 2024.
39. Meena, M., Swapnil, P., Divyanshu, K., Kumar, S., Harish, Tripathi, Y.N., Upadhyay, R.S., PGPR-mediated induction of systemic resistance and physiochemical alterations in plants against the pathogens: Current perspectives. *J. Basic Microbiol.*, 60, 10, 828–861, 2020.
40. Mehmood, N., Saeed, M., Zafarullah, S., Hyder, S., Rizvi, Z.F., Gondal, A.S., Kupe, M., Multifaceted impacts of plant-beneficial pseudomonas spp. in managing various plant diseases and crop yield improvement. *ACS Omega*, *8*, 25, 22296–22315, 2023.
41. Molnár, Z., Solomon, W., Mutum, L., Janda, T., Understanding the mechanisms of Fe deficiency in the rhizosphere to promote plant resilience. *Plants*, *12*, 10, 1945, 2023.

42. Mosier, S., Córdova, S.C., Robertson, G.P., Restoring soil fertility on degraded lands to meet food, fuel, and climate security needs *via* perennialization. *Front. Sustainable Food Syst.*, 5, 706142, 2021.
43. Nithyapriya, S., Sundaram, L., Eswaran, S.U.D., Perveen, K., Alshaikh, N.A., Sayyed, R.Z., Mastinu, A., Purification and Characterization of Desferrioxamine B of Pseudomonas fluorescens and Its Application to Improve Oil Content, Nutrient Uptake, and Plant Growth in Peanuts. *Microb. Ecol.*, *87*, 1, 1–13, 2024.
44. Njeru, E.M. and Koskey, G., Using beneficial microorganisms to promote sustainable crop production and resilience of smallholder agroecosystems to changing climate, in: *Climate Change and Resilient Food Systems: Issues, Challenges, and Way Forward*, pp. 287–314, 2021.
45. Pattnaik, S., Mohapatra, B., Gupta, A., Plant growth-promoting microbe mediated uptake of essential nutrients (Fe, P, K) for crop stress management: Microbe–soil–plant continuum. *Front. Agron.*, *3*, 689972, 2021.
46. Pecoraro, L., Wang, X., Shah, D., Song, X., Kumar, V., Shakoor, A., Rani, R., Biosynthesis pathways, transport mechanisms and biotechnological applications of fungal siderophores. *J. Fungi*, *8*, 1, 21, 2021.
47. Prabhakar, P.K., Bacterial siderophores and their potential applications: a review. *Curr. Mol. Pharmacol.*, *13*, 4, 295–305, 2020.
48. Rabani, M.S., Hameed, I., Gupta, M.K., Wani, B.A., Fayaz, M., Hussain, H., Ahad, M.B., Introduction of biofertilizers in agriculture with emphasis on nitrogen fixers and phosphate solubilizers, in: *Microbiomes for the Management of Agricultural Sustainability*, pp. 71–93, Springer Nature Switzerland, Cham, 2023.
49. Rahal, S., Menaa, B., Chekireb, D., Screening of heavy metal-resistant rhizobial and non-rhizobial microflora isolated from Trifolium sp. growing in mining areas. *Environ. Monit. Assess.*, *196*, 3, 283, 2024.
50. Richa, Metabolite-based Bioformulation: Next Generation of Biofertilizers, in: *Metabolomics, Proteomics and Gene Editing Approaches in Biofertilizer Industry: Volume II*, pp. 53–81, Springer Nature Singapore, Singapore, 2024.
51. Saha, L., Tiwari, J., Bauddh, K., Ma, Y., Recent developments in microbe–plant-based bioremediation for tackling heavy metal-polluted soils. *Front. Microbiol.*, *12*, 731723, 2021.
52. Saikanth, K., Singh, B.V., Sachan, D.S., Singh, B., Advancing sustainable agriculture: a comprehensive review for optimizing food production and environmental conservation. *Int. J. Plant Soil Sci.*, *35*, 16, 417–425, 2023.
53. Sarker, A., Ansary, M.W.R., Hossain, M.N., Islam, T., Prospect and challenges for sustainable management of climate change-associated stresses to soil and plant health by beneficial rhizobacteria. *Stresses*, *1*, 4, 200–222, 2021.
54. Schalk, I.J., Bacterial siderophores: diversity, uptake pathways and applications. *Nat. Rev. Microbiol.*, 1–17, 2024.
55. Schouten, H.J. and Bai, Y., Obtaining disease resistance by changing S-genes. *Prophyta Annu.*, *2017*, 44–45, 2017.

56. Seenivasagan, R. and Babalola, O.O., Utilization of microbial consortia as biofertilizers and biopesticides for the production of feasible agricultural product. *Biology*, *10*, 11, 1111, 2021.
57. Sepehri, M. and Khatabi, B., Combination of siderophore-producing bacteria and Piriformospora indica provides an efficient approach to improve cadmium tolerance in alfalfa. *Microb. Ecol.*, *81*, 717–730, 2021.
58. Shaikh, T.A., Rasool, T., Lone, F.R., Towards leveraging the role of machine learning and artificial intelligence in precision agriculture and smart farming. *Comput. Electron. Agric.*, *198*, 107119, 2022.
59. Sharma, R., Jasrotia, T., Kumar, R., Kumar, R., Umar, A., Alharthi, F.A., Al-Zaqri, N., An insight into the mechanism of 'symbiotic-bioremoval'of heavy metal ions from synthetic and industrial samples using bacterial consortium. *Environ. Technol. Innovation*, *21*, 101302, 2021.
60. Singh, D., Thapa, S., Geat, N., Mehriya, M.L., Rajawat, M.V.S., Biofertilizers: Mechanisms and application, in: *Biofertilizers*, pp. 151–166, Woodhead Publishing, Cambridge, Cambridgeshire, United Kingdom, 2021.
61. Singh, P., Chauhan, P.K., Upadhyay, S.K., Singh, R.K., Dwivedi, P., Wang, J., Jiang, M., Mechanistic insights and potential use of siderophores producing microbes in rhizosphere for mitigation of stress in plants grown in degraded land. *Front. Microbiol.*, *13*, 898979, 2022.
62. Sivaram, A.K., Abinandan, S., Chen, C., Venkateswartlu, K., Megharaj, M., Microbial inoculant carriers: Soil health improvement and moisture retention in sustainable agriculture. *Adv. Agron.*, *180*, 35–91, 2023.
63. Soares, E.V., Perspective on the biotechnological production of bacterial siderophores and their use. *Appl. Microbiol. Biotechnol.*, *106*, 11, 3985–4004, 2022.
64. Soares, M.P. and Weiss, G., The Iron age of host–microbe interactions. *EMBO Rep.*, *16*, 11, 1482–1500, 2015.
65. Srikanth, P., Sivakumar, D., Sharma, A., Kaushik, N., Recent developments in omics techniques for improving plant abiotic stress using microbes. *Int. J. Environ. Sci. Technol.*, 1–24, 2024.
66. Suchithra, M.R., Muniswami, D.M., Sri, M.S., Usha, R., Rasheeq, A.A., Preethi, B.A., Dineshkumar, R., Effectiveness of green microalgae as biostimulants and biofertilizer through foliar spray and soil drench method for tomato cultivation. *S. Afr. J. Bot.*, *146*, 740–750, 2022.
67. Sun, X., Zhang, C., Bei, S., Wang, G., Geisen, S., Bedoussac, L., Zhang, J., High bacterial diversity and siderophore-producing bacteria collectively suppress Fusarium oxysporum in maize/faba bean intercropping. *Front. Microbiol.*, *13*, 972587, 2022.
68. Swarnalatha, G.V., Goudar, V., Reddy, E. C. R. G. S., Al Tawaha, A.R.M., Sayyed, R.Z., Siderophores and their applications in sustainable management of plant diseases, in: *Secondary Metabolites and Volatiles of PGPR in Plant-Growth Promotion*, pp. 289–302, 2022.

69. Thakur, B. and Kaur, S., The role of artificial intelligence in biofertilizer development, in: *Metabolomics, Proteomics and Gene Editing Approaches in Biofertilizer Industry: Volume II*, pp. 157–176, Springer Nature Singapore, Singapore, 2024.
70. Timofeeva, A.M., Galyamova, M.R., Sedykh, S.E., Bacterial siderophores: Classification, biosynthesis, perspectives of use in agriculture. *Plants*, *11*, 22, 3065, 2022.
71. Tiwari, A.K., Moond, V., Choudhari, R.J., Badekhan, A., Tejasree, P., Baral, K., Bharti, R., Optimizing Bio-fertilizers to Address Food Security and Advance Nutritional Sustainability. *J. Exp. Agric. Int.*, *45*, 12, 241–255, 2023.
72. Wang, Y., Zhang, G., Huang, Y., Guo, M., Song, J., Zhang, T., Liu, H., A potential biofertilizer—siderophilic bacteria isolated from the rhizosphere of Paris polyphylla var. yunnanensis. *Front. Microbiol.*, *13*, 870413, 2022.
73. Wiche, O., Dittrich, C., Pourret, O., Monei, N., Heim, J., Lambers, H., Relationships between carboxylate-based nutrient-acquisition strategies, phosphorus-nutritional status and rare earth element accumulation in plants. *Plant Soil*, *489*, 1, 645–666, 2023.
74. Wu, D., Wang, W., Yao, Y., Li, H., Wang, Q., Niu, B., Microbial interactions within beneficial consortia promote soil health. *Sci. Total Environ.*, 165801, 2023.
75. Yadav, A. and Yadav, K., Challenges and Opportunities in Biofertilizer Commercialization. *SVOA Microbiol.*, *5*, 1, 01–14, 2024.
76. Zhu, J., Chen, T., Ju, Y., Dai, J., Zhuge, X., Transmission Dynamics and Novel treatments of high risk carbapenem-resistant Klebsiella pneumoniae: the Lens of One Health. *Pharmaceuticals*, *17*, 9, 1206, 2024.
77. Zulfiqar, U., Ayub, A., Hussain, S., Ahmad, M., Rehman, A., Ishfaq, M., Yong, J.W.H., Iron biofortification in cereal crops: Recent progress and prospects. *Food Energy Secur.*, *13*, 4, e547, 2024.

8

Siderophores in Biocontrol of Insects: Suppressing Insects and Enhancing Plant Defense

Bimal Kumar Sahoo[1], Ansh Raj[2], Ipsita Samal[2*], Deepak Kumar Mahanta[3] and Tanmaya Kumar Bhoi[4]

[1]*Department of Agricultural Entomology, Centre for Plant Protection Studies, Tamil Nadu Agricultural University, Coimbatore, India*
[2]*ICAR-National Research Centre on Litchi, Mushahari, Muzaffarpur, Bihar, India*
[3]*Forest Entomology Discipline, Forest Protection Division, Indian Council of Forestry Research and Education (ICFRE)-Forest Research Institute (ICFRE-FRI), Dehradun, Uttarakhand, India*
[4]*Forest Protection Division, ICFRE-Arid Forest Research Institute (ICFRE-AFRI), Jodhpur, Rajasthan, India*

Abstract
Siderophores are compounds with a strong affinity for binding iron, which are released by various microorganisms, and perform an essential role in plant health and defense mechanisms. This chapter explores the dual functionality of siderophores in biocontrol strategies, focusing on their ability to suppress insect pests while simultaneously enhancing plant defenses against pathogens. By facilitating iron acquisition in nutrient-deficient soils, siderophores not only improve plant growth but also trigger systemic immune responses that bolster resistance to phytopathogens. Recent studies indicate that siderophore-producing bacteria can inhibit the proliferation of harmful insects and pathogens, thereby serving as effective biocontrol agents. The chapter delves into the mechanisms through which siderophores mediate these effects, including their role in inducing systemic resistance (ISR) in plants and disrupting the iron acquisition strategies of insect pests. Additionally, the potential for integrating siderophore applications into sustainable agricultural practices is discussed, highlighting their eco-friendly nature compared to synthetic pesticides. The implications for enhancing crop resilience

**Corresponding author*: happyipsu29@gmail.com

Debasis Mitra, Ayush Madan and Anju Rani (eds.) Microbial Siderophores for Agricultural Productivity: Perspectives of Use in Agriculture, (183–206) © 2026 Scrivener Publishing LLC

and productivity in the face of biotic stressors are significant, positioning siderophores as a promising tool in modern agricultural biocontrol strategies. Overall, this chapter emphasizes the importance of understanding the multifaceted roles of siderophores in plant–insect interactions and their potential applications in promoting sustainable agriculture through enhanced biocontrol mechanisms.

***Keywords*:** Siderophores, insect pests, reactive oxidative stress, PGPR, ISR

8.1 Introduction

Agricultural crops are threatened by insect pests by inflicting direct damage by impeding plant growth during vegetative and reproductive stages or indirect damage by spreading several diseases as vectors, resulting in substantial economic losses, ultimately jeopardizing food security worldwide [1–4]. Traditionally, chemical pesticides have been employed to control insect pests to protect the crops, but this approach harms the agroecosystem [5]; biomagnification [6, 7], development of resistance, and secondary pest outbreaks complicate pest management efforts even further [8]. Chronic exposure to these harmful chemicals at lower doses can trigger several detrimental impacts on human health like immune suppression, reproductive deformities, hormonal imbalances, and even deadly diseases like cancer [9, 10]. Furthermore, soil fertility can be degraded by the application of these pesticides which ultimately results in disruption of the soil microflora composition. Additionally, the water bodies can also be contaminated as these chemicals leach into the environment [11]. So, effective alternative control measures for managing these notorious insect pests are thus imperative [12]. The use of biocontrol agents can be a viable alternative to these synthetic chemical pesticides as these agents can effectively control insect pests without posing detrimental impacts to human health as well as the ecosystem [13–15]. These biocontrol agents include numerous microorganisms like bacteria and fungi [16] as well as insect antagonists like predators and parasitoids [17]. Apart from the production of antibiotics and lytic enzymes, these biocontrol agents scavenge iron by producing siderophores, limiting the iron acquisition by insect pests [18, 19]. Furthermore, antagonistic insects harbor pathogenic bacteria in their gut that sequester iron through the production of siderophores, thereby limiting the availability of iron to the host insect and inhibiting its growth [17].

Pathogenic microorganisms as well as insect pests require iron for their growth and development, and they obtain this iron from the host they infect and colonize [20]. Iron (Fe) ranks as the fourth most prevalent metal in the Earth's crust, following oxygen, silicon, and aluminum [21, 22]. It serves important functions in oxidation–reduction reactions [23]. Iron is also directly involved in several metabolic processes in plants such as photosynthesis, chlorophyll synthesis, and respiration-related pathways such as oxidative phosphorylation, the electron transport chain, and tricarboxylic acid cycle as well as nitrogen fixation and antioxidant defense [24, 25]. It also plays a major role in the synthesis of various important compounds like nucleic acids, heme proteins, siderophores, cofactors, vitamins, secondary metabolites, antibiotics, enzymes, signaling molecules, toxins, and pigments [24]. Fe is a micronutrient, and Fe insufficiency is caused by reduced Fe bioavailability rather than low total Fe concentrations [26]; to overcome these constraints, different microbes like fungi and bacteria as well as plants have been shown to absorb Fe utilizing siderophores [27].

These siderophores are iron-chelating chemicals that bind particularly to iron, resulting in limiting the quantity of absorbable iron for both phytopathogens and insect pests [28]. In this way, due to the deprivation of essential iron nutrients, the growth of plant pathogens and insect pests is suppressed [19, 29]. At the same time, iron is supplied to plants in iron-limited conditions to promote plant growth and development as well as enhance the plant defense system against phytopathogens and insect pests [30]. This chapter explores the possible role of siderophores as a biocontrol agent by suppressing the natural pest population as well as enhancing plant defense.

8.2 Siderophore and Their Mode of Action

The term "siderophore" is derived from the Greek word "siderophore", meaning "iron carrier" [33]. Siderophore, a small Fe (III)-binding molecule with lower molecular mass, acts as a biogenic chelator with exceptional affinity and selectivity for Fe complexes [31]. Many bacteria, yeasts, fungi, and plants make and secrete siderophores for Fe (III) chelation because they have mechanisms to create Fe ions in iron-deficient circumstances [32]. Neilands [34] first isolated the crystalline form of siderophores, which laid the foundation for extensive research on these compounds in the future [11]. Subsequently, Klopper *et al.* [35] presented initial evidence supporting the role of siderophore-producing plant growth-promoting rhizobacteria (PGPR) as biocontrol agents. The most abundant form of iron in the environment is the ferric form which is predominantly unavailable to plants and microorganisms at specific pH conditions (7.35–7.40).

To address this challenge, microorganisms produce siderophores that strongly bind to ferric iron. The ferric iron-siderophore complexes are then transported into the cytosol, where the ferric iron is converted to its ferrous form, making it available for the microorganism to use [36].

Siderophores exhibit a wide range of chemical diversity, including hydroxamate, hydroxylate, carboxylate-type, peptide-based, catecholate, phenolate, and mixed-type siderophores [37]. Some of the siderophores also show their action as elicitors for ISR in plants. In this, pyoverdines or pseudobactines, secreted by *Pseudomonas*, were one of the first PGPR metabolites studied for ISR elicitors in 1990. The different types of siderophores secreted by different microorganisms are achromobactin, citrate, ferrichrome, desferrioxamine-E, enterobactin, pyochelin, yersiniabactin, aerobactin, and pyoverdine [38, 39]. Siderophores exhibit a wide range of chemical diversity, including carboxylate-type, hydroxamate, catecholate, phenolate, and mixed-type varieties [37]. Sideropres can also play a role as elicitors for inducing ISR in plants—for example, siderophores like pyoverdines or pseudobactines, secreted by *Pseudomonas*, were among the first PGPR products studied as ISR elicitors in 1990. Various microorganisms secrete different types of siderophores, such as citrate, aerobactin, ferrichrome, desferrioxamine-E, enterobactin, achromobactin, yersiniabactin, pyoverdine, and pyochelin [38, 39].

Microbes employ several unique mechanisms to absorb iron from the environment, one of which involves the production of Fe^{3+}-chelating molecules called siderophores [40]. From the surroundings, siderophores are capable of binding even trace amounts of Fe^{3+}, enabling iron uptake by microorganisms while limiting the precipitation and loss of bioavailability [41].

Ferric ion-depleted conditions significantly reduce the hydrophobicity of microbial surfaces by altering the composition of surface proteins, ultimately hindering biofilm formation and eventually microbial death. As a result, during the scarcity of Fe^{+3} in the environment, microorganisms manufacture and release siderophores to cope with the Fe-deficient habitat [42]. During their transportation into the microbes, these Fe^{+3} coordinate with siderophores to form a hexadentate octahedral complex [44]. Subsequently, within the microorganisms, the Fe^{2+}–siderophore complexes break apart, releasing the siderophores, which are then recycled for further Fe^{3+} uptake [45]. Siderophores increase the concentration of accessible Fe^{3+} by solubilizing insoluble iron particles, which directly impacts the availability of Fe in the surrounding habitat, influencing the diversity and dynamics of microbial populations [46]. Furthermore, this process affects the access of microbes to Fe^{3+} that are unable to produce siderophores, thereby regulating Fe levels and other biogeochemical cycles [46].

8.3 Insect Pathogenic Microorganisms Producing Siderophores

Microbes, particularly bacteria and fungi, play a vital role in pest control by producing siderophores, iron-chelating molecules that deprive pests and pathogens of essential iron (Table 8.1).

8.3.1 Bacteria

Entomopathogenic bacteria are regarded as effective biocontrol agents, capable of promoting plant growth and development and providing an eco-friendly alternative to synthetic insecticides [5]. These groups of bacteria include *Bacillus*, *Serratia*, *Xenorhabdus*, and *Photorhabdus*. Bacterial species like *Photorhabdus* and *Xenorhabdus* are entomopathogenic bacteria

Table 8.1 Siderophores produced by different entomopathogens.

Sl. no.	Organism	Siderophore	Type	References
1	*Beauveria bassiana*	Ferrichrome	Hydroxamate	[20]
2	*Metarhizium anisopliae*	Ferrichrome	Hydroxamate	
3	*Xenorhabdus nematophila*	Avaroferrin	Hydroxamate	
4	*Xenorhabdus doucetiae*	Putrebactin	Hydroxamate	
5	*Photorhabdus luminescens*	Aerobactin	Hydroxamate	
6	*Pseudomonas fluorescens*	Pyoverdine	Hydroxamate	[119]
7	*Pseudomonas putida*	Pyoverdine	Hydroxamate	
8	*Bacillus thuringiensis*	Bacillibactin	Hydroxamate	
9	*Bacillus subtilis*	Bacillibactin	Hydroxamate	[120]
10	*Serratia marcescens*	Aerobactin	Hydroxamate	
11	*Aspergillus flavus*	Aurofusarin	Hydroxamate	
12	*Metarhiziumrileyi*	Oosporein	Hydroxamate	

associated with entomopathogenic nematode and are indirectly involved in producing virulence against insect pests by synthesizing a range of metabolites and protein toxins [47].

Insect pathogen belonging to the genus *Serratia* isolated from entomopathogenic nematode *Caenorhabditis briggsae* had several enzymatic activities such as gelatinase, DNase, protease, and chitinase activity in addition to siderophore synthesis [48, 49]. The most well-known *Bacillus* species for producing siderophores include *Bacillus licheniformis, Bacillus anthracis, Bacillus velezensis, Bacillus thuringiensis, Bacillus cereus, Bacillus halodenitrificans, Bacillus atrophaeus, Bacillus mojavensis, Bacillus pumilus*, and *Bacillus subtilis* [50–53]. Similarly, *Bacillus thuringiensis* can produce siderophores like bacillibactin [54–57]. A study conducted by Williams *et al.* [58] revealed that though *B. thuringiensis* was not able to produce siderophore petrobactin, it produced protocatechuic acid or 3,4-dihydroxybenzoic acid which was the iron-chelating component of petrobactin.

Other siderophores include photobactin, isolated from *Photorhabdus luminescens* NC19 [59]. Similarly, certain hydroxamate siderophores like avaroferrin, ochrobactin C, aerobactin, and putrebactin are vital for the virulence of nematode endosymbionts like *Xenorhabdus* and *Photorhabdus* species, out of which aerobactin plays an important role in pathogenicity toward *Galleria mellonella* [20]. *Pseudomonas entomophila* produces two siderophores, pyoverdine and pseudomonine, which have also been characterized in *Pseudomonas fluorescens* [60]. In addition to pyoverdine, several other secondary siderophores have been identified in *fluorescent pseudomonads*, including thioquinolobactin, pyochelin, *pseudomonine, corrugatin*, and ornicorrugatin [61–66]. Several factors influence the effectiveness of siderophore-producing bacteria as biocontrol agents, including their iron-binding affinity, amount of siderophore produced, stability of the Fe–siderophore complex, the availability of receptors for siderophore activation, and the interactions between the siderophore-producing bacteria and the target insect [28].

8.3.2 Fungi

Nomuraea, Beauveria, Metarhizium, Paecilomyces, Hirsutella, and *Verticillium* are the major fungal genera utilized as bioinsecticides [67]. Entomopathogenic fungi infect the insect pests through two different mechanisms: direct and indirect infection processes. By direct infection, entomopathogenic fungi enter the interior of the insect body through natural orifices or by degrading the insect's cuticle by secretion of enzymes [43]. By indirect mechanism, these fungi can also produce siderophores to acquire iron nutrients from the insect hemolymph, limiting the iron nutrient for insect growth and development and thus resulting in the ultimate death of insects or by secreting toxic or repellent/antifeedant secondary metabolites and volatiles for attracting natural enemies or activating plant defense [67, 68].

Endophytic insect pathogenic fungi have the potential to act as biocontrol agents; they assist plants to respond to different biotic and abiotic factors like iron deficiency [69]. Fungi possess two different types of mechanisms for iron uptake: non-ribosomal synthesized Fe^{3+} chelators (siderophore-mediated iron acquisition, SIA) and reductive iron assimilation, RIA (ferric reductase-mediated conversion of insoluble Fe^{3+} to soluble Fe^{2+}) [70]. Siderophores are the compounds secreted by these fungi which may help to improve the solubilization of Fe in soil, which, in turn, is easily accessible to plants *via* elevated microbial activity [71]. Unlike bacteria, fungi are poor competitors for nutrients with plants [72].

The fungus *Rhizopus microsporus* var. *rhizopodiformis* is the source of the siderophore rhizoferrin, which was first isolated and characterized [82]. This molecule was classified under carboxylate type which is synthesized by linking two citric acid molecules to 1,4-diaminobutane through a pair of amide bonds [83]. Ferricrocin, a siderophore, has been identified as a virulence factor in fungi that target insect hosts. It is commonly found in *Aspergillus fumigatus*, *A. nidulans*, *B. bassiana*, *Metarhizium robertsii*, and *M. grisea* [73–78]. It is vital for virulence against insects, oxidative stress resistance, conidiation, and conidial germination [74, 76, 78]. Similarly, *Metarhizium brunneum* and *B. bassiana* could enhance the iron concentration both in dry matter and substrate by exudating iron-binding siderophores [69]. Under iron-limiting conditions, the entomopathogenic fungi *M. anisopliae* and *M. robertsii* synthesize a mixture of extracellular glycosylated and non-glycosylated coprogen-type siderophores, along with Nα-dimethylcoprogen and dimerumic acid where metachelin A and metachelin B were the main components of siderophores [27]. However, the results from Donzelli *et al.* [70] revealed no detection of metachelin A and its associative derivatives in the samples from the mrsidD mutant D3 under iron-depleted conditions. Barra-Bucarei *et al.* [79] suggested that *B. bassiana* was an efficient siderophore producer and was effective in the control of whitefly in tomato crops under greenhouse conditions. Similarly, *B. bassiana* synthesizes the well-known intracellular siderophores tenellin and ferricrocin when exposed to iron-rich conditions [80]. Recently, *Trichoderma hamatum* has been documented as an entomopathogen, and the siderophore rhizoferrin is responsible for its insecticidal properties [81].

8.4 Molecular Mechanisms of Siderophores Activity against Insect Pests

The synthesis of siderophores in microorganisms is tightly regulated by specific genes. These genes are often located in clusters within the microbial genome and are controlled by iron-responsive regulators—for example, the exbD gene encodes a part of the TonB complex (comprising ExbD, ExbB, and TonB,), which is essential for siderophore-mediated uptake of ferric iron (Fe^{3+}) in numerous bacteria [84]. The exbD gene also plays a crucial role for pathogenicity and symbiosis in *Photorhabdus temperata* K122 [85]. Based on this, Watson *et al.* [86] investigated the virulence potential of *Photorhabdus luminescens* mutants with targeted deletions in *exbD*, *feoABC*, and *yfeABCD* against *Manduca sexta* and found that exbD mutant was unable to obtain Fe^{3+} in iron-limited media, which resulted in avirulence to the insect. The removal of the *exbD* gene, which encodes a component of the TonB complex, resulted in an impairment of pathogenicity in this bacterial strain, along with *Photorhabdus temperata* [85]. However, the yfeABCD-encoded transporter for divalent cations is crucial for iron uptake (*via* the TonB complex and the Yfe transporter) and the virulence of *Photorhabdus luminescens* to *Manduca sexta*. Another study conducted by Li *et al.* [29, 87] reported that deletion of NrSidA and MaCFEM1 resulted in loss of virulence of *N. rileyi* and *M. acridium* against *Spodoptera litura* and *Locusta migratoria*, respectively.

Two different types of biosynthesis gene cluster (BGC) arrangements such as aerobactin BGC (hydroxamate I) and desferrioxamine BGC (hydroxamate II) were involved in siderophore production in *Xenorhabdus* and *Photorhabdus*. However, some stains of *Xenorhabdus* and *Photorhabdus* were unable to produce these gene clusters. In those stains, genes that regulate the production of enzymes linked to 2,3-dihydroxybenzoic acid (2,3-DHB) or 2-hydroxybenzoic acid (2-HB, salicylic acid) contributing to iron binding mechanisms were responsible for siderophore production [20]. The transcriptome analysis of *Photorhabdus temperata* during its contamination of the *Rhizotrogus majalis* host insect demonstrated that the *pchC* gene, responsible for the production of pyochelin-type siderophores, was upregulated [88, 89]. The virulence potential of bacteria depends on the presence of different ion transporters for the ferrous ion in the siderophore uptake system. In *Photorhabdus luminescens*, Fe (II) intake mechanism YfeABCD is essential for complete pathogenicity in the larvae of *Manduca sexta* and also contributes to the recolonization process of the nematode [86].

Beyond insect–pathogenic bacteria, entomopathogenic fungi also play a crucial role in insect pest control by producing siderophores under iron-limiting conditions. The biosynthesis of these siderophores is driven by specific genes, which are key to various fungal processes, including vegetative growth, stress resistance, development, fungal physiology, host interactions, and virulence—for example, in the white muscardine fungus *Beauveria bassiana*, among the genes encoding enzymes for siderophore biosynthesis (SID), BbSidA is essential for synthesizing the majority of siderophores, while SidC and SidD are critically involved in the biosynthesis of ferricrocin initially and later fusarinine C. The polyketide synthase genes PKS14 and PKS15 that are widely preserved among entomopathogenic fungi contribute significantly to the virulence of insects. When these genes were overexpressed in *B. bassiana*, and the metabolites in the overexpressed strains were compared to knockout strains, several metabolites were detected, including compounds such as bassianolide, enniatin A, beauvericins, and ferricrocin, the last of which functions as a siderophore. Consequently, PKS14 and PKS15, together with their related proteins or elements of their gene clusters, play a critical role in insect pathogenesis and the secondary metabolic processes of *B. bassiana* [90].

García-Espinoza *et al.* [69] investigated the different ferric reductase enzyme gene levels in cucumber and melon plants subjected to iron-depleted environments responsible for siderophore production. It was found that in cucumber plants, iron-deficient plants inoculated with *Metarhizium brunneum* showed an elevated reductase activity by the upregulation of different genes FRO1, FRO2, FRO3, FRO4, IRT1, FIT, and HA1 in iron-deficient conditions compared to iron-sufficient ones [69]. *Metarhizium robertsii* produces two recognized iron-chelating compounds: metachelins, which scavenge iron from extracellular sources, and ferricrocin, which is involved in intracellular iron acquisition. Null mutants of the gene (mrsidD and mrsidC) involved in siderophore synthesis and iron uptake transcriptional repressor (mrsreA) exhibited a reduced rate of germination, altered sensitivity to hydrogen peroxide, and variable capacity to mitigate the effects of iron chelation under iron-deficient conditions. Under iron-provided conditions, *Metarhizium robertsii* elevated the levels of insect iron-binding proteins, such as transferrins, while infecting *Spodoptera exigua* [91].

8.5 Interaction of Siderophores with Fungal and Insect Physiology

Siderophores produced by entomopathogens sequester iron, limiting its availability to insects and weakening their immune systems while enhancing their pathogenicity by supporting enzymatic activity, modulating immune responses, influencing insect behavior, including impaired grooming, and facilitating fungal infection (Figure 8.1).

Siderophore production is essential for the virulence of entomopathogenic fungus *Nomuraea rileyi* against *S. litura* as iron is an indispensable element for the production of microsclerotia [29]. Besides microsclerotia production, siderophore-mediated iron uptake is also indispensable for the morphogenesis of pigmented microsclerotia, resistance to oxidative stress, spore formation, phase transition, and conidial germination of *N. rileyi*. Iron also plays a crucial role in the detoxification of reactive oxidative stress (ROS) in nutrient toxicity in fungus for proper nutrient mobilization. Based on this fact, loss of siderophore biosynthesis limited the intracellular iron, resulting in a deficiency of ROS detoxification during iron toxicity. The NrSidA mutants exhibited higher transcript levels of NrSOD at both the initiation and formation stages of microsclerotia, leading to an increase in ROS [29].

Li *et al.* [29] compared the virulent activity of wild types (WT), high iron permease activity (NrFtrA), and mutants of siderophores synthesis by deletion of genes (NrSidA) of *N. rileyi* against *S. litura*, and they found that NrFtrA and WT conidia successfully bypassed the insect immune defenses and transitioned to yeast-like budding, forming multiple hyphal bodies 72 h after fungal spore injection, leading to larval mortality before pupation.

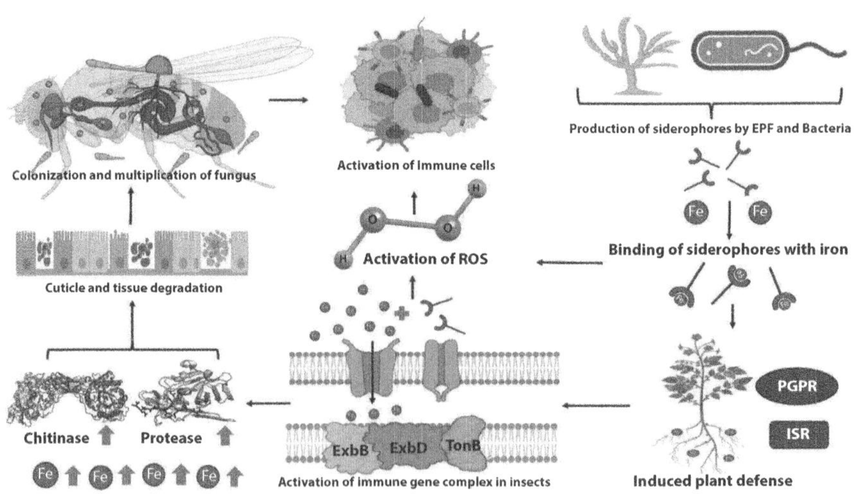

Figure 8.1 Schematic representation of siderophore-mediated pest control: interactions between entomopathogens, insects, and plants.

However, NrSidA spores were consistently attacked and encapsulated by insect hemocytes, with minimal yeast-like cell formation observed, leading to an inability to establish infection in *Spodoptera litura*. Similarly, in *A. fumigatus*, *sidA* conidia failed to germinate *in vivo*, resulting in a complete loss of virulence due to the absence of SidA [92].

In insects, iron is predominantly stored intracellularly or bound to ferritin [93]. Furthermore, insects utilize iron withholding as a form of nutritional immunity to restrict the availability of iron to invading microbes [94, 95]. MaCFEM1 is a protein that binds iron, allowing *M. acridum* to obtain iron in iron-limited environments. *M. acridum* faces challenges in the infection process due to the presence of symbiotic gut bacteria in migratory locusts. According to the report by Li *et al.* [87], *M. acridum* infected the migratory locust by the over-expression of MaCFEM1, which allowed *M. acridum* to compete for iron with symbiotic gut bacteria, resulting in a monopoly for iron resources by fungi. Furthermore, decreased proliferation of gut bacteria impaired the defense mechanism and immunity of host insects against fungi, making them vulnerable to fungal infection. During *M. acridum* (MaCFEM1) infection, certain bacterial symbionts of the host multiply uncontrollably, becoming opportunistic pathogens and substantially hastening the host's death.

8.6 Siderophores in Plant Growth, Development, and Defense Mechanisms

Bacteria that produce siderophores can enhance plant growth while offering protection against fungal infections and insect pest damage [96]. Several mechanisms by which plants acquire iron from microbial chelators have been suggested. One mechanism involves microbial siderophores chelating iron from the soil and exchanging their ligands with phytosiderophores [97]. Another mechanism involves microbial chelating agents binding soil iron and then exchanging their ligands with phytosiderophores [97].

Entomopathogenic fungi are capable of colonizing plant organs endophytically, including those in rhizosphere, without eliciting pathogenic responses in the plant. These fungi specifically target insect pests and can offer systemic protection to the plant, ultimately enhancing plant health [98–100]. Studies have shown that entomopathogenic fungi are capable of colonizing a diverse array of cultivated plants, including species like tomatoes, beans, corn, and coffee [101–103]. Barra-Bucarei *et al.* [79] reported that inoculating tomato roots with indigenous endophytic isolates of

B. bassiana improved phosphate mobilization and increased the production of iron siderophores, resulting in increased plant height and biomass, including both above-ground and below-ground dry matter. Additionally, it decreased the count of whitefly eggs and nymphs on tomato leaves. Similarly, entomopathogenic fungal endophytes have been shown to promote growth in various crops, including tomato, wheat, corn, cotton, maize, and beans [104–106]. Furthermore, Powell *et al.* [107] and Lopez and Sword [106] also observed a decrease in the survival rate of *Helicoverpa zea* larvae when *B. bassiana* was inoculated into tomato and cotton plants, respectively. The growth-promoting activity mediated by endophytes in plants can be attributed to the enhanced production of phytohormones, growth regulators (metabolites), siderophores, and phosphate solubilization, which make essential nutrients more available for plant growth and development [108, 109].

Fungal endophytes secrete siderophores that not only bind iron but also facilitate the uptake of other important nutrients like Zn, Mn, and Cu. Furthermore, the production of organic acids by these fungi helps plants absorb nutrients from the rhizosphere [27, 110]. Recently, a siderophore "rhizoferrin" produced by *Trichoderma hamatum* and other fungi of the order Mucorales (division Zygomycota) has been crucial for their growth and development, without posing any threat to plants or humans [111]. Similarly, the inoculation of sesame plants with the catecholate siderophore bacillibactin, secreted by *Bacillus subtilis*, resulted in 17%, 24%, 31%, and 19% increase in leaf, shoot, root, and overall plant length, respectively. This treatment also led to 11% increase in iron content, 47% increase in oil content, and 50% and 42% increase in chlorophyll and carotenoid levels, respectively. Additionally, the fresh weight was increased by 16%, and the dry weight improved by 24% compared to control plants [112].

Entomopathogens can also combat harmful insect pests indirectly by enhancing the plant's defense mechanisms through increased growth and development. Recent studies have demonstrated that entomopathogenic fungi are essential for enhancing the plant's ability to absorb nutrients [110, 113], and plants cultivated with fungal biocontrol agents show improved growth and increased productivity [114]—for example, inoculating plants with *Isaria farinosa* and *B. bassiana* has a substantial impact on the growth and development of key crops like sunflower, sorghum, wheat, and tomato [79, 115]. Similarly, *M. brunneum* could potentially alleviate Fe deficiency chlorosis effects in sorghum plants cultivated in calcareous soil. It can also increase the number of inflorescence and plant height of sunflowers under calcareous and non-calcareous soil conditions [110].

8.7 Siderophore-Mediated Insect Defense against Entomopathogens

The production of siderophores and outer membrane receptor proteins (OMRPs) by *Pseudomonas* sp. and *Brachybacterium* sp. has been shown to protect the host insect from entomopathogens and regulate other gut bacterial populations that may induce disease in the host insect without manifesting clinical signs [116]. The siderophore pyoverdine chelates iron in an inaccessible form which will be unavailable to entomopathogens, thereby inhibiting the pathogen's ability to acquire soluble iron in the gut environment, which is already iron-limited [59, 117]. Thus, the production of siderophores by insect gut bacteria in response to exogenous siderophores (from invading microorganisms such as entomopathogenic fungi or *Bacillus thuringiensis* (Bt)) in iron-limited circumstances in the host gut can help the host insect alleviate the virulence of invading microorganisms, either through cross-feeding or by proactively sequestering available nutrients [118].

8.8 Challenges and Future Scope

Siderophores, produced by microorganisms, have the potential for biocontrol in pest management by bringing the pest population below economic injury level and increasing plant defense. However, several challenges hinder the success rate of these biocontrol products, such as host specificity, interaction with other biocontrol agents, and environmental factors. However, siderophore-based commercial products as biocontrol agents are still lacking in the current global market due to some technical hindrances like scaling production, effective delivery mechanisms, and understanding the proper mechanism of action. Moreover, environmental concerns include potential disruptions to soil microbiomes, environmental degradation of siderophores, and the risk of pests developing resistance over time. Other challenges in this area are public acceptance and economic feasibility, which also hinder widespread adoption.

Additionally, gaps in knowledge, including unclear mechanisms of action and the long-term efficacy of siderophores, require further research. Addressing these issues is crucial to realize the full potential of siderophores in sustainable pest management. Future research and development aimed at utilizing the capabilities of siderophore-producing microorganisms through targeted genetic engineering techniques could lead to the development of efficient biocontrol agents with improved effectivity and a wider target range. Addressing these research gaps is essential for advancing siderophore-based biological pest control strategies and ensuring their successful adoption in sustainable agricultural practices. By optimizing siderophore production and application, these microorganisms can offer an eco-friendly substitute for chemical pesticides, fostering environmental sustainability while enhancing agricultural output. Siderophores have the potential to play an essential role in sustainable pest management strategies, reducing reliance on harmful chemicals and supporting long-term agricultural sustainability.

8.9 Conclusion

Agricultural crops are seriously threatened by insect pests, which result in losses both before and after harvest. New approaches to sustainable farming and pest control require an understanding of siderophore synthesis mechanisms and how they affect plant growth and development. As long as siderophore-based insect antagonists are safe for people, using them to manage insect pests and protect crops from diseases can be both efficient and ecologically benign. In addition to reducing the negative impacts of synthetic agrochemicals, we may gradually return agricultural lands to their natural state by implementing biocontrol agents that generate siderophores. Biocontrol agents that produce siderophores do not cause resistance in the target species, in contrast to chemical pesticides. However, in order to fully utilize siderophores, a thorough comprehension of the mode of action and the elements that affect their effectiveness is necessary. When selecting biocontrol agents, factors like resource competition, strong root colonizing activity, the interactions between different species in the rhizosphere, and the stability of iron–siderophore complexes should be carefully understood. The application of microbial siderophores has great potential to transform contemporary agriculture and promote a more robust and balanced ecosystem with more study and prudent application.

AI disclosure statement: The author acknowledges the use of AI-based tools for minor language editing or grammatical correction. However, all intellectual contributions, analyses, and conclusions presented in this chapter are entirely the author's own.

References

1. Savary, S., Willocquet, L., Pethybridge, S.J., Esker, P., McRoberts, N., Nelson, A., The global burden of pathogens and pests on major food crops. *Nat. Ecol. Evol.*, 3, 3, 430–439, 2019.
2. Basir, F.A., Banerjee, A., Ray, S., Role of farming awareness in crop pest management—A mathematical model. *J. Theor. Biol.*, 461, 59–67, 2019.
3. Gurung, K., Wertheim, B., Falcao Salles, J., The microbiome of pest insects: It is not just bacteria. *Entomol. Exp. Appl.*, 167, 3, 156–170, 2019.
4. Liu, J. and Wang, X., Plant diseases and pests detection based on deep learning: A review. *Plant Methods*, 17, 1–18, 2021.
5. Sonawane, M.S., Chaudhary, R.D., Shouche, Y.S., Sayyed, R.Z., Insect gut bacteria: A novel source for siderophore production. *Proc. Natl. Acad. Sci., India, Sect. B Biol. Sci.*, 88, 567–572, 2016.
6. Tongo, I., Onokpasa, A., Emerure, F., Balogun, P.T., Enuneku, A.A., Erhunmwunse, N., Ezemonye, L., Levels, bioaccumulation and biomagnification of pesticide residues in a tropical freshwater food web. *Int. J. Environ. Sci. Technol.*, 1–16, 2022.
7. Chormare, R. and Kumar, M.A., Environmental health and risk assessment metrics with special mention to biotransfer, bioaccumulation, and biomagnification of environmental pollutants. *Chemosphere*, 302, 134836, 2022.
8. Hawkins, N.J., Bass, C., Dixon, A., Neve, P., The evolutionary origins of pesticide resistance. *Biol. Rev.*, 94, 1, 135–155, 2019.
9. Ayilara, M.S., Adeleke, B.S., Akinola, S.A., Fayose, C.A., Adeyemi, U.T., Gbadegesin, L.A., Babalola, O.O., Biopesticides as a promising alternative to synthetic pesticides: A case for microbial pesticides, phytopesticides, and nanobiopesticides. *Front. Microbiol.*, 14, 1040901, 2023.
10. Kharrazian, D., Exposure to environmental toxins and autoimmune conditions. *Intgr. Medicine A Clinician's J.*, 20, 2, 20, 2021.
11. Chincholkar, S.B., Chaudhari, B.L., Talegaonkar, S.K., Kothari, R.M., Microbial iron chelators: A sustainable tool for the biocontrol of plant diseases, in: *Biocontrol Potential and Its Exploitation in Sustainable Agriculture: Crop Diseases, Weeds, and Nematodes*, pp. 49–70, 2000.
12. Headrick, D., The future of organic insect pest management: Be a better entomologist or pay for someone who is. *Insects*, 12, 2, 140, 2021.
13. Bonato, M., Martin, E.A., Cord, A.F., Seppelt, R., Beckmann, M., Strauch, M., Applying generic landscape-scale models of natural pest control to real data: Associations between crops, pests, and biocontrol agents make the difference. *Agric. Ecosyst. Environ.*, 342, 108215, 2023.
14. Ratto, F., Bruce, T., Chipabika, G., Mwamakamba, S., Mkandawire, R., Khan, Z., Sait, S.M., Biological control interventions and botanical pesticides for insect pests of crops in sub-Saharan Africa: A mapping review. *Front. Sustainable Food Syst.*, 6, 883975, 2022.

15. Pandit, M.A., Kumar, J., Gulati, S., Bhandari, N., Mehta, P., Katyal, R., Kaur, J., Major biological control strategies for plant pathogens. *Pathogens*, 11, 2, 273, 2022.
16. Tatung, M. and Deb, C.R., Isolation, characterization, and investigation on potential multi-trait plant growth promoting rhizobacteria from wild banana (*Musa itinerans*) rhizospheric soil. *J. Pure Appl. Microbiol.*, 17, 3, 1578–1590, 2023.
17. Nadarasah, G. and Stavrinides, J., Quantitative evaluation of the host-colonizing capabilities of the enteric bacterium *Pantoea* using plant and insect hosts. *Microbiology*, 160, 3, 602–615, 2014.
18. Tariq, M., Khan, A., Asif, M., Khan, F., Ansari, T., Shariq, M., Siddiqui, M.A., Biological control: A sustainable and practical approach for plant disease management. *Acta Agric. Scand. Sect. B—Soil Plant Sci.*, 70, 6, 507–524, 2020.
19. Wang, X., Yang, R., Zafar, J., Peng, C., Zhang, X., Hong, Y., Xu, X., Symbiotic and antagonistic functions of the bacterium *Burkholderia cepacia* BsNLG8, from the *Nilaparvata lugens* (Stal). *Agriculture*, 12, 12, 2106, 2022.
20. Hirschmann, M., Grundmann, F., Bode, H.B., Identification and occurrence of the hydroxamate siderophores aerobactin, putrebactin, avaroferrin, and ochrobactin C as virulence factors from entomopathogenic bacteria. *Environ. Microbiol.*, 19, 10, 4080–4090, 2017.
21. Huber, D.L., Synthesis, properties, and applications of iron nanoparticles. *Small*, 1, 482–501, 2005.
22. Gamit, D. and Tank, S., Effect of siderophore producing microorganism on plant growth of *Cajanus cajan* (Pigeon pea). *Int. J. Res. Pure Appl. Microbiol.*, 4, 20, 2014.
23. Ahsan, S.M., Injamum-Ul-Hoque, M., Das, A.K., Rahman, M.M., Mollah, M.M.I., Paul, N.C., Choi, H.W., Plant–entomopathogenic fungi interaction: Recent progress and future prospects on endophytism-mediated growth promotion and biocontrol. *Plants*, 13, 10, 1420, 2024.
24. Messenger, A.J. and Barclay, R., Bacteria, iron and pathogenicity. *Biochem. Educ.*, 11, 54–63, 1983.
25. Fardeau, S., Audic, N., Sasaki, A., Bacterial iron uptake: A promising solution against multidrug resistant bacteria, in: *Science Against Microbial Pathogens: Communicating Current Research and Technological Advances*, Formatex Research Center, Norristown, PA, USA, 2011.
26. Kraemer, S.M., Iron oxide dissolution and solubility in the presence of siderophores. *Aquat. Sci.*, 66, 3–18, 2004.
27. Krasnoff, S.B., Keresztes, I., Donzelli, B.G.G., Gibson, D.M., Metachelins, mannosylated and N-oxidized coprogen-type siderophores from *Metarhizium robertsii*. *J. Nat. Prod.*, 77, 1685–1692, 2014, C.R.
28. Deb, and Tatung, M., Siderophore producing bacteria as biocontrol agent against phytopathogens for a better environment: A review. *S. Afr. J. Bot.*, 165, 153–162, 2024.

29. Li, Y., Wang, Z., Liu, X., Song, Z., Li, R., Shao, C., Yin, Y., Siderophore biosynthesis but not reductive iron assimilation is essential for the dimorphic fungus *Nomuraea rileyi* conidiation, dimorphism transition, resistance to oxidative stress, pigmented microsclerotium formation, and virulence. *Front. Microbiol.*, 7, 931, 2016.
30. Crowley, D.E., Microbial siderophores in the plant rhizosphere, in: *Iron Nutrition in Plants and Rhizospheric Microorganisms*, pp. 169–198, Springer, Berlin/Heidelberg, Germany, 2006.
31. Yadav, A.N., Singh, J., Rastegari, A.A., Yadav, N., Plant microbiomes for sustainable agriculture: Current research and future challenges, in: *Plant Microbiomes for Sustainable Agriculture, Sustainable Development and Biodiversity*, vol. 25, A.N. Yadav, J. Singh, A.A. Rastegari, N. Yadav (Eds.), Springer, Cham, 2020.
32. Timofeeva, A.M., Galyamova, M.R., Sedykh, S.E., Bacterial siderophores: Classification, biosynthesis, perspectives of use in agriculture. *Plants*, 11, 22, 3065, 2022.
33. Neilands, J.B., Siderophores: Structure and function of microbial iron transport compounds. *J. Biol. Chem.*, 270, 45, 26723–26726, 1995.
34. Neilands, J.B., A crystalline organo-iron pigment from a rust fungus (*Ustilago sphaerogena*). *J. Am. Chem. Soc.*, 74, 19, 4846–4847, 1952.
35. Klopper, J.W., Leong, J., Teintze, M., Schroth, M.N., Enhanced plant growth by siderophores produced by plant growth-promoting rhizobacteria. *Nature*, 286, 885–886, 1980.
36. Saha, M., Sarkar, S., Sarkar, B., Sharma, B.K., Bhattacharjee, S., Tribedi, P., Microbial siderophores and their potential applications: A review. *Environ. Sci. Pollut. Res.*, 23, 3984–3999, 2016.
37. Kour, D., Negi, R., Khan, S.S., Kumar, S., Kaur, S., Kaur, T., Yadav, A.N., Microbes mediated induced systemic response in plants: A review. *Plant Stress*, 100334, 2023.
38. Ghosh, S.K., Bera, T., Chakrabarty, A.M., Microbial siderophore–A boon to agricultural sciences. *Biol. Control*, 144, 104214, 2020.
39. Del Carmen Orozco-Mosqueda, M., Fadiji, A.E., Babalola, O.O., Santoyo, G., Bacterial elicitors of the plant immune system: an overview and the way forward. *Plant Stress*, 100138, 2023.
40. Sheng, M., Jia, H., Zhang, G., Zeng, L., Zhang, T., Long, Y., Lan, J., Hu, Z., Zeng, Z., Wang, B., Siderophore production by rhizosphere biological control bacteria *Brevibacillus brevis* GZDF3 of *Pinellia ternata* and its antifungal effects on *Candida albicans*. *J. Microbiol. Biotechnol.*, 30, 689–699, 2020.
41. Galica, T., Borbone, N., Mareš, J., Kust, A., Caso, A., Esposito, G., Hájek, J., Reháková, K., Urajová, P., Costantino, V., Cyanochelins, an overlooked class of widely distributed. *Appl. Environ. Microbiol.*, 87, e0312820, 2021.
42. Barakat, H., Qureshi, K.A., Alsohim, A.S., Rehan, M., The purified siderophore from *Streptomyces tricolor* HM10 accelerates recovery from iron-deficiency-induced anemia in rats. *Molecules*, 27, 4010, 2022.

43. Liu, D., Smagghe, G., Liu, T.X., Interactions between entomopathogenic fungi and insects and prospects with glycans. *J. Fungi*, 9, 5, 575, 2023.
44. Teta, R., Esposito, G., Kundu, K., Stornaiuolo, M., Scarpato, S., Pollio, A., Costantino, V., A glimpse at siderophores production by *Anabaena flos-aquae* UTEX 1444. *Mar. Drugs*, 20, 0256, 2022.
45. Sultana, S., Alam, S., Karim, M.M., Screening of siderophore-producing salt-tolerant rhizobacteria suitable for supporting plant growth in saline soils with iron limitation. *J. Agric. Food Res.*, 4, 100150, 2021.
46. Park, J., Durham, B.P., Key, R.S., Groussman, R.D., Pinedo-Gonzalez, P., Hawco, N.J., John, S.G., Carlson, M.C.G., Lindell, D., Juranek, L., *et al.*, Siderophore production and utilization by microbes in the North Pacific Ocean, bioRxiv, 2022.
47. Goodrich-Blair, H. and Clarke, D.J., Mutualism and pathogenesis in *Xenorhabdus* and *Photorhabdus*: two roads to the same destination. *Mol. Microbiol.*, 64, 260–268, 2007.
48. Petersen, L.M. and Tisa, L.S., Influence of temperature on the physiology and virulence of the insect pathogen *Serratia* sp. strain SCBI. *Appl. Environ. Microbiol.*, 78, 24, 8840–8844, 2012.
49. Grimont, F. and Grimont, P., The genus Serratia, in: *Prokaryotes*, vol. 6, pp. 219–244, 2006.
50. Sherpa, M.T., Bag, N., Das, S., Haokip, P., Sharma, L., Isolation and characterization of plant growth promoting rhizobacteria isolated from organically grown high yielding pole type native pea (*Pisum sativum* L.) variety Dentami of Sikkim, India. *Curr. Res. Microb. Sci.*, 20, 2, 100068, 2021.
51. Shen, N., Li, S., Li, S., Zhang, S., Jiang, M., The siderophore-producing bacterium *Bacillus simensis* Gxun-6, has an antifungal activity against *Fusarium oxysporum* and promotes the growth of banana. *Egypt. J. Biol. Pest Control*, 32, 34, 2022.
52. Hu, Q.P. and Xu, J.G., A simple double-layered chrome azurol S agar (SD-CASA) plate assay to optimize the production of siderophores by a potential biocontrol agent *Bacillus*. *Afr. J. Microbiol. Res.*, 5, 25, 4321–4327, 2011.
53. Patil, S., Bheemaraddi, M.C., Shivannavar, C.T., Gaddad, S.M., Biocontrol activity of siderophore producing *Bacillus subtilis* CTS-G24 against wilt and dry root rot causing fungi in chickpea. *IOSR J. Agric. Vet. Sci. (IOSR-JAVS)*, 7, 9, 63–68, 2014, doi: 10.9790/2380 07916368.
54. Wilson, M.K., Abergel, R.J., Raymond, K.N., Arceneaux, J.E., Byers, B.R., Siderophores of *Bacillus anthracis*, *Bacillus cereus*, and *Bacillus thuringiensis*. *Biochem. Biophys. Res. Commun.*, 348, 1, 320–325, 2006.
55. Ali, B., Hafeez, A., Ahmad, S., Javed, M.A., Sumaira, S., Afridi, M.S., Selim, S., *Bacillus thuringiensis* PM25 ameliorates oxidative damage of salinity stress in maize *via* regulating growth, leaf pigments, antioxidant defense system, and stress responsive gene expression. *Front. Plant Sci.*, 13, 921668, 2022.

56. Raddadi, N., Cherif, A., Ouzari, H., Marzorati, M., Brusetti, L., Boudabous, A., Daffonchio, D., *Bacillus thuringiensis* beyond insect biocontrol: plant growth promotion and biosafety of polyvalent strains. *Ann. Microbiol.*, 57, 481–494, 2007.
57. Chaabouni, I., Guesmi, A., Cherif, A., Secondary metabolites of Bacillus: potentials in biotechnology, in: *Bacillus thuringiensis Biotechnology*, E. Sansinenea (Ed.), pp. 347–366, Springer, Dordrecht, 2012.
58. Williams, K.M., Martin, W.E., Smith, J., Williams, B.S., Garner, B.L., Production of protocatechuic acid in *Bacillus thuringiensis* ATCC33679. *Int. J. Mol. Sci.*, 13, 3, 3765–3772, 2012.
59. Ciche, T.A., Blackburn, M., Carney, J.R., Ensign, C., Photobactin: a catechol siderophore produced by *Photorhabdus luminescens*, an entomopathogen mutually associated with *Heterorhabditis bacteriophora* NC1 nematodes. *Appl. Environ. Microbiol.*, 69, 4706–4713, 2003.
60. Matthijs, S., Laus, G., Meyer, J.M., Abbaspour-Tehrani, K., Schäfer, M., Budzikiewicz, H., Cornelis, P., Siderophore-mediated iron acquisition in the entomopathogenic bacterium *Pseudomonas entomophila* L48 and its close relative *Pseudomonas putida* KT2440. *Biometals*, 22, 951–964, 2009.
61. Cox, C.D. and Graham, R., Isolation of an iron-binding compound from *Pseudomonas aeruginosa. J. Bacteriol.*, 137, 357–364, 1979.
62. Risse, D., Beiderbeck, H., Taraz, K., Budzikiewicz, H., Gustine, D., Corrugatin, a lipopeptide siderophore from *Pseudomonas corrugata. Z Naturforsch*, 53c, 295–304, 1998.
63. Mercado-Blanco, J., van der Drift, K.M.G.M., Olsson, P.E., Thomas Oates, J.E., van Loon, L.C., Bakker, P.A.H.M., Analysis of the pmsCEAB gene cluster involved in the biosynthesis of salicylic acid and the siderophore pseudomonine in the biocontrol strain *Pseudomonas fluorescens* WCS374. *J. Bacteriol.*, 183, 1909–1920, 2001.
64. Cornelis, P. and Matthijs, S., Diversity of siderophore-mediated iron uptake systems in fluorescent pseudomonads: not only pyoverdines. *Environ. Microbiol.*, 4, 787–798, 2002.
65. Matthijs, S., Abbaspour Tehrani, K., Laus, G., Jackson, R.W., Cooper, R.M., Cornelis, P., Thioquinolobactin, a *Pseudomonas* siderophore with antifungal and anti-*Pythium* activity. *Environ. Microbiol.*, 9, 425–434, 2007.
66. Matthijs, S., Budzikiewicz, H., Schäfer, M., Whatelet, B., Cornelis, P., Ornicorrugatin, a new siderophore from *Pseudomonas fluorescens* AF76. *Z Naturforsch*, 63, 8–12, 2008.
67. Sinha, K.K., Choudhary, A.K., Kumari, P., Entomopathogenic fungi, in: *Ecofriendly Pest Management for Food Security*, Omkar (Ed.), pp. 475–505, Academic Press, Cambridge, 2016.
68. Poveda, J., Beneficial effects of microbial volatile organic compounds (MVOCs) in plants. *Appl. Soil Ecol.*, 168, 104118, 2021, doi: 10.1016/j.apsoil.2021.104118.

69. García-Espinoza, F., Quesada-Moraga, E., García del Rosal, M.J., Yousef-Yousef, M., Entomopathogenic fungi-mediated solubilization and induction of Fe related genes in melon and cucumber plants. *J. Fungi*, 9, 2, 258, 2023.
70. Donzelli, B.G.G., Gibson, D.M., Krasnoff, S.B., Intracellular siderophore but not extracellular siderophore is required for full virulence in *Metarhizium robertsii. Fungal Genet. Biol.*, 82, 56–68, 2015.
71. Jin, C.W., Ye, Y.Q., Zheng, S.J., An underground tale: Contribution of microbial activity to plant iron acquisition *via* ecological processes. *Ann. Bot.*, 113, 7–18, 2014.
72. Dijkstra, F.A., Carrillo, Y., Pendall, E., Morgan, J.A., Rhizosphere priming: A nutrient perspective. *Front. Microbiol.*, 4, 216, 2013.
73. Wallner, A., Blatzer, M., Schrettl, M., Sarg, B., Lindner, H., Haas, H., Ferricrocin, a siderophore involved in intra- and transcellular iron distribution in *Aspergillus fumigatus. Appl. Environ. Microbiol.*, 75, 4194–4196, 2009.
74. Eisendle, M., Schrettl, M., Kragl, C., Müller, D., Illmer, P., Haas, H., The intracellular siderophore ferricrocin is involved in iron storage, oxidative-stress resistance, germination, and sexual development in *Aspergillus nidulans. Eukaryotic Cell*, 5, 1596–1603, 2006.
75. Gibson, D.M., Donzelli, B.G., Krasnoff, S.B., Keyhani, N.O., Discovering the secondary metabolite potential encoded within entomopathogenic fungi. *Nat. Prod. Rep.*, 31, 1287–1305, 2014.
76. Hof, C., Eisfeld, K., Welzel, K., Antelo, L., Foster, A.J., Anke, H., Ferricrocin synthesis in *Magnaporthe grisea* and its role in pathogenicity in rice. *Mol. Plant Pathol*, 8, 163–172, 2007.
77. Jirakkakul, J., Wichienchote, N., Likhitrattanapisal, S., Ingsriswang, S., Yoocha, T., Tangphatsornruang, S., Wasuwan, R., Cheevadhanarak, S., Tanticharoen, M., Amnuaykanjanasin, A., Iron homeostasis in the absence of ferricrocin and its consequences in fungal development and insect virulence in *Beauveria bassiana. Sci. Rep.*, 11, 19624, 2021.
78. Schrettl, M., Bignell, E., Kragl, C., Sabiha, Y., Loss, O., Eisendle, M., Wallner, A., Arst Jr., H.N., Haynes, K., Haas, H., Distinct roles for intra- and extracellular siderophores during *Aspergillus fumigatus* infection. *PLoS Pathog.*, 3, 1195–1207, 2007.
79. Barra-Bucarei, L., González, M.G., Iglesias, A.F., Aguayo, G.S., Peñalosa, M.G., Vera, P.V., Beauveria bassiana multifunction as an endophyte: Growth promotion and biological control of *Trialeurodes vaporariorum* (Westwood) (Hemiptera: Aleyrodidae) in tomato. *Insects*, 11, 9, 591, 2020.
80. Jirakkakul, J., Cheevadhanarak, S., Punya, J., Chutrakul, C., Senachak, J., Buajarern, T., Tanticharoen, M., Amnuaykanjanasin, A., FEMS Microbiol. Lett. *FEMS Microbiol. Lett.*, 362, 1–8, 2015.
81. Lana, M., Simón, O., Velasco, P., Rodríguez, V.M., Caballero, P., Poveda, J., First study on the root endophytic fungus *Trichoderma hamatum* as an entomopathogen: Development of a fungal bioinsecticide against cotton leafworm (*Spodoptera littoralis*). *Microbiol. Res.*, 270, 127334, 2023.

82. Drechsel, H., Metzger, J., Freund, S., Jung, G., Boelaert, J.R., Winkelmann, G., Rhizoferrina: A novel siderophore from the fungus *Rhizopus microsporus* var. *rhizopodiformis*. *Biol. Met.*, 4, 238–243, 1991.
83. Ghorai, A.K., Patsa, R., Jash, S., Dutta, S., Microbial secondary metabolites and their role in stress management of plants, in: *Biocontrol Agents and Secondary Metabolites*, S. Jogaiah (Ed.), pp. 283–319, Woodhead Publishing, Sawston, Cambridge, UK, 2021.
84. Braun, V. and Braun, M., Active transport of iron and siderophore antibiotics. *Curr. Opin. Microbiol.*, 5, 194–201, 2002.
85. Watson, R.J., Joyce, S.A., Spencer, G.V., Clarke, D.J., The exbD gene of *Photorhabdus temperata* is required for full virulence in insects and symbiosis with the nematode *Heterorhabditis*. *Mol. Microbiol.*, 56, 763–773, 2005.
86. Watson, R.J., Millichap, P., Joyce, S.A., Reynolds, S., Clarke, D.J., The role of iron uptake in pathogenicity and symbiosis in *Photorhabdus luminescens* TT01. *BMC Microbiol.*, 10, 177, 2010.
87. Li, J., Li, J., Cao, L., Chen, Q., Ding, D., Kang, L., An iron-binding protein of entomopathogenic fungus suppresses the proliferation of host symbiotic bacteria. *Microbiome*, 12, 1, 202, 2024.
88. Eppinger, M., Rosovitz, M.J., Fricke, W.F., Rasko, D.A., Kokorina, G., Fayolle, C., Lindler, L.E., Carniel, E., Ravel, J., The complete genome sequence of *Yersinia pseudotuberculosis* IP31758, the causative agent of Far East scarlet-like fever. *PLos Genet.*, 3, e142, 2007.
89. An, R., Sreevatsan, S., Grewal, P.S., Comparative *in vivo* gene expression of the closely related bacteria *Photorhabdus temperata* and *Xenorhabdus koppenhoeferi* upon infection of the same insect host, *Rhizotrogus majalis*. *BMC Genomics*, 10, 1–16, 2009.
90. Toopaang, W., Panyawicha, K., Srisuksam, C., Hsu, W.C., Lin, C.C., Tanticharoen, M., *et al.*, Metabolomic analysis demonstrates the impacts of polyketide synthases PKS14 and PKS15 on the production of beauvericins, bassianolide, enniatin A, and ferricrocin in entomopathogen *Beauveria bassiana*. *Metabolites*, 13, 3, 425, 2023.
91. Donzelli, B.G.G. and Krasnoff, S.B., *Genetics and Molecular Biology of Entomopathogenic Fungi*, 1st ed., vol. 94, R.J.S. Leger and B. Lovett (Eds.), pp. 365–436, Academic Press, Amsterdam, Elsevier/Academic Press, Cambridge, MA, 2016.
92. Schrettl, M., Bignell, E., Kragl, C., Joechl, C., Rogers, T., Arst Jr., H.N., *et al.*, Siderophore biosynthesis but not reductive iron assimilation is essential for *Aspergillus fumigatus* virulence. *J. Exp. Med.*, 200, 1213–1219, 2004, doi: 10.1084/jem.20041242.
93. Gorman, M.J., Iron homeostasis in insects. *Annu. Rev. Entomol.*, 68, 51–67, 2023.
94. Hood, M.I. and Skaar, E.P., Nutritional immunity: transition metals at the pathogen-host interface. *Nat. Rev. Microbiol.*, 10, 8, 525–537, 2012.

95. Iatsenko, I., Marra, A., Boquete, J.P., Pena, J., Lemaitre, B., Iron sequestration by transferrin 1 mediates nutritional immunity in *Drosophila melanogaster*. *PNAS*, 117, 13, 7317–7325, 2020.
96. Sarvepalli, M., Velidandi, A., Korrapati, N., Optimization of Siderophore Production in Three Marine Bacterial Isolates along with Their Heavy-Metal Chelation and Seed Germination Potential Determination. *Microorganisms*, 11, 12, 2873, 2023.
97. Ahmed, E. and Holmström, S.J.M., Siderophores in Environmental Research: Roles and Applications. *Microb. Biotechnol.*, 7, 196–208, 2014.
98. Quesada-Moraga, E., Entomopathogenic Fungi as Endophytes: Their Broader Contribution to IPM and Crop Production. *Biocontrol Sci. Technol.*, 30, 864–877, 2020.
99. Branine, M., Bazzicalupo, A., Branco, S., Biology and Applications of Endophytic Insect-Pathogenic Fungi. *PLoS Pathog.*, 15, e1007831, 2019.
100. Vega, F.E., The Use of Fungal Entomopathogens as Endophytes in Biological Control: A Review. *Mycologia*, 110, 4–30, 2018.
101. García, J.E., Posadas, J.B., Perticari, A., Lecuona, R.E., Metarhizium anisopliae (Metschniko) Sorokin promotes growth and has endophytic activity in tomato plants. *Adv. Biol. Res.*, 5, 22–27, 2011.
102. Gathage, J.W., Lagat, Z.O., Fiaboe, K.K.M., Akutse, K.S., Ekesi, S., Maniania, N.K., Prospects of fungal endophytes in the control of *Liriomyza* leafminer flies in common bean *Phaseolus vulgaris* under field conditions. *BioControl*, 61, 741–753, 2016.
103. Posada, F., Aime, M.C., Peterson, S.W., Rehner, S.A., Vega, F.E., Inoculation of coffee plants with the fungal entomopathogen *Beauveria bassiana* (Ascomycota: Hypocreales). *Mycol. Res.*, 111, 748–757, 2007.
104. Sasan, R.K. and Bidochka, M.J., The insect-pathogenic fungus *Metarhizium robertsii* (Clavicipitaceae) is also an endophyte that stimulates plant root development. *Am. J. Bot.*, 99, 101–107, 2012.
105. Tall, S. and Meyling, N.V., Probiotics for plants? Growth promotion by the entomopathogenic fungus *Beauveria bassiana* depends on nutrient availability. *Microb. Ecol.*, 76, 1002–1008, 2018.
106. Lopez, D.C. and Sword, G.A., The endophytic fungal entomopathogens *Beauveria bassiana* and *Purpureocillium lilacinum* enhance the growth of cultivated cotton (*Gossypium hirsutum*) and negatively affect survival of the cotton bollworm (*Helicoverpa zea*). *Biol. Control*, 89, 53–60, 2015.
107. Powell, W.A., Klingeman, W.E., Ownley, B.H., Gwinn, K.D., Evidence of endophytic *Beauveria bassiana* in seed-treated tomato plants acting as a systemic entomopathogen to larval *Helicoverpa zea* (Lepidoptera: Noctuidae). *J. Entomol. Sci.*, 44, 391–396, 2009.
108. Waqas, M., Khan, A.L., Kamran, M., Hamayun, M., Kang, S.M., Kim, Y.H., Lee, I.J., Endophytic fungi produce gibberellins and indoleacetic acid and promote host-plant growth during stress. *Molecules*, 17, 10754–10773, 2012.

109. Khan, A.L. and Lee, I.J., Endophytic *Penicillium funiculosum* LHL06 secretes gibberellin that reprograms *Glycine max* L. growth during copper stress. *BMC Plant Biol.*, 13, 86, 2013.
110. Raya-Díaz, S., Sánchez-Rodríguez, A.R., Segura-Fernández, J.M., del Campillo, M.D.C., Quesada-Moraga, E., Entomopathogenic fungi-based mechanisms for improved Fe nutrition in sorghum plants grown on calcareous substrates. *PLoS One*, 12, e0185903, 2017.
111. Skríba, A., Patil, R.H., Hubáček, P., Dobias, R., Palyzová, A., Maresová, H., Rhizoferrin glycosylation in *Rhizopus microsporus*. *J. Fungi*, 6, 89, 2020.
112. Nithyapriya, S., Lalitha, S., Sayyed, R.Z., Reddy, M.S., Dailin, D.J., El Enshasy, H.A., Herlambang, S., Production, purification, and characterization of bacillibactin siderophore of *Bacillus subtilis* and its application for improvement in plant growth and oil content in sesame. *Sustainability*, 13, 10, 5394, 2021.
113. Sánchez-Rodríguez, A.R., Raya-Díaz, S., Zamarreño, AM., García-Mina, J.M., del Campillo, M.C., Quesada-Moraga, E., An endophytic *Beauveria bassiana* strain increases spike production in bread and durum wheat plants and effectively controls cotton leafworm (*Spodoptera littoralis*) larvae. *Biol. Control*, 116, 90–102, 2018.
114. Barelli, L., Moonjely, S., Behie, S.W., Bidochka, M.J., Fungi with multifunctional lifestyles: Endophytic insect pathogenic fungi. *Plant Mol. Biol.*, 90, 657–664, 2016.
115. Sánchez-Rodríguez, A.R., Barrón, V., del Campillo, M.C., Quesada-Moraga, E., The entomopathogenic fungus *Metarhizium brunneum*: A tool for alleviating Fe chlorosis. *Plant Soil*, 406, 295–310, 2016.
116. Indiragandhi, P., Anandham, R., Madhaiyan, M., Kim, G.H., Sa, T., Cross-utilization and expression of outer membrane receptor proteins for siderophore uptake by Diamondback moth *Plutella xylostella* (Lepidoptera: Plutellidae) gut bacteria. *FEMS Microbiol. Lett.*, 289, 1, 27–33, 2008.
117. Lim, H.S., Jung, M.L., Sang, D.K., A plant growth-promoting *Pseudomonas fluorescens* GL20: mechanism for disease suppression, outer membrane receptors for ferric siderophore, and genetic improvement for increased biocontrol efficacy. *J. Microbiol. Biotechnol.*, 12, 2, 249–257, 2002.
118. Joshi, F., Archana, G., Desai, A., Siderophore cross-utilization amongst rhizospheric bacteria and the role of their differential affinities for Fe^{3+} on growth stimulation under iron-limited conditions. *Curr. Microbiol.*, 53, 141–147, 2006.
119. Litwin, A., Nowak, M., Różalska, S., Entomopathogenic fungi: unconventional applications. *Rev. Environ. Sci. Bio/Technol.*, 19, 23–42, 2020.
120. Khan, S. and Lang, M., A Comprehensive Review on the Roles of Metals Mediating Insect–Microbial Pathogen Interactions. *Metabolites*, 13, 839, 2023.

9

Formulation and Delivery Systems for Siderophore-Based Products

S. Neelufar Shama[1]*, K.R. Prasanna[2], Joshna Avula[3] and Divya Suneesh[4]

[1]*Department of Pharmacognosy, Scient Institute of Pharmacy, Ibrahimpatnam, Telangana, India*
[2]*Department of Pharmacognosy, Hillside College of Pharmacy and Research Centre, Bengaluru, Karnataka, India*
[3]*Department of Pharmacognosy, Sri Padmavati School of Pharmacy, Tirupati, Andhra Pradesh, India*
[4]*Department of Pharmacology, Hillside College of Pharmacy and Research Centre, Bengaluru, Karnataka, India*

Abstract

Siderophore-based products represent a groundbreaking advancement in sustainable agriculture, addressing critical challenges such as nutrient deficiencies, pathogen infestations, and environmental contamination. These products tap the specificities of microbial siderophores that are encapsulated into niche formulations, such as seed coatings, granular fertilizers, liquid sprays, and nanotechnology-based delivery systems. Other agrochemicals differ from this class of siderophore-based formulation since these are environment-friendly and have effects on iron bioavailability improvement, restoration of soil health, and pathogen suppression without synthetic fertilizer and pesticide use. Seed coatings stimulate germination and promote healthy root growth; granular fertilizers liberate nutrients slowly into the soil, which may improve iron-deficient soils. Advanced liquid formulations and applications of nanotechnology allow for the formulation of precise delivery mechanisms that ensure bioavailability under harsh conditions, such as alkaline soils or dry regions. This might best be represented in more applied ways, such as using siderophore-based sprays to combat iron chlorosis in citrus cultures or granular products for cleaning up polluted soils. There is still much to work on in terms of scalability, environmental applicability, and take-up among farmers. This chapter gives a comprehensive review of the progress made,

**Corresponding author*: neelufarshama@gmail.com

Debasis Mitra, Ayush Madan and Anju Rani (eds.) Microbial Siderophores for Agricultural Productivity: Perspectives of Use in Agriculture, (207–232) © 2026 Scrivener Publishing LLC

methods of preparation, and practical applications of siderophore-based products, focusing their important roles in promoting sustainable, resilient, and environmental friendly agricultural practices.

Keywords: Siderophore-based products, sustainable agriculture, nutrient bioavailability, eco-friendly formulations, pathogen suppression, precision delivery systems

9.1 Introduction

Siderophore-based products are a significant innovation within the scope of sustainable agriculture. Such products will offer an ecofriendly solution to problems associated with the accessibility of nutrients and health of plants generally. Siderophores are natural compounds produced by bacteria, fungi, and plants for the chelation and transportation of iron in assimilation into the plant. Therefore, for agricultural systems where available soil iron is limited—for instance, in alkaline and calcareous soils—siderophore-based products supply an avenue to enhance its bioavailability, thus having sound plant growth and protecting it from deficiency [1]. These products are really surfacing as a mandatory input in integrated nutrient management and plant disease control mechanisms over traditional chemical-based fertilizers and pesticides.

With regard to this, one of the primary drivers that fuel the development and application of siderophore-based products has been the rising demand for sustainable and organic agricultural methods. Furthermore, these have also drawn people's attraction because they are deemed to be more environment-friendly, in view of the overall trend to rely on fewer synthetic chemical applications. Unlike chemical-based fertilizers that cause some nutrient imbalances and damage to the ecosystems, the siderophore-based products work in harmony with natural soil processes. These products enhance the soil vitality by encouraging good microorganisms and improving the cycling processes of nutrients. By doing so, they bring down reliance on destructive chemical pesticides [2, 3]. Due to this, they promote crop productivity while significantly contributing toward the advancement of the sustainability of agricultural methodologies.

These products have proven particularly useful in a variety of different agricultural applications, from improved nutrition and biocontrol agents to soil beneficial organisms that reduce harmful pathogens within the soil. They find wider applicability beyond crop-specific areas into specialty crops, horticulture, and sustainable landscape creation. Improved formulations as well as delivery technologies also allow for encapsulated formulations or use of slow-release technologies and can be further enhanced. Such products, based on siderophores, would thus ensure the production of high-yielding crops, in terms of quality and resistance to environmental stresses, due to an efficient and continuous supply of available nutrients to plants. Consequently, the scope of siderophore-based products expands by revolutionizing modern techniques of agriculture to contribute toward food security in the world [4, 5].

9.2 Advantages of Siderophore-Based Products

a. **Eco-friendly:** One of the major advantages of siderophore-based products is that it is eco-friendly. While the synthetic agrochemicals used in farming settle down in water and soil and hence present various kinds of hazards to the environment, the siderophore-based compounds are biogenic and non-toxic because they are the byproducts of the micro-organism of bacteria or fungi. These organisms will decompose organic material without creating any harmful products. This, in turn, reduces their input to environmental pollution and therefore enhances the quality of soil or water, and thus these are helpful in sustainable agricultural practices [1, 2].
b. **Enhanced nutrient efficiency:** Siderophore-based products increase the iron intake of plants in a highly bioavailable form. Iron, although it is present in soil in plenty, often is not available to plants in most soils, especially those of higher pH or calcareous in nature. Siderophores chelate the iron and form a readily plant-taken complex. This helps reduce the dependency on traditional high-input fertilizers, which often show low efficiency and may cause nutritional imbalances. Siderophore-based products play a significant role in maximizing natural nutrient uptake in soil, thus contributing to the attainment of more sustainable agriculture with decreased environmental impacts [6].
c. **Pathogen suppression:** Besides increasing the availability of nutrients, some siderophore-producing microorganisms possess biocontrol characteristics that also suppress soil-borne pathogens. These siderophores can remove iron from pathogenic microorganisms, and therefore they cannot grow

or infect the plants because it is the most essential nutrient required for their growth. This natural mode of pathogen suppression reduces the reliance on chemical pesticides, thus providing an integrated approach in pest management. Because most of the beneficial microbes produced *via* siderophore-based products compete with harmful pathogenic microbes, it develops a healthier soil ecosystem.

d. **Compatibility:** The siderophore-based products are relatively compatible with the sustainable agricultural systems, the case including organic farming. They could simply introduce them to the existing farming structures without negatively interfering with already positively pre-sent beneficial microorganisms. Therefore, they enhance soil health by encouraging the development of beneficial microorganisms and improving the nutrient cycling processes. These properties make them an extremely effective choice for farming operators interested in improving crop production with minimum disturbances to environmental balance and least dependence on chemical resources. The addition of siderophore-formulated preparations in organic and integrated pest management systems strengthens the concept of sustainable agriculture further [7].

9.3 Mechanisms of Action

Siderophore-based products work on the natural characteristics of siderophores which enable plant growth, facilitating nutrient uptake efficiently with no diseases. Products in the market differ from one another; however, they will contribute toward sustainable agriculture in all of their potential positive impacts. Discussed below are the primary mechanisms of action.

9.3.1 Iron Chelation and Nutrient Uptake

Siderophores are specialized molecules that bind tightly to iron, one of the essential nutrients required for plant growth. Iron in many agricultural soils is found to be in unavailable forms, particularly those with high pH or calcareous conditions. Siderophores act as chelators to form soluble iron complexes, which are readily absorbed by the plant roots (Figure 9.1) [1]. This process increases the bioavailability of iron in the rhizosphere

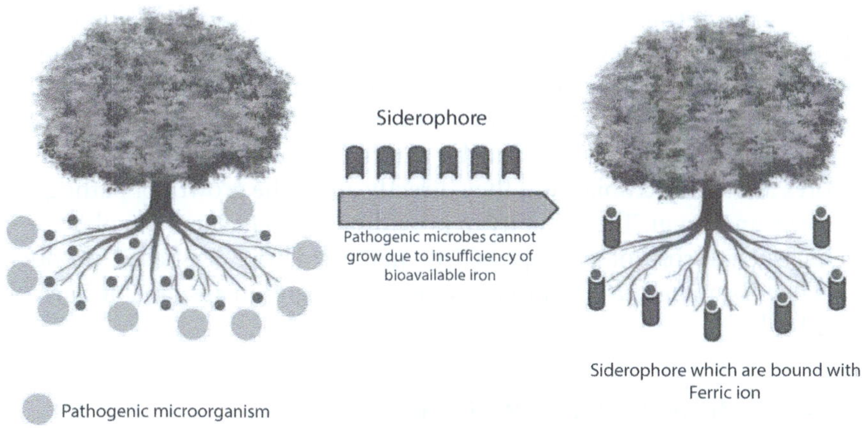

Figure 9.1 Siderophores acting as chelators to form soluble iron complexes.

so that the plants will obtain the critical micronutrient for processes like photosynthesis, respiration, and chlorophyll synthesis. This consequently reduces the application of synthetic fertilizers in farming and increases sustainability and efficiency in agriculture [8].

9.3.2 Microbial Interactions and Soil Health

Microorganisms that synthesize siderophores such as some species of *Pseudomonas* and *Bacillus* provide nutrient uptake while promoting healthy microbial activity in the soil. Siderophores released by the microorganisms effectively compete for iron that is available from pathogens and are harmful, limiting their growth by several folds [3]. This is known as nutrient limitation, where competition for the nutrient is established with pathogens through deprivation; thus, the potential to trigger disease is weakened. More than that, siderophores promote the growth of beneficial soil microorganisms, which facilitates further the process of decomposition and nutrient cycling. Siderophore-based products improve the health and fertility of soil by promoting a healthy microbial community [9].

9.3.3 Biocontrol of Pathogens

Besides their function in improving nutrient uptake, some siderophore-producing microorganisms also have innate biocontrol capabilities. These microbes can control the growth of plant pathogens in several ways, among which is the chelation of iron that results from siderophore-mediated iron sequestration. Pathogenic organisms, especially those that require iron to be virulent, will be deprived of this essential nutrient as long as it is bound by siderophores [2]. This mechanism of competitive exclusion restricts the formation of infections in plants by pathogens. In addition, some siderophores are antimicrobial, and they directly affect the growth of pathogenic bacteria and fungi. Their duality in providing nutrient supply to plants and being suppressors of pathogens will make siderophore-based products valuable for integrated pest management strategies in sustainable agriculture [10].

9.3.4 Induced Systemic Resistance (ISR)

It has also been discovered that siderophore-producing microbes can induce systemic resistance in plants. This is through the colonization of the roots of plants and then the release of siderophores that trigger the immune response in the plant, making it more ready to fight against subsequent attacks by pathogens. This phenomenon, referred to as induced systemic resistance (ISR), resembles the action of conventional chemical fungicides but does not carry the adverse environmental consequences. Consequent to being siderophore-derived products, they contribute both to plant nutritional needs and enhance innate plant defenses against many types of biotic stresses. They represent, therefore, an important resource for nutrient and disease management within sustainable methodologies of agriculture [11].

9.3.5 Synergy with Other Fertilizers and Biocontrol Agents

Siderophore-based products are most often used in combination with many organic fertilizers and biocontrol agents. These products can team up with composts, organic amendments, and other biological agents to enhance soil fertility and plant health through improvement in nutrient uptake and stimulation of microbial activity. This is a synergistic effect which enhances plant growth, yield, and resistance to various environmental stresses such as drought and salinity. The congruence of siderophore-based products with a variety of sustainable agricultural practices renders them highly adaptable within a variety of farming systems [12].

9.4 Types of Siderophore-Based Products

a. Biofertilizers

Biofertilizers formulated with siderophores can be used as an alternative to synthetic fertilizers. Among the bacteria used for biofertilizer formulations, *Pseudomonas fluorescens* and *Bacillus subtilis* are most commonly used because they enhance iron availability which, in turn, promotes plant growth. For instance, rice and wheat crops which were grown in iron-deficient soils and supplied with biofertilizers enriched with siderophores yielded enhanced biomass and also gave a high yield.

b. Biopesticides

Microbes that produce siderophores in biopesticides help by binding iron and also have antifungal and antibacterial effects. For instance, products from *Trichoderma harzianum* can reduce harmful soil pathogens and help plant roots grow [1]. This dual ability makes them perfect for integrated pest and nutrient management systems.

c. Seed coatings

Siderophore-based seed coatings have enormous potential for improving seed germination as well as promoting early growth of the plant. These coatings allow for a sustained and localized availability of iron in the soil, even in situations where the soil is iron deficient. For instance, seeds coated with *Pseudomonas fluorescens* have been found beneficial in germination of maize and wheat, and also increased root length under stress conditions [9].

d. Foliar sprays

Siderophore-rich foliar sprays are a practical approach to ameliorating iron chlorosis in horticultural crops. Foliar sprays with siderophore solutions on citrus and grape plants have resulted in an increase in chlorophyll content, decrease in leaf yellowing, and improved quality of fruits [13].

e. Compost enhancers

Siderophore-enhanced compost formulations increase nutrient cycling and decomposition efficiency. The main advantage in these formulations is that it combines microbial siderophores and organic compost and hence promotes the faster release of nutrients for better yields of crops such as tomatoes and lettuce [14].

A list of various siderophore-based products and delivery systems along with their agricultural applications are given in Table 9.1.

Table 9.1 Examples of siderophore-based products and delivery systems with agricultural applications.

Delivery system	Formulation example	Application area	Target crop/ soil	Benefits/ applications	References
Seed coating	*Pseudomonas fluorescens*-based coating	Seed treatment	Wheat, corn	Enhanced germination, improved iron uptake, and root growth in iron-deficient soils	[3]
Granular fertilizer	Siderophore-enriched biofertilizer	Soil amendment	Rice, barley	Slow-release nutrient supply, enhances soil fertility, and improves crop growth	[4]
Foliar spray	Aqueous siderophore suspension	Direct foliar application	Citrus, tomato, grapes	Mitigates iron chlorosis, promotes chlorophyll synthesis, and improves fruit quality	[4]
Nanoparticle delivery	Siderophore-loaded polymeric nanoparticles	Precision delivery of nutrients	Maize, lettuce	Controlled release, targeted delivery to roots under drought or nutrient deficiency stress	[9]
Bioinoculants (liquid biofertilizers)	*Azospirillum brasilense* + siderophore	Liquid fertilizer application	Rice, soybean	Provides immediate bioavailable nutrients, enhances growth under iron-limited conditions	[8]

(*Continued*)

Table 9.1 Examples of siderophore-based products and delivery systems with agricultural applications. (*Continued*)

Delivery system	Formulation example	Application area	Target crop/ soil	Benefits/ applications	References
Soil inoculation for disease control	Siderophore-producing *Pseudomonas* strain	Soil inoculation for disease control	Cotton, wheat	Promotes plant growth, suppresses soil-borne pathogens, and enhances soil fertility	[8]
Liquid biofertilizer	*Bacillus megaterium*-based siderophore	Liquid fertilizer application	Rice, soybean	Provides immediate bioavailable nutrients, enhances growth under iron-limited conditions	[10]
Microbial inoculants (root inoculation)	*Trichoderma harzianum* + siderophore	Root inoculation	Strawberry, tomato	Enhances root growth, provides pathogen suppression, and improves nutrient uptake	[15]
Biodegradable granules	Siderophore-enriched chitosan pellets	Controlled release in soil	Various crops	Sustainable nutrient release, improves soil health and iron availability	[20]
Soil inoculants	Siderophore-producing *Bacillus* sp.	Soil inoculation for disease control	Cotton, wheat	Promotes plant growth, suppresses soil-borne pathogens, and enhances soil fertility	[21]

(*Continued*)

Table 9.1 Examples of siderophore-based products and delivery systems with agricultural applications. (*Continued*)

Delivery system	Formulation example	Application area	Target crop/ soil	Benefits/ applications	References
Iron-chelating fertilizers	Siderophore–iron complex (e.g., *Bacillus* sp.)	Fertilizer application	Fruit trees (mango, citrus)	Provides essential micronutrients, prevents iron deficiency, and improves fruit yield	[21]
Siderophore-enhanced compost (foliar spray)	*Bacillus subtilis* siderophore formulation	Compost enrichment	Vegetables tomato, lettuce)	Enhances compost decomposition, promotes nutrient cycling, and improves soil health	[25]
Sprayable gel systems	Siderophore-based hydrogel sprays	Foliar application for growth promotion	Grapes, olive trees	Controlled release *via* gel formulation, reduces nutrient leaching, and enhances plant growth	[27]
Coated fertilizers	Siderophore-encapsulated micronutrient granules	Soil and crop-based application	Soybean, corn, wheat	Provides long-term nutrient release and improves iron and micronutrient availability	[28]
Microencapsulated fertilizers	Siderophore-encapsulated NPK fertilizers	Fertilization for balanced nutrient supply	Vegetable crops (carrot, lettuce)	Ensures controlled release of nutrients and enhances uptake under stress conditions	[29]

(*Continued*)

Table 9.1 Examples of siderophore-based products and delivery systems with agricultural applications. (*Continued*)

Delivery system	Formulation example	Application area	Target crop/ soil	Benefits/ applications	References
Granular fertilizer	Siderophore-enriched biofertilizer	Soil amendment	Rice, barley	Slow-release formulation for sustained nutrient availability, improving crop growth	[29]
Iron-chelating biostimulant	Siderophore-based formulation (e.g., *Bacillus* sp.)	Crop health promotion	Fruit trees (citrus, mango)	Enhances photosynthetic activity and fruit yield, preventing micronutrient deficiencies	[30]

9.5 Key Principles in Formulating Siderophore-Based Products

9.5.1 Microbial Selection and Optimization

Typically, the formulation starts with the selection of robust siderophore-producing microorganisms, including *Pseudomonas fluorescens*, *Bacillus subtilis*, and *Azotobacter chroococcum*, grown under optimized conditions to ensure optimal siderophore production. Some carbon sources such as glucose or glycerol or certain nitrogen sources such as ammonium sulfate significantly affect the yields of the siderophores [1].

9.5.2 Extraction and Purification

Siderophores require isolation by solvent extraction, ultrafiltration, and chromatography techniques to ensure that they are highly pure compounds. Purification removes impurities that may interfere with the performance of the formulated product. Very pure siderophores, such as ferrioxamines and pyoverdines, have been very effective in solving iron deficiencies in plants [15].

9.5.3 Stabilization

Siderophores are sensitive to other environmental factors like pH, temperature, and ultraviolet radiation. The work in the formulation is focused more on stabilizing those compounds by encapsulation techniques involving polymeric coatings, liposomes, or hydrogels. Alginate-based granules have been specifically designed to prevent the degradation of these siderophores while constantly releasing it in the soil [3].

9.5.4 Delivery Systems

Successful utilization of siderophore-based products depends heavily on effective delivery systems. Some of the delivery systems that are commonly employed are granular formulations, liquid sprays, seed coatings, and nanoparticles. Each system has an application, such as in soil-borne pathogens or foliar iron uptake [4].

9.6 Importance of Formulating Siderophore-Based Products

9.6.1 Enhancing Bioavailability and Stability

Siderophores often are challenged by degradation under environmental stress and low mobility in soil. Thus, formulation strategies will be essential to ensure stability and ensure targeted delivery. For example, encapsulation of siderophores in polymeric carriers like alginate or chitosan ensures extended activity in alkaline soils, which have naturally low bioavailability of iron. This is particularly relevant in regions such as South Asia, where iron chlorosis severely affects crop yield.

9.6.2 Broadening Applications in Agriculture

Siderophores beyond iron chelation include being involved in disease suppression and abiotic stress tolerance while fortifying nutrient acquisition. Formulation tailored to these various applications best utilizes siderophores for these diverse uses. In this context, bio-fertilizers are enriched in siderophore-producing microorganisms and foliar spray applications targeting citrus and maize crops that have iron-limiting conditions.

9.6.3 Enabling Cost-Effective Delivery

Agricultural practices require effective and inexpensive methods. The manufacture of siderophores as stable, easily applied products like granules or liquid sprays ensures that they are available to the farmer. The most recent breakthroughs in microbial fermentation and synthetic biology have greatly reduced the cost of siderophore production, allowing them to enter mainstream agricultural use [16].

9.7 Innovative Siderophore-Based Products in the Market

i. IronMax Pro
This granular formulation has ferrioxamine-type siderophores enriched for targeted application in iron-deficient soils, which improves nutrient uptake within crops such as rice and maize while enhancing yields by as much as 25% at field trials [17].

ii. GreenBoost FS
With content rich in pyoverdines, GreenBoost FS brings low chlorosis levels in critical crops like citrus and grape that happen in horticultural ones. It further boosts photogenic efficiency and quality [18].

iii. RootShield SC
This seed coating that employs microorganisms contains *Bacillus subtilis* and siderophore compounds, which improve the germination and early growth stages of both wheat and barley. It is also highly utilized in precision agriculture systems [19].

iv. AgriSorb NanoFe
A nanoparticle-based formulation incorporating ferrichrome siderophores, AgriSorb NanoFe provides an iron delivery mechanism with controlled release. It shows good result in alkaline soils where conventional chelates does not work effectively [20].

9.8 Applications of Siderophore-Based Products

Siderophore-based products have been increasingly applied to different forms of agriculture to improve crop health, nutrient uptake, and control of plant diseases. These products have many case studies that highlight their practical uses, which can be considered a model in sustainable farming. Some of the most important case studies and practical applications for siderophore-based agricultural products include the following:

i. Use of siderophore-producing *Pseudomonas* spp. in crop protection
One of the most notable applications of siderophore-based products in agriculture is by using the *Pseudomonas* spp. bacteria that produce pyoverdine and pyochelin as examples. The siderophores that it produces play a key role in acquiring iron and also possess biocontrol properties. In India, *Pseudomonas fluorescens* was applied as biocontrol to soil borne plant pathogens like *Fusarium* spp. and *Rhizoctonia solani* causing damages to cotton, wheat, or chickpea crop during a field trial. Disease incidence is lowered when it competes with pathogenic microbes for iron and, in turn, keeps healthy plants; this caused a significant yield from crops than for the non-treated plot. This research reveals the double-edged function of siderophore-derived products by promoting nutrient accessibility while concurrently offering natural disease suppression [2].

ii. Siderophore-based fertilizer for iron deficiency in rice
Iron deficiency is a widespread problem in rice cultivation, especially in alkaline and calcareous soils where iron availability to the plant is significantly reduced. A study carried out in China targeted a biofertilizer using a siderophore as a tool for this problem. The biofertilizer is comprised of *Bacillus* spp. that can produce siderophores and was applied to low-iron rice fields. The siderophore-based fertilizer enhanced the chelation of iron in the soil; hence, its availability in rice plants increased. This biofertilizer resulted in a higher growth rate, high chlorophyll content, and higher grain yield in rice plants grown using this biofertilizer. The use of this biofertilizer replaced synthetic iron fertilizers, leading to cost savings and less harm to the environment, which proved the possibility of using siderophore-based products for the improvement of nutrient use efficiency in rice cultivation [21].

iii. Siderophore-based products for drought tolerance in crops
In drought-prone areas, the uptake of nutrients by crops becomes more critical. A report in the Mediterranean region reviewed a siderophore-based biostimulant formulated to enhance the drought tolerance of wheat. This biostimulant, which contained *Pseudomonas* spp., was then applied to wheat plants at drought stress. The results further showed that

the siderophore-based product led to increased iron acquisition, eventually enhancing the photosynthetic activity and root development in the wheat plants. Overall, the plants were then more resistant to drought stress, maintaining a higher biomass as well as grain yield relative to the untreated plants. This case is an example of how siderophore-based products can be used in improving plant stress tolerance toward sustainable agricultural practices in water-scarce areas [22].

iv. Siderophore-based biocontrol in tomato plants

Tomato plants are easily attacked by several fungal pathogens such as *Fusarium oxysporum* and *Verticillium dahliae* that cause wilt diseases. In Spain, the efficiency of a siderophore-producing *Bacillus subtilis* strain as a biocontrol agent against these pathogens was tested. The growth of the pathogens was reduced due to the presence of the siderophores in the bacteria, which competitively occupied the iron absorbed from the soil. This shows that when siderophore-based biocontrol is applied, there will be a healthier tomato plant and lesser wilt symptoms compared to the control. This case, therefore, elucidates the effectiveness of siderophore-based products, not only aiding in plant growth but also effective against the pathogen, hence an alternative mode of action for chemical-based fungicides in tomato cultivation [23].

v. Siderophore-based products for improving soil fertility in organic farming

The availability of micronutrients, such as iron, often stands as the limiting factor, especially in soils that possess high pH values or insufficient quantities of organic matter, with regard to organic farming systems. An experiment in Brazil dealt with the application of siderophore-based product, wherein *Pseudomonas putida* was added for an increase in soil fertility to promote growth of crops involved in organic farming systems. This intervention was applied to crops such as tomato, lettuce, and cucumber grown in organic soil. To promote iron absorption in plants, siderophore application was expected to increase both growth and yield in the said crops. This product promotes microbial diversity in soils and thereby increases their long-term vitality. It is in this regard that siderophore-based products were introduced into organic farming systems in this case study, aimed at optimizing nutrient availability and crop production sustainability [24].

9.9 Delivery Systems for Siderophore-Based Products

9.9.1 Importance of Delivery Systems

Delivery systems play many different roles in siderophore-based products, namely:

1. Stability enhancement: Protects siderophores against environmental degradation from UV radiation, temperature fluctuations, or microbial hydrolysis.
2. Targeted application: Facilitates delivery to specific plant parts such as roots or leaves or to soil zones that are in most need of iron.
3. Sustained release: It facilitates the slow and controlled release of siderophores, thus lowering the need for repeated applications and decreasing nutrient loss.
4. Compatibility with farming practices: Suitable to prevailing farm operations such as seed treatments, foliar sprays, or fertigation systems.

9.9.1.1 Delivery Systems for Siderophore-Based Products

Optimization of effectiveness in agricultural contexts through efficient administration of siderophore-based products is essential. Effective delivery means that the products will reach the target plants and soil under desirable conditions, thereby enhancing absorption, encouraging plant growth, and providing biocontrol against diseases caused by pathogens. Thus, various delivery systems have already been developed for siderophore-based products, consisting of traditional formulations and the advanced technology to target given conditions in soil and the plant. Some of the main delivery systems for siderophore-based products are described briefly in the following subsections.

9.9.1.2 Liquid Formulations

Liquid formulations of siderophore-based products include one of the major types of delivery used widely in agriculture. In general, the products are either irrigated or applied topically onto the leaves through foliage and seed treatments. Usage through liquid formulations allows rapid intake not only by the plants but also by soil microbes where a full concentration of siderophores is delivered in a totally dissolved form that is ready for uptake by plant mechanisms. Liquid formulations have significant uses in foliar spraying where it allows for the instant deposit of siderophores onto the leaves of a plant where they can immediately complex with iron and encourage its absorption. These formulations also go into the soil by the use of drip irrigation or fertigation systems. Such a system ensures adequate dispersion of siderophores throughout the root zone that compete with the soil-borne pathogens for iron supplies. Such products as Sequester, acting as a chelating liquid iron solution, have been developed to enhance the yield of crops in agriculture, including rice and wheat, where poor availability of iron is seen in the soils. Apart from that, these help to reduce the deficiency of iron and support healthy crop development without supplementary chemical fertilizers [25].

9.9.1.3 Granular Formulations

Solid granular formulations have direct applications of siderophore-based products directly applied to the soil. Siderophores coated on inert carriers, like clay or peat or any other organic matter, constitute the basically prepared products. Granular formulations are most common for their field applications concerning crops, especially maize, rice, and cotton. The action of the granular product is such as the slow-release formulation whereby siderophores slowly become available at different times to the plant. These products are usually applied during planting or as a side dressing that will provide long-term availability of iron in the soil. The form is especially effective in promoting microbial activities in the rhizosphere, enabling siderophores to out-compete pathogenic microbes for access to iron and reduce disease pressure. The BioControl Granules are one of the most acknowledged granular formulations containing the siderophore-producing *Pseudomonas* spp. These products are significantly used as a soil-borne pathogen suppressor that can possibly cause diseases in crops; these include tomatoes and cucumbers, which, in particular, are subject to infection by diseases including *Fusarium* wilt [26].

9.9.1.4 Seed Coatings and Pellets

Seed coating and pelletization are advanced technologies that deliver siderophore-based products directly to the target location. Here siderophores are blended with polymers or clay-based materials, and the seed is coated with it before sowing. The siderophore from the seed coating is released gradually as the seed germinates, thus providing plants with nutrients throughout their entire life cycle from germination. The delivery mechanism in question is advantageous in ensuring that crops receive an adequate supply of siderophores during the early growth phases, a time at which nutrient availability in the soil is most critical. Furthermore, seed coatings can provide a protective benefit against pathogens by introducing beneficial microbial populations into the soil precisely at the time of planting. Seed treatments were developed to coat crops, including maize, wheat, and soybean with siderophores to improve seedling germination and maximum nutrient uptake. These formulations are very potent in iron-limited soil where readily available bioavailable iron is critical for growth. Commercial products include PlantGuard Seed Coating that contains siderophores to assist seed germination and overall plant health [27].

9.9.1.5 *Encapsulation and Controlled-Release Systems*

Encapsulation is a delivery system where siderophores are incorporated into materials that include polymers or even lipid-based carriers. Siderophores are consequently released slowly from the material, thus providing a controlled and lengthy supply of iron to plants. This reduces the amount of applications and prolongs the duration that siderophores are present in the soil, hence increasing their efficacy. Controlled-release formulations represent the best option for maintaining a steady supply of siderophores throughout the growing season, thus minimizing the need for repeated applications. Furthermore, encapsulation protects the siderophores from environmental degradation such as UV radiation and microbial degradation, thus prolonging their activity. This system has shown to be very effective in large-scale agricultural use, especially for crops grown in nutrient-poor soils. The controlled-release siderophore formulations have been tested in field crops such as rice and maize where continuous nutrient supply is needed to maximize growth. For instance, encapsulated formulations of siderophore-producing *Bacillus* spp. have shown increased iron acquisition as well as decreased pathogen incidences during field evaluations; this offers long-term effects for crops [16].

9.9.1.6 *Nanotechnology-Based Delivery Systems*

One of the promising delivery systems of siderophore-based products has been nanotechnology. Here nanoparticles are prepared to encapsulate the siderophores. Nanoparticles can be engineered to have controlled release properties of the siderophores. Nanoparticles enhance the stability, solubility, and bioavailability of siderophores, thereby facilitating targeted delivery to specific plant tissues or soil environments. Nanotechnology-based delivery systems especially work well in precision agriculture whereby delivery can be tailor-made depending on the needs of different plants. The delivery systems ensure effectiveness in the delivery process and minimize waste and potential adverse effects on the environment due to waste. Nanoparticles prevent the degradation of siderophores in the soil as they enhance their effect in the soil. Products like nano-Fe use nanoparticles to deliver siderophores to improve iron uptake where crops are grown in depleted soils. Such products show an increase in crop production and resistance to soil-borne crop diseases in tomatoes, potatoes, and rice [28].

9.9.1.7 Hydrogel-Based Systems

Hydrogel-based delivery systems provide a new approach to the controlled release of siderophore-based products. Hydrogels are polymers that absorb water and have the capacity to hold large amounts of water and nutrients, which they can gradually release into the environment. These systems are very beneficial in areas where there is a scarcity of water since they help maintain the level of moisture and ensure nutrient delivery to plants. Such systems of hydrogels are frequently utilized in the cases of dryland agriculture as well as soils where the retention of water is poor; such systems ensure gradual siderophores over a considerable duration for providing adequate supplies of bioavailable iron, thus saving much water and flow into the plants. Likewise, an irrigation event seldom occurs such that nutrition reaches crops. There have been formulations involving siderophore-producing microbes in hydrogels, namely, *Pseudomonas* spp., to be applied in arid conditions. These hydrogels help to increase the growth and enhance the nutrient uptake of plant as well as increase tolerance toward drought stress. Another such formulation is HydroBio hydrogel, which besides improving plant growth in drought condition, has also made available more iron to the plants from soil [29].

9.9.1.8 Biodegradable Polymers

Another promising candidate for the controlled and sustained-release delivery systems of siderophore-based products is through biodegradable polymers. Natural sources such as starch, chitosan, or PLA are used as the materials for these types of biodegradable polymers, thus degrading readily in soils over time due to minimizing the environmental burdens of farm practices. As a non-pollutive delivery strategy, biodegradable polymer-based delivery strategies are deemed to be relatively sustainable by delivering siderophores toward crops. These systems provide a slow, constant release of siderophores into the soil to promote long-term plant-available iron and decrease applications. Additionally, these technologies are considered to enhance the general health of the soil because of increased beneficial microbial production within the rhizosphere. *Bacillus* spp. encapsulated in biodegradable polymers have been tested in the field to make siderophores more available to plants by simultaneously lowering the environmental impact generated by synthetic fertilizers. Products such as EcoFe use biodegradable polymers to encase siderophore-producing microbes and therefore provide long-term advantages to crops like wheat, maize, and soybeans [14].

9.10 Challenges and Future Directions

Siderophore-derived products have become highly recognized as efficient tools in sustainable agriculture that can provide a natural way to enhance nutrient uptake in addition to providing protection against pathogenic microorganisms for the plants. The approach also promotes the ways of friendly farming. However, many challenges are still seen in connection with the development and distribution of these products. These need to be addressed so that maximum effectiveness and scalability of such solutions can be ensured within the agricultural environment.

9.10.1 Challenges in Delivery Systems

9.10.1.1 Stability and Shelf Life of Siderophores

One of the primary challenges regarding the longer-term stability of siderophores lies with their vulnerability to agrochemical formulations' degradation. As such, siderophores are organic molecules and generally responsive to environment conditions—temperature fluctuations, pH variances, as well as UV lighting which breaks over time, due to which chelate-iron ability will decrease and generally lead to decreasing overall performance capabilities in products produced using these siderophores. For instance, in liquid formulations that contain siderophores, they may tend to degrade faster than when in solid formulation, making their shelf life short and possibly requiring frequent reapplication. Researchers have successfully synthesized encapsulation technologies such as including siderophores within degradable polymers or hydrogels that are made with materials protecting them against other stressors from the environment. It will protect the siderophores and simultaneously ensure time-controlled release, thereby granting continuous availability of iron to plants on frequent applications. For instance, Rojas-Sánchez *et al.* examined the encapsulation of siderophores in nanoparticles and biodegradable polymers, enhancing their shelf life and efficacy in the action against various environmental stresses [30].

9.10.1.2 Controlled and Sustained Release

The efficacy of siderophores depends upon the attainment of controlled and prolonged release. Products emanating from siderophores often suffer from the issue of rapid release, which only temporarily provides iron available for uptake by plants. Fast mechanisms of release may rush in easy access to nutrients, but they mostly end up in losses and are used more than once. On the other hand, slow mechanisms of releasing nutrients can depict extended action, but issues of release usually occur during critical moments for plant uptake. For better controlled release, smart delivery systems that can respond to environmental factors like moisture or temperature may be used. Smart polymers or nanoparticles that can release siderophores according to the nutrient requirements of the plant or environmental conditions will be useful in this regard. Hydrogel-based systems, such as those discussed by Jampílek and Králová, provide controlled release of siderophores based on soil moisture levels, which is efficient and targeted delivery of nutrients [31].

9.10.1.3 Soil and Environmental Compatibility

The effectiveness of siderophore-based products can be affected by the soil type and environmental conditions, including the content of organic matter and the pH in soils. This also includes microbial activity that establishes the efficiency through which siderophores can transfer iron to plants. Siderophores may have a reduced chelating ability of iron in calcareous or alkaline soils because competing ions inhibit them; therefore, this reduces the availability of iron for plants. It depends, on top of this, to what extent the microorganisms composed or are active in this soil might degrade or, as the opposite, intensify functionality. Customized formulations for specific types of soil are currently under study to overcome these challenges. For example, siderophore-based products could be formulated specifically for certain soil conditions such as pH adjustment or the addition of co-factors to enhance iron solubility. Moreover, applying microbial inoculants with siderophore-based products enhances the effectiveness of the former because compatibility with the native soil microbiome is ensured.

An understanding of soil microbial dynamics and the development of site-specific applications improve the general effectiveness of siderophore products in different soil environments [4].

9.10.1.4 Cost and Scalability

Siderophore-based products assist in sustainable farming, but producing them is still highly expensive, especially in large quantities. Microbial fermentation for the production of siderophores is relatively expensive, and scaling up such processes for business applications is not easy. Also, the better delivery methods like encapsulation or hydrogel-based formulas increase production costs. New biotechnology developments, such as genetic engineering and synthetic biology, may provide cheaper ways to make siderophores in large quantities. For example, scientists can design microbes that produce more siderophores for less money. Furthermore, with research on better delivery systems, the production of such products in larger quantities could help decrease costs and make them more accessible to farmers. For instance, Rojas-Sánchez *et al.*, state that genetically engineered microorganisms will increase the production of siderophores for industries [30].

9.10.1.5 Environmental Impact and Residues

Siderophore-based products are, in general, viewed as environmentally friendly; however, attention to their long-term impact should be paid to both microbial resistance and residues in the soil and water. On large-scale application of siderophore-based products, their possible effects on the microbiomes of soils are still open to speculation—for instance, the development of resistance among microbes to the siderophore-producing strains might adversely affect the long-term sustainability of these products. Therefore, long-term ecological experiments will evaluate what is achieved when this effect arising through repeated application is internalized at the level of the individual experiments. To develop a critical understanding on both relationships developed with soil microbic resources and potential appearance, or rather management, of resistance as well as the prospective risks for the environment whatever when combined with strategy or strategies by IPs is likely. According to Herrmann and Lesueur, continuation of the research into the dynamics of the microbe–plant–soil system might ensure lesser risks associated with the development of siderophore-based products [31, 32].

9.11 Future Directions

The future development of siderophore-based products is a function of overcoming the aforementioned challenges *via* continuous research, technological growth, and formulation of agriculture-friendly methodologies. Several areas likely to be considered for such future developments are as follows:

i. **Integration with other biocontrol agents:** The potential could lie in combining siderophores with other biocontrol agents like PGPR or even biofungicides to present multifunctional formulations that might not only provide the nutrients the plants need but would also protect them from such plant pathogens.
ii. **Precision agriculture:** Using precision agriculture technologies—the sensors and drones, *inter alia*—farmers may also be able to apply siderophore-based products optimally so as to realize nutrient delivery without any waste end.
iii. **Biodegradable and smart delivery systems:** These studies will be aimed at biodegradable and "smart" delivery systems which would respond to environmental cues for better siderophore-based product efficiency that requires minimal reapplications while ensuring that plant benefits are persisting in nature.

9.12 Conclusion

Siderophore-based products mark a significant advancement in the direction of sustainable agriculture and offer a comprehensive approach for enhancing plant nutrition and health while reducing the dependency on synthetic fertilizers and agrochemicals. Its potential to effectively deliver bioavailable iron, thereby enhancing nutrient uptake and the development of plants, renders these siderophores very crucial in contemporary agricultural practice. Moreover, as biocontrol agents, siderophores not only facilitate health in plants by the suppression of pathogens but also provide significant roles for strengthening the general resilience of crops to environmental stresses. Inclusion of sophisticated delivery systems, such as biodegradable polymers, hydrogels, and encapsulation technologies ensures the stabilization and regulated release of siderophores, which overcomes crucial product efficacy and shelf-life-related challenges. Such developments in the products could, therefore, have an important impact on increasing the use efficiency of nutrients, minimizing environmental pollution, and driving a change toward more sustainable agriculture.

Despite all of these efforts, major barriers to the widespread use of siderophore-based products remain: mainly, cost-effective production methods, better delivery mechanisms, and overcoming soil-specific compatibility issues. Future research should be focused on overcoming these barriers, above all, by developing precision agriculture technologies, promoting microbial synergy, and implementing smart delivery systems that adjust the rate of siderophore release to the needs of the plants. Henceforth, siderophore-related products will become a standard for sustainable agriculture, the lack of which can no longer let people go hungry without damaging the planet. There will be a tremendous chance toward achieving the world's needs to make the earth sound at that time when such an end product is used correctly within the current production as well as management systems associated with agribusiness industries.

AI disclosure statement: The author confirms that no artificial intelligence (AI)-generated content or AI-assisted writing tools were used in the preparation of this chapter (including summaries and the abstract). All text, analysis, and conclusions are the result of the author's original work.

References

1. Sharma, A. and Johri, B.N., Growth promoting influence of siderophore-producing Pseudomonas strains GRP3A and PRS9 in maize (*Zea mays* L.) under iron limiting conditions. *Microbiol. Res.*, 158, 3, 243–8, 2003.
2. Ahmed, E. and Holmström, S.J., Siderophores in environmental research: roles and application. *Microb. Biotechnol.*, 7, 3, 196–208, 2014.
3. Fukushima, T., Allred, B.E., Sia, A.K., Nichiporuk, R., Andersen, U.N., Raymond, K.N., Grampositive siderophore-shuttle with iron-exchange from Fe-siderophore to apo-siderophore by Bacillus cereus YxeB. *Proc. Natl. Acad. Sci. U.S.A.*, 110, 13821–13826, 2013.
4. Mathew, A., Eberl, L., Carlier, A.L., A novel siderophore-independent strategy of iron uptake in the genus Burkholderia. *Mol. Microbiol.*, 91, 4, 05–820, 2014.
5. Radzki, W., Manero, F.G., Algar, E., García, J.L., García-Villaraco, A., Solano, B.R., Bacterial siderophores efficiently provide iron to iron-starved tomato plants in hydroponics culture. *Antonie van Leeuwenhoek*, 104, 3, 321–330, 2013.

6. Sayyed, R., Gangurde, N., Patel, P., Joshi, S., Chincholkar, S., Siderophore production by Alcaligenes faecalis and its application for growth promotion in *Arachis hypogaea. Indian J. Biotechnol.*, 9, 302–307, 2010.
7. Shanmugaiah, V., Nithya, K., Harikrishnan, H., Jayaprakashvel, M., Balasubramanian, N., Biocontrol mechanisms of siderophores against bacterial plant pathogens, in: *Sustainable Approaches to Controlling Plant Pathogenic Bacteria*, pp. 182–205, CRC Press, Taylor and Francis Group, New York, 2015.
8. Dimkpa, C.O., Svatos, A., Dabrowska, P., Schmidt, A., Boland, W., Kothe, E., Involvement of siderophores in the reduction of metal-induced inhibition of auxin synthesis in Streptomyces spp. *Chemosphere*, 74, 19–25, 2008.
9. Aguado-Santacruz, G.A.A., Moreno-Gomez, B.A., Jimenez-Francisco, B.B., Garcia-Moya, E.B., PreciadoOrtiz, R.E., Impact of the microbial siderophores and phytosiderophores on the iron assimilation by plants: a synthesis. *Rev. Fitotec. Mex.*, 35, 9–21, 2008.
10. Ahmed, E. and Holmstrom, S.J.M., Siderophores in environmental research: roles and applications. *Microb. Biotechnol.*, 7, 196–208, 2014.
11. Rajkumar, M., Ae, N., Prasad, M.N.V., Freitas, H., Potential of siderophore-producing bacteria for improving heavy metal phytoextraction. *Trends Biotechnol.*, 28, 142–149, 2010.
12. Raymond, K. and Dertz, E.M., Biochemical and physical properties of siderophores, in: *Iron transport in Bacteria*, J.M. Crosa, A.M. Mey, S.M. Pyne (Eds.), pp. 1–16, ASM, Washington, DC, 2004.
13. Solano, R.B., Garcıa, J.A.L., Garcia-Villaraco, A., Algar, E., Garcia-Cristobal, J., Manero, F.J.G., Siderophore and chitinase producing isolates from the rhizosphere of *Nicotiana glauca* Graham enhance growth and induce systemic resistance in *Solanum lycopersicum* L. *Plant Soil*, 334, 189–197, 2010.
14. Sharma, M., Mishra, V., Rau, N., Sharma, R.S., Increased iron-stress resilience of maize through inoculation of siderophore-producing *Arthrobacter globiformis* from mine. *J. Basic Microbiol.*, 56, 7, 719–735, 2016.
15. Shen, Y., Zhang, Y., Liu *et al*, C., *Micromonospora zeae* sp. nov., a novel endophytic actinomycete isolated from corn root (*Zea mays* L.). *J. Antibiot (Tokyo)*, 67, 11, 739–743, 2014.
16. Karmakar, P., Sharma, D., Das, P., *et al.*, Phosphate solubilizing capacity and siderophore production by Arthroderma cuniculi Dawson isolated from rhizospheric soil. *Res J. Life Sci. Bioinform. Pharmac. Chem. Sci.*, 4, 3, 330–336, 2018.
17. Winkelmann, G., A search for glomuferrin: a potential siderophore of arbuscular mycorrhizal fungi of the genus Glomus. *BioMetals*, 30, 4, 559–564, 2017.
18. Prashant, S., Makarand, R., Bhushan, C., Sudhir, C., Siderophoregenic Acineotobacter calcoaceticus isolated from wheat rhizosphere with strong PGPR activity. *Malays. J. Microbiol*, 5, 6–12, 2009.

19. Jonhson, G.V., Lopez, A., La Valle Foster, N., Reduction and transport of Fe from siderophores. *Plant Soil*, 241, 27–33, 2001.
20. De Villegas, M.E.D., Villa, P., Frias, A., Evaluation of siderophores production by Pseudomonas aeruginosa. *Microbiol.*, 44, 112–117, 2002.
21. Huston, W.M., Potter, A.J., Jennings, M.P., Rello, J., Hauser, A.R., McEwan, A.G., Survey of ferroxidase expression and siderophore production in clinical isolates of Pseudomonas aeruginosa. *J. Clin. Micriobiol.*, 42, 2806–2809, 2004.
22. Sayyed, R.Z., Badgujar, M.D., Sonawane, H.M., Mhaske, M.M., Chincholkar, S.B., Production of microbial iron chelators (siderophores) by fluorescent Pseudomonads. *Indian J. Biotechnol.*, 4, 484–490, 2005.
23. De Freitas, J.R., Banerjee, M.R., Germida, J.J., Phosphate-solubilizing rhizobacteria enhance the growth and yield but not phosphorus uptake of canola (*Brassica napus L.*). *Biol. Fertil. Soils*, 24, 358–364, 1997.
24. Hayat, R., Ali, S., Amara, U., Khalid, R., Ahmed, I., Soil beneficial bacteria and their role in plant growth promotion: a review. *Ann. Microbiol.*, 60, 579–598, 2010.
25. Hinsinger, P., Bioavailability of soil inorganic P in the rhizosphere as affected by root-induced chemical changes: a review. *Plant Soil*, 237, 173–195, 2001.
26. Leeman, M., Ouden, FMD., Van Pelt, J.A., Dirkx, F.P.M., Steijl, H., Bakker, P.A.H.M., Schippers, B., Iron availability affects induction of systemic resistance to Fusarium wilt of radish by *Pseudomonas fluorescens*. *Phytopathology*, 86, 149–155, 1996.
27. Lugtenberg, B. and Kamilova, F., Plant-growth-promoting rhizobacteria. *Annu. Rev. Microbiol.*, 63, 541–556, 2009.
28. Boukhalfa, H. and Crumbliss, A.L., Chemical aspects of siderophores mediated iron transport. *Biometals*, 15, 325–339, 2002.
29. Boukhalfa, H., Lack, J., Reilly, S.D., Hersman, L., Neu, M.P., Siderophore production and facilitated uptake of iron and plutonium in *P. putida*. *AIP Conf. Proc.*, 673, 343–344, 2003.
30. Rojas-Sánchez, B., Guzmán-Guzmán, P., Morales-Cedeño, L.R., Orozco-Mosqueda, MdC., Saucedo-Martínez, B.C., Sánchez-Yáñez, J.M., *et al.*, Bioencapsulation of microbial inoculants: mechanisms, formulation types and application techniques. *Appl Biosci.*, 1, 2, 198–220, 2022.
31. Jampílek, J. and Králová, K., Nanomaterials for delivery of nutrients and growth-promoting compounds to plants, in: *Nanotechnology: An Agricultural Paradigm*, R. Prasad, M. Kumar, and V. Kumar (eds.), pp. 177–226, Springer, Singapore, 2017.
32. Herrmann, L. and Lesueur, D., Challenges of formulation and quality of biofertilizers for successful inoculation. *Appl. Microbiol. Biotechnol.*, 97, 8859–8873, 2013.

10

Genetic Engineering of Microbes for Enhanced Siderophore Production

Rajdip Goswami[1], Bhaskar Pal[2], Arunava Ghosh[3], Kushal Roychoudhuri[4], Pinki Biswas[5], Biswajit Basu[1] and Saikat Santra[6*]

[1]*Department of Pharmaceutical Technology, School of Health and Medical Sciences, Adamas University, Barasat, Kolkata (WB), India*
[2]*Department of Pharmacology, Charaktala College of Pharmacy, Charaktala, Mothabari, Malda, West Bengal, India*
[3]*Bengal School of Technology, Sugandha, Hooghly, (WB), India*
[4]*Department of Pharmaceutics, Charaktala College of Pharmacy, Charaktala, Gangaprosad, Mothabari, Malda, West Bengal, India*
[5]*Department of Pharmaceutics, JRSET College of Pharmacy, Panchpota, Chakdah, Nadia, India*
[6]*Department of Pharmaceutics, JRSET College of Pharmacy, Panchpota, Chakdah, Nadia, India*

Abstract
Siderophores, small high-affinity iron-chelating compounds produced by microorganisms, play a pivotal role in plant growth promotion and disease suppression in sustainable agriculture. With the growing demand for bio-based solutions, the genetic engineering of microbes has emerged as a potent tool for enhancing siderophore production and functionality. This chapter delves into the advancements in genetic engineering techniques used to optimize siderophore biosynthesis, such as CRISPR-Cas systems, metabolic pathway modifications, and synthetic biology approaches. We explore how genetic alterations can enhance the efficiency of siderophore production in both natural and engineered microbial strains, promoting

**Corresponding author*: utpalsantra98@gmail.com; ORCID: 0000-0002-9398-7945
Rajdip Goswami: ORCID: 0000-0003-3296-8799
Bhaskar Pal: ORCID: 0009-0007-6299-4871
Arunava Ghosh: OCRID: 0009-0001-4823-7053
Kushal Roychoudhuri: ORCID: 0009-0006-8854-795X
Pinki Biswas: ORCID: 0009-0008-5419-5102
Biswajit Basu: ORCID: 0000-0001-8442-810X

Debasis Mitra, Ayush Madan and Anju Rani (eds.) Microbial Siderophores for Agricultural Productivity: Perspectives of Use in Agriculture, (233–268) © 2026 Scrivener Publishing LLC

their use in biofertilizers and biocontrol agents. The chapter will also highlight successful case studies of genetically modified microbes in field trials and their potential commercial applications and address the challenges faced, such as regulatory concerns and biosafety. This research is expected to provide deeper insights into the future potential of microbial genetic engineering to improve agricultural productivity through enhanced siderophore production.

***Keywords*:** Siderophores, genetic engineering, CRISPR-Cas9, metabolic engineering, synthetic biology, microbial biosynthesis, biofertilizers, biocontrol

10.1 Introduction

Microbial genetic engineering is one of the potential approaches to increase the synthesis of siderophores—small organic molecules produced by microorganisms to sequester iron from the surrounding. Iron is an essential nutrient for plants and various other organisms but is rarely in a chemically usable form especially in agricultural soils which are often low in nutrients [1]. The liberation of siderophores allows it to chelate with iron (Fe^{3+}) and turns the nutrient into a usable form by plants [109]. This process enhances plant growth, crop health, and yield, and therefore siderophores should be encouraged as important in agriculture. Their work also becomes very critical in such areas as exhaustion of soil nutrients, and they call for the least use of chemical fertilizers that are bad for the environment [2]. Siderophores serve not only in the access of plant nutrients but also in the modulation of microbial populations in the rhizosphere [3]. It promotes positive symbiotic relationships between plant and microbes, enhancing on the quality and stability of the soil. Improving the ability of microbes to produce siderophores naturally is, therefore, an important area of interest in the improvement of sustainable farming. However, the approaches used to enhance siderophore production have several drawbacks, including poor efficiency and inability to address large-scale production. To tackle these problems, researchers are now going back to genetic engineering to design better solutions [2, 4].

Siderophore enhancement is achievable by microbial DNA using some techniques that include gene editing and metabolic engineering. Because scientists can learn what genes are involved in making and controlling siderophores, they can create microbial strains that make more siderophores or better ones at getting iron. These genetic modifications can result in microbes that are fit to the existing agricultural requirements. This is in terms of enhancing the uptake of nutrients by plants under varying soil types [5].

This approach has a lot of advantages for agriculture because microbes can be designed to increase soil fertility as well as improve yields and decrease reliance on chemicals. They also help contribute to sustainable farming by funding environmentally sound farming practices that conserve our soil and water [6].

With each passing year in the field of biotechnology, it is becoming increasingly apparent that engineered microbes to meet current and future world agriculture needs can be developed. It means that these microbes can greatly contribute to the availability of food production, increased productivity of farming, and sustainable farming. Further research and technology development may play a critical role in improving the sustainability of future farming systems and achieving healthier plant growth and improved soil productivity [1].

10.2 Understanding Siderophore Biosynthesis Pathways

Siderophore production is a necessity in iron-limited areas and genetically related activities such as NRPS and NIS to facilitate the synthesis of siderophores from amino acids or non-peptidic structures [7]. These are released into the environment to chelate or coordinate with Fe (III) ions, and the resultant complexes are picked up by transport systems from microbes [8]. Genetic manipulation assists this course by either optimizing enzymes or developing new siderophores with more effectiveness in iron binding. Understanding of natural pathways makes engineered strains for use in agriculture and bioremediation; this is why these systems must be studied and are broadly classified into hydroxamates, catechols, carboxylates, and mixed types, as shown in Figure 10.1 [9].

10.2.1 Crucial Enzymes for Siderophore Production

Microbes produce specific small molecules known as siderophores to solubilize and consequently chelate iron from its surrounding. This process depends on various enzymes involved in siderophore biosynthesis, which follows two main pathways: the non-ribosomal peptide synthetase-dependent pathway and the non-ribosomal peptide synthetase-independent (NIS) pathway [11]. In the NRPS pathway, the large multifunctional enzyme complexes called non-ribosomal peptide synthetases

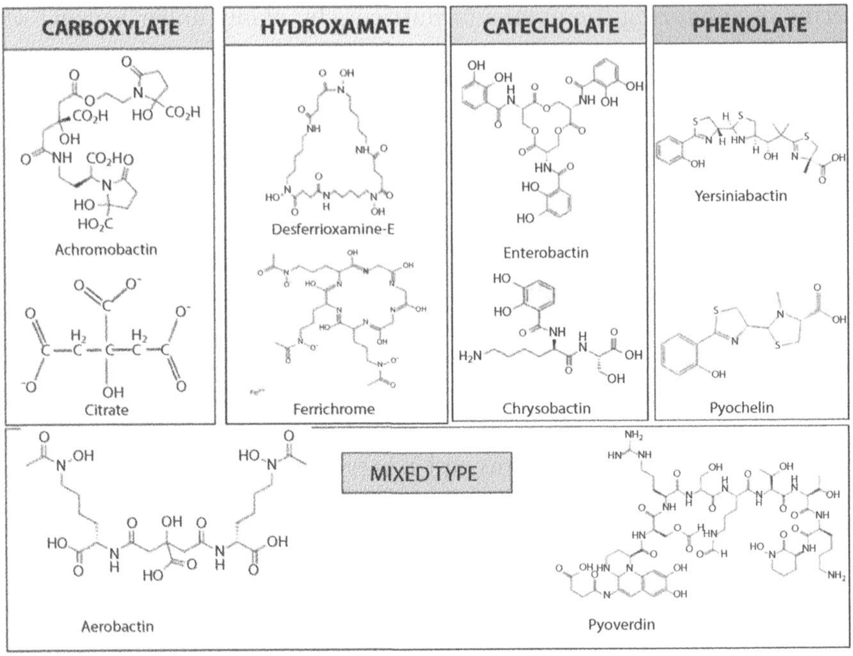

Figure 10.1 Types of siderophores (hydroxamates, catechols, carboxylates, and mixed types) with their respective structures and roles in microbial iron acquisition and agricultural applications [10].

(NRPSs) are involved in the stepwise incorporation of amino acids for the assembly of peptide siderophores. Some of the most important domains in NRPS adenylation, peptidyl carrier protein, and condensation domains regulate the selection, activation, and installation of the amino acids, while the NIS pathway is in charged with the synthesis of nonpeptidic siderophores. It greatly depends on the catalytic activities, for example, monooxygenases and ligases. These enzymes are critical in the synthesis of the complicated architectures specific to nonpeptidic siderophores [12].

The study of and interference with enzymes of both pathways have led to tremendous improvements in genetic engineering. Several strategies including the upregulation of biosynthetic genes and use of new strategies in gene modification have been used to increase the yield of siderophores in microbial strains. Out of these innovations, the ones that are plausible for further enhancement of siderophore synthesis are for medicinal use, agricultural use, and bioremediation [13].

10.2.2 Control of Siderophore Synthesis Pathways

The biosynthesis of siderophores is regulated by a host of environmental and genetic factors. This regulation relies on iron availability where the ferric uptake regulator (Fur), a central transcription factor, represses siderophore biosynthetic gene expression in iron-loaded conditions [7, 14]. On the contrary, under iron starvation, derepression of these genes promotes increased siderophore production to promote iron acquisition. Moreover, two component systems (TCS) are used to modulate siderophore production in response to different stimuli, such as osmotic stress and nutrient configuration [15]. These regulatory networks can be targeted by genetic engineering strategies to promote siderophore production by creating knockout strains deficient of repressors or by overexpressing positive regulators. Knowledge of these regulatory mechanisms increases our understanding of microbial iron homeostasis and allows us to design microbial strains to produce higher yields of siderophores in industrial applications [16, 17].

10.2.3 Natural Diversity in Microbial Siderophore Synthesis

Information about natural variability in siderophore production among microbial species emphasizes the evolutionary adaptations used by various organisms to survive in ferum-limited environments [18]. The species specificity of siderophore biosynthesis gene clusters assures their diversity in chemical form from catecholates and hydroxamates. Specifically, *Pseudomonas* species synthesize predominantly pyoverdine and pyochelin, whereas other microbes customize their siderophores to their ecological niche [7]. This variability is important in microbial interactions in communities as it determines the competition for iron and survival strategies [19, 20]. This diversity offers genetic engineering approaches an opportunity to introduce genes from high-producing strains to less-efficient producers to produce desirable siderophores more efficiently. In addition, an understanding of the ecological interactions and selective pressures that drive siderophore evolution may lead to the development of conservation strategies or biotechnological applications in which to harness the natural microbial capabilities for enhanced production [21, 22].

10.3 Genetic Engineering Techniques for Siderophore Enhancement

Siderophore production by genetic engineering is another method used in advancing technology that focuses on improving the biosynthesis of the microorganisms. Because siderophores are essential for iron which is scarce in iron-limited environments, they find a wide application in farming, healthcare, and industry [23–25]. The use of gene overexpression, in which important biosynthetic genes are intentionally activated, is one of the most widely used techniques. This is usually done by putting in front of these genes powerful promoters that stimulate enhanced transcription speed. For example, upregulation of genes including entA, entB, and entC has been demonstrated to highly increase the synthesis of enterobactin, the most efficient siderophore. Multiple-copy plasmids are again utilized to enhance the production and guarantee the robust synthesis of siderophores [2].

The other revolutionary method is the CRISPR-Cas 9 gene editing technology that enables targeted changes on the gene as well. Thus, through this functional genomic approach, scientists are able to either increase the biosynthesis, alter the regulatory mechanisms, or inhibit the genes that negatively control the biosynthesis of siderophore [26]. For instance, editing of promoter regions or transcriptional regulators enhances the tractability of biosynthetic pathways to the environment, while knocking out repressor genes eliminates negative regulation altogether [27, 28].

Along with the abovementioned strategies, the improvements in the metabolic control provide an opportunity to enhance the entire metabolic pathways. These strategies enhance siderophore production, efficacy, and specificity, thereby providing immense value for use in the bioremediation of contaminated sites, plant growth promotion, and the creation of better antimicrobial approaches [29–31].

10.3.1 CRISPR-Cas9-Mediated Genetic Editing

CRISPR-Cas9-mediated genetic editing is a revolutionary tool used to enhance siderophore production by enabling precise and efficient modifications in microbial genomes. This system relies on a programmable RNA molecule (guide RNA or gRNA) to direct the Cas9 nuclease to a specific DNA sequence, where it introduces double-strand breaks. The cell then repairs these breaks through either non-homologous end joining (NHEJ) or homology-directed repair (HDR), allowing for targeted gene knockouts, insertions, or edits [32–35] using CRISPR-Cas9 to modify genes related to siderophore production, as shown in Figure 10.2.

10.3.1.1 Siderophore Enhancement

Certain genes negatively regulate siderophore biosynthesis by suppressing the expression of pathway enzymes. CRISPR-Cas9 can be used to knock out these repressor genes, leading to increased siderophore production. For example, removing regulatory proteins that downregulate the *ent* operon in *E. coli* can enhance enterobactin synthesis [36, 37].

10.3.1.2 Activation of Silent Pathways

Many microorganisms have latent or low-activity siderophore biosynthetic pathways. CRISPR-Cas9 can be employed to activate these pathways by modifying regulatory elements, such as promoters or ribosome binding sites, to ensure the strong expression of biosynthetic genes [38].

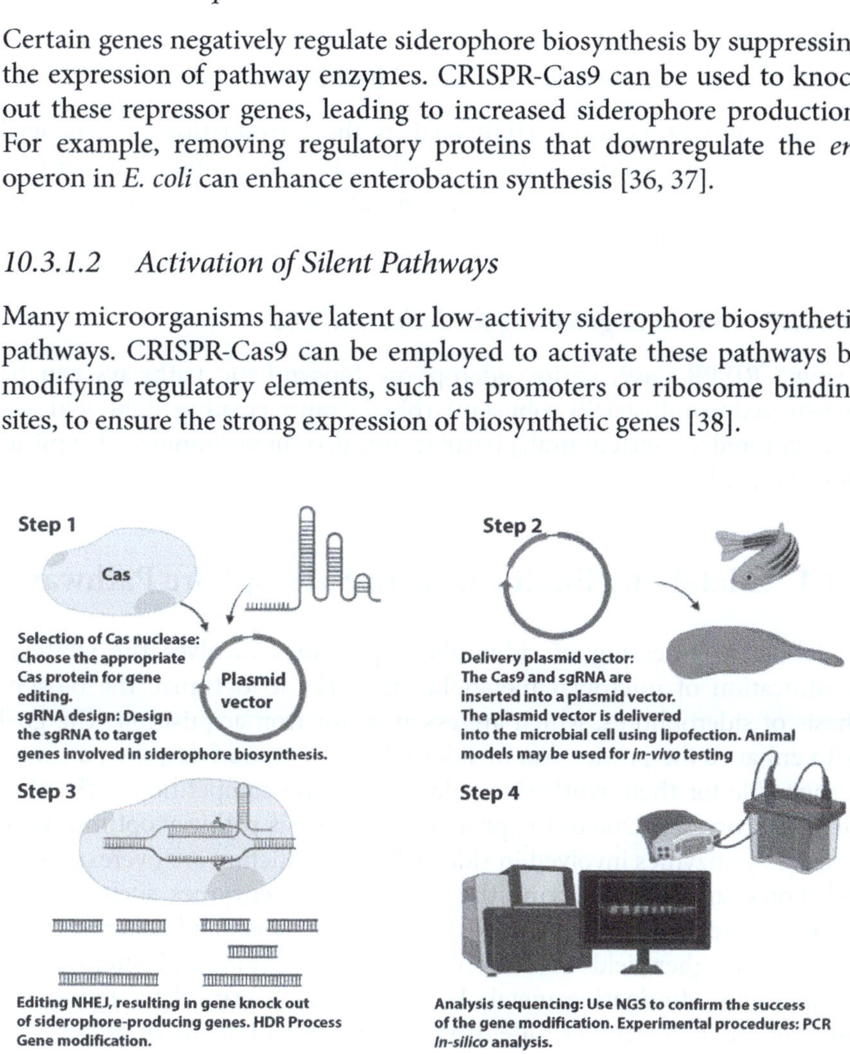

Figure 10.2 CRISPR-Cas9 mechanism for gene editing, showcasing its application in modifying genes responsible for siderophore production to improve microbial efficiency in agriculture [43].

10.3.1.3 Pathway Optimization

Cas9 can be used in combination with CRISPR activation (CRISPRa) systems to upregulate multiple genes in siderophore pathways simultaneously. This is especially useful for increasing the efficiency of complex pathways that involve several enzymes [39].

10.3.1.4 Introduction of Beneficial Mutations

Site-specific editing using HDR enables the introduction of beneficial mutations in biosynthetic enzymes, enhancing their catalytic activity or substrate specificity. This approach can improve the efficiency of siderophore synthesis [40].

10.3.1.5 Strain Engineering for Industrial Use

Using CRISPR-Cas9, entire siderophore biosynthetic pathways can be transferred to industrially robust microbial strains, creating high-yield systems tailored for agricultural, pharmaceutical, or biotechnological applications [41, 42].

10.4 Metabolic Engineering of Siderophore Pathways

Metabolic engineering of siderophore pathways involves the strategic modification of microbial metabolic networks to optimize the biosynthesis of siderophores, which are essential for iron acquisition. The goal is to enhance the production of siderophores by modifying the pathways responsible for their synthesis while minimizing competition with other cellular processes. One of the primary strategies is pathway optimization, where key enzymes involved in siderophore production are overexpressed [8]. For example, amplifying the expression of enzymes such as non-ribosomal peptide synthetases (NRPS) in *Pseudomonas* and *Escherichia coli* can lead to higher yields of siderophores like pyoverdine and enterobactin. Additionally, eliminating metabolic bottlenecks is critical for increasing production efficiency [44]. Identifying steps that limit the overall rate of siderophore biosynthesis and modifying the expression of corresponding enzymes can help increase flux through the biosynthetic pathway. This could involve enhancing the availability of precursor molecules like amino acids, iron, or CoA derivatives, which are essential for siderophore synthesis [9]. Another powerful tool in metabolic engineering is the redirection of carbon flux. By shifting the metabolic focus of a microorganism towards siderophore production and away from competing pathways, such as carbon or nitrogen metabolism, overall yield can be enhanced [7]. Moreover, synthetic biology approaches allow the integration of foreign biosynthetic genes from other organisms, enabling the production of novel or high-yield siderophores [45]. Overall, metabolic engineering of siderophore pathways enables the development of microbial strains with optimized production capabilities, which is beneficial for applications in agriculture, bioremediation, and pharmaceuticals [46].

10.5 Synthetic Biology Approaches for Siderophore Optimization

Synthetic biology approaches for siderophore optimization leverage advanced techniques in genetic engineering, metabolic pathway design, and system biology to create microorganisms with enhanced or novel siderophore production capabilities [47]. These strategies aim to improve the yield, efficiency, and functionality of siderophores for a variety of applications, from agriculture to medicine. Synthetic biology allows the precise modification of genetic and metabolic networks, facilitating the creation of tailored biosynthetic systems [48].

10.5.1 Pathway Construction and Engineering

Synthetic biology enables the design and construction of synthetic pathways for siderophore biosynthesis. By introducing genes from different organisms into a host strain, researchers can create novel siderophores or optimize existing pathways [49]. For example, introducing genes involved in siderophore biosynthesis from *Pseudomonas* into *E. coli* or yeast can lead to the production of siderophores that are not naturally synthesized by these hosts. This modular approach allows for the creation of customized biosynthetic pathways with enhanced efficiency or unique properties [50].

10.5.2 Optimization of Biosynthetic Gene Clusters

In synthetic biology, biosynthetic gene clusters—groups of genes that work together to produce a specific compound—can be optimized for siderophore production [51]. By using synthetic promoters, ribosome binding sites, or transcription factors, the expression of these gene clusters can be finely tuned to enhance siderophore output. Additionally, the insertion of artificial biosynthetic pathways can introduce new functions or improve the efficiency of the existing biosynthetic process, such as increasing the affinity of the siderophore for iron [52].

10.5.3 Metabolic Flux Control

Synthetic biology approaches also allow the optimization of metabolic flux towards siderophore production by constructing genetic circuits that regulate the flow of carbon, nitrogen, and other precursors into the siderophore biosynthesis pathway. This can be achieved by overexpressing genes that provide precursors or by knocking down competing pathways. By manipulating the overall metabolic network, it is possible to maximize the production of siderophores and minimize the synthesis of by-products that may divert resources away from siderophore biosynthesis [53].

10.5.4 Engineering of Regulatory Networks

Synthetic biology provides tools to modify regulatory networks that control siderophore production. By utilizing synthetic promoters, CRISPR interference (CRISPRi), or RNA-based regulatory systems, the expression of siderophore genes can be precisely controlled. For example, using inducible promoters that activate siderophore production under specific conditions (e.g., low iron availability) ensures efficient production in response to environmental cues. Additionally, synthetic biology enables the development of feedback regulation systems that optimize siderophore production based on the metabolic state of the microorganism [54].

10.5.5 Development of Hybrid and Chimeric Siderophores

One of the most innovative synthetic biology approaches is the design of hybrid or chimeric siderophores. By combining elements from different siderophore classes (e.g., catecholate and hydroxamate), researchers can create novel siderophores with enhanced iron-binding properties or greater stability under various environmental conditions [55]. These hybrid siderophores can be tailored for specific applications, such as improving iron uptake in plants or targeting pathogens more effectively.

10.5.6 Synthetic Consortia for Siderophore Production

Synthetic biology also enables the development of synthetic microbial consortia, where multiple engineered strains work together to optimize siderophore production. For instance, one strain may focus on producing the siderophore, while another may be responsible for utilizing the siderophore to acquire iron more efficiently. This approach mimics natural microbial communities but with the added benefit of engineered control over each strain's contribution, leading to optimized siderophore production in complex environments [56].

10.6 Enhancing Siderophore Production in Different Microbes

Iron-chelating siderophores are essential second metabolites used by microbes to acquire iron from their surroundings and are of great importance for plant well-being and pathogen elimination. There is great potential for using siderophore-producing microorganisms through genetic engineering and metabolic elicitation in sustainable agriculture. To this end, the research studies in this dissertation have provided for a better understanding of how to enhance the efficacy and yield of such iron-chelating compounds in bacteria and fungi for the development of next-generation biofertilizers and biocontrol agents for addressing emerging issues in modern farming practices. This section describes the main microbial opportunists and novel approaches to enhance siderophore synthesis [2].

10.6.1 Bacterial Species

Siderophores are produced excessively by bacteria and contain a great importance in sustainable agriculture because of their ability to foster plant growth, increase nutrient accessibility, and inhibit the growth of pathogenic organisms [57]. The most known species for their well-known siderophores are *Pseudomonas* spp., which produce pyoverdine and pseudobactin. Some of the approaches include engineering biosynthetic genes and have shown to improve yields through optimization of genetic engineering techniques to foster plant health and defensive capabilities against pathogens [58]. Like any other bacteria, *Bacillus subtilis* synthesizes its own siderophores called bacillibactin and petrobactin, which are crucial for nutrient recycling and soil health, and genetic manipulations such as the removal of inhibitory genes have enhanced the overall iron-chelating potential of these siderophores.

Other indispensable bacteria are *Escherichia coli* that is widely applied in illustrating the siderophore pathway. EntA, EntB, and EntC genes

Table 10.1 Bacterial species and their engineered siderophores: Strategies and applications in agriculture.

Bacterial species	Siderophore produced	Role in agriculture	Genetic engineering strategies	Benefits of engineering	Reference
Pseudomonas spp.	Pyoverdine, pseudobactin	Plant growth promotion, biocontrol of pathogens	Overexpression of pyoverdine biosynthetic genes	Increased plant growth and pathogen suppression	[62]
Bacillus subtilis	Bacillibactin, petrobactin	Biofertilizer, improved nutrient cycling	Deletion of genes inhibiting siderophore biosynthesis	Boosted iron chelation efficiency	[63]
Escherichia coli	Enterobactin	Model organism for siderophore pathway research	Overexpression of *entA*, *entB*, and *entC* genes	High siderophore yield for experimental and agricultural uses	[64]
Azotobacter spp.	Azotobactin	Nitrogen fixation, plant growth promotion	Introduction of synthetic pathways for hybrid siderophores	Multi-functional bacteria for iron uptake and nitrogen fixation	[65]
Rhizobium spp.	Rhizobactin	Symbiotic nitrogen fixation in legumes	Engineering to co-produce siderophores and nodulation factors	Improved symbiosis and crop yield in legumes	[66]
Streptomyces spp.	Desferrioxamine B	Soil health improvement, suppression of plant pathogens	Overexpression of desferrioxamine biosynthetic genes	Enhanced biocontrol properties	[67]

overexpression is characterized by high levels of siderophore production for practical and farming purposes. Some bacteria like *Azotobacter* spp. and *Rhizobium* spp. are highly beneficial for nitrogen fixation and plant symbiotic relationships. The engineering approaches developed incorporated mere siderophore hybrid pathways and co-production of nodulation substances, all of which increase its versatility and crop output capabilities

[59, 60]. In the same way, desferrioxamine B is produced by *Streptomyces* spp., which was genetically modified to enhance biocontrol function, hence contributing to soil health and plants [61].

As stated in Table 10.1, these bacterial species and their recombinant siderophores illustrate how genetic manipulations can amplify superior microbial roles in agriculture innovation. These microbes increase siderophore production that can be used as bio-fertilizers, bio-controllers, and models for effective and sustainable farming.

10.6.2 Fungal Species

Siderophore species are far-reaching in the life of fungi noteworthy for their contribution to agriculture as nutrient cycling and pathogen control and biocontrol agents. Ferrichrome and hydroxamate, the siderophores produced by *Aspergillus niger*, isolated from soil are important for nutrient cycling and plant uptake of iron [68]. Being in soil area, they developed various genetic engineering techniques, including siderophore biosynthetic gene overexpression to enhance *P. fluorescens* soil enrichment and to improve plant growth. For example, *Aspergillus fumigatus* produces a siderophore called triacetylfusarinine C, which is well recognized for pathogen suppression. This siderophore has been enhanced for biocontrol by knocking out metabolic pathways that otherwise compete for nutritional resources [69].

Another embraces *Trichoderma harzianum*, which secretes coprogen that helps in managing pathogenic organisms in the soil. Interestingly, CRISPR modifications to improve siderophore production unlocked this fungus' ability to improve pathogen control and soil iron acquisition for sustainable agriculture. The soil and plant growth promotion by *Penicillium* spp. include fusarinines and hydroxamates. Biotechnological intervention in these species has led to an increase in iron chelation through manipulation of regulatory signals and thus an improvement in soil fertility and plant health [70].

Fusarinines synthesized by *Fusarium oxysporum* play an important role in plant pathogen suppression. By producing such hybrid genes to increase the range of siderophores, they perform as dual organisms showing both antimicrobial property and nutrient solubilization. Of all the known fungi, *Rhizopus oryzae* which produces a phyto-iron called rhizoferrin is involved in enhancing plant iron acquisition. Through metabolic engineering, it has enhanced and augmented the level of siderophore production, thus enhancing the availability of iron in the soil. Lastly, *Beauveria bassiana*, an insect-killing fungus and an entomopathogen, synthesizes and secretes

Table 10.2 Genetically engineered fungal strains for siderophore production.

Fungal species	Siderophore produced	Role in agriculture	Genetic engineering strategies	Outcomes	Reference
Aspergillus niger	Ferrichrome, hydroxamate	Soil nutrient cycling, iron uptake	Overexpression of siderophore biosynthetic genes	Enhanced soil enrichment and plant growth	[10]
Aspergillus fumigatus	Triacetylfusarinine C	Biocontrol and pathogen suppression	Gene knockout of competing metabolic pathways	Higher siderophore yields	[72]
Trichoderma harzianum	Coprogen	Biocontrol of soil-borne pathogens	CRISPR modification to increase siderophore synthesis	Improved pathogen suppression and iron availability	[73]
Penicillium spp.	Fusarinines, hydroxamates	Soil conditioning, plant growth promotion	Genetic manipulation of regulatory pathways	Enhanced iron chelation	[74]
Fusarium oxysporum	Fusarinines	Plant pathogen control	Hybrid gene expression for broader siderophore range	Dual function in pathogen suppression and nutrient enhancement	[75]
Rhizopus oryzae	Rhizoferrin	Iron uptake in plants	Metabolic engineering to boost siderophore secretion	Increased bioavailability of iron in soils	[76]
Beauveria bassiana	Ferricrocin	Biocontrol of insect pests	Gene editing for combined insecticidal and siderophore functions	Multi-functional agricultural application	[77]

ferricrocin in soil and phytosphere, genetically modified to have both insecticidal and siderophore properties, which is beneficial for biocontrol and agriculture [71].

As indicated in Table 10.2, these fungal species illustrate the relevance of siderophores in agriculture and the versatility of their functions. Genetic engineering has been a useful technology to increase siderophore synthesis, improve biocontrol potential, and influence plant health.

10.7 Application of Engineered Microbes in Agriculture

10.7.1 Use of Biofertilizer

Rapidly increasing agricultural demands coupled with numerous environmental issues have necessitated the adoption of a new-fangled sustainable farming era *via* inclusion of biofertilizers and engineered microbes. Such technologies provide promising solutions to achieve higher crop productivity, maintaining soil health and reducing detrimental impact on our planet due to existing conventional farming practices [78].

10.7.1.1 Enhancement of Soil Fertility

Biofertilizers are natural fertilizers which are in the form of powder or liquid and contain a good number of living microorganisms which help in the growth and development of plants by increasing the nutrient availability to the host plants. They enrich soil fertility in different ways.

10.7.1.2 Nitrogen Fixation

Certain bacteria such as *Rhizobium* and *Azotobacter* convert atmospheric nitrogen to forms that can be absorbed and utilized by plants. This is particularly beneficial for leguminous crops which enter into a symbiotic relationship with these bacteria [79].

10.7.1.3 Phosphorus Solubilization

Fungi (*Mycorrhizae*) and bacteria (e.g., *Bacillus* spp.) can solubilize inorganic phosphorus, making it available to plants. Since phosphorus is often a limiting nutrient in agricultural soils, this is an important function [80].

10.7.1.4 Organic Matter Decomposition

Biofertilizers help decompose organic matter. Organic matter decomposition occurs when biofertilizer releases enzymes and metabolites into the soil/rhizosphere. These enzymes, in turn, catalyze other enzyme reactions necessary for the rapid degradation of complex molecules in organic residues. During this process, mineral nutrients are released as obtained from biodegradation [81].

10.7.2 Synergistic Benefits of the Biological Hybrid Biofertilizers and Engineered Microbes

The combination of biofertilizers and engineered microbe technology for plant growth should have several advantages, which would increase markedly the productivity in agriculture.

10.7.2.1 Nutrient Efficiency and Crop Yield

By integrating biofertilizers with engineered microbes, farmers can improve the efficiency of nutrient use. Engineered strains can release nutrients in a more available form, which are then taken up by the plants. The result is higher crop yields and reduced fertilizer costs as less synthetic fertilizer needs to be applied [82].

10.7.2.2 Soil Health and Microbial Diversity

Biofertilizers help in maintaining a healthy microbial ecosystem of soil. Engineered microbes can improve this diversity. A diverse microbial community helps in improving soil health and nutrient cycling; improved soil structure and disease resistance help in making agriculture sustainable for long duration. Positive-impact-machinery-reduced biofertilizers are used as an alternative to the harmful chemical fertilizers that are widely available [83].

10.7.2.3 Pathogen Reduction Strategies

The use of bioengineered microorganisms, especially when applied along with biofertilizers, is gaining attention in biocontrol methods. Beneficial microorganisms are known to give pathogens a run for their resources and space, while engineered microorganisms are known to possess certain agents, which are antagonistic toward the pathogens. This decreases the dependence on agrochemicals and encourages the use of IPM strategies [84].

10.7.2.4 Abilities and Resistance to Stress

The introduction of engineered microbes can also aid the plants in dealing with environmental stresses such as salinity, temperature extremes, and drought. Such strains, for instance, may secrete biomolecules of osmoregulatory nature in the plants' tissues to retain more water. Such engineered microbes, when used alongside biofertilizers, enhance plant resilience, as a whole improving their growth under difficult conditions [85].

10.7.2.5 Concern Over the Environment

The application of biofertilizers and bioengineered microorganisms minimizes the use of artificial chemical fertilizers and insecticides, thereby promoting environmentally friendly farming techniques. The limitations imposed on the chemical applications help curtail the nutrient poured into the water bodies that create eutrophication and other related problems [86].

10.7.2.6 Protection of the Climate by Carbon Sequestration

Some engineered microbes are constructed with the capability to increase soil carbon sequestration to assist in climate change. Biofertilizers help in this process by maintaining the health of the soil organic matter and encouraging more microbial activity which helps to store carbon.

The application of biofertilizers along with added microorganisms creates a prospective biological approach to achieve sustainable agriculture. This combination can mitigate most of the problems found in contemporary agriculture by boosting the supply of nutrients, enhancing existing soils, and developing stress-resistant crops. Research is advancing at a rapid pace; therefore, the wider adoption of these technologies will transform agricultural practices for good, making them more efficient, sustainable, and environmentally friendly. This integration not only enhances the sustenance of the human population but also the wellbeing of Mother Nature [87].

10.7.2.7 Role in Biocontrol Against Plant Pathogens

Agribusiness has seen a shocking upsurge in the use of chemical pesticides, raising issues of ecological balance, health hazards, and the possibility of evolving strains of pests that are resistant to prescribed insecticides. Consequently, this has led to an expeditious finding of other options for pest management. One such option which shows great potential is the use of engineered microbes for the control of plant disease organisms. This not only helps in the effective management of the diseases but also boosts the practice of agricultural sustainability.

This solution involves the use of living agents to mitigate the effect of pest or plant disease populations. This may include natural enemies, such as predators and parasitoids, or beneficial microbes that can outcompete, inhibit, or kill pathogens. Out of all of them, engineered microbes are proven effective because they can be designed for a particular purpose [87].

Microorganisms, including bacteria and fungi, can be designed in such a way that they possess or exhibit characteristic traits which make them more effective in the control of plant diseases. These traits may include the ability to produce different types of antimicrobial agents, the capacity to efficiently colonize plant surfaces and tissues, or the resistance to abiotic stress [88]. Most parts of their contribution to biocontrol are highlighted in the discussion below.

10.7.3 Producing Antimicrobial Agents

Microbes that can be modified can also be made to synthesize certain metabolites which are bacteriostatic in nature toward the pests. For instance, *Pseudomonas fluorescens* strains have been altered to synthesis fungicidal phenazines and 2,4-diacetylphloroglucinol compounds that inhibit the growth of soil-borne fungi. Through the use of genetic engineering to increase the secretion activity of these compounds, transgenic microorganisms can serve to improve the prevention of diseases such as root rot or damping-off [89, 90].

10.7.3.1 Competitive Exclusion

Genetically modified microorganisms can overpower pathogens in the bid for nutrients and space. This is achieved by covering the plant root zone or the leaf surfaces with beneficial microorganisms which pose a threat to the pathogenic microorganisms. For instance, genetically modified strains of *Bacillus* sp. can be constructed in such a way that they bind more to the roots and other plant surfaces, thereby inhibiting the growth of pathogens. This mechanism of competitive exclusion helps to lower the occurrence and severity of the disease in question [91].

10.7.3.2 Induced Systemic Resistance

Certain microorganisms that have been modified can activate the defenses of plants, thereby inducing systemic resistance (ISR). Plants responding to beneficial microbes initiate their immune defense, rendering the plants stronger against subsequent attacks from pests. *Trichoderma* species, for example, can induce severe ISR to the plants genetically altered to improve their amplitude of signaling pathways and to enable the plants in repelling infections more effectively [92].

10.7.3.3 Enhanced Stress Tolerance

Moreover, the genetically modified microorganisms are effective in promoting the enhancement of stress resilience in crops which can worsen the vulnerability of diseases. For instance, some strains can be engineered to express osmoprotectants that aid in moisture retention in plants under drought stress, hence alleviating the stress that predisposes them to diseases. The ability to increase tolerance to stress and at the same time manage the pests is key to progressive crop production [93, 94].

10.7.3.4 Biodegradation of Pathogen Residues

Biodegradable microbes can also decrease pathogenic remains in soils or in the plant materials, thus helping to reduce pathogen pressure as well as avoid a reinfection. For instance, such engineered fungi that are capable of degrading toxins generated by plant pathogens can also be utilized in cleaning up contaminated soils for replanting purposes. One of the techniques that can be employed for sustainable agriculture is the use of designed microbes for the biocontrol of plant-disease-causing pathogens [95]. These natural biologicals help in plant defense induction, the establishment of competition with pathogens, and improving the health of the soil, thereby reducing the use of agrochemicals and enhancing the growth of crops and their yields. With this, research and development in this field is quickening, and in the not ignorable near future, engineered microbes will become indispensable to the way agriculture is practiced and diseases of plants managed. This kind of innovation encourages not only the farmers but also the environment and the agriculture as a whole [96].

10.8 Field Trials and Real-World Applications

The use of modified microorganisms in agriculture is highly promising since it can increase crop yields, regenerate soils, and foster more sustainable systems across the globe. With the progress of these technologies, however, field trials and real-world usages become essential for assessing their merit and practicability in a broad range of agricultural applications. In this article, we examine different field trials and real-world applications of engineered microbes in agriculture so as to address advantages, disadvantages, and the prospects of this technologies [97].

10.8.1 Application to Real-Life Situations

The application of genetically engineered microbes (GEMs) in real-world agricultural environments requires extensive testing and evaluation under field conditions. These field trials help determine not only the effectiveness of siderophore-producing microbes in enhancing plant growth but also their environmental stability, competitive ability, and interactions with native microbial populations, as shown in Table 10.3.

10.8.1.1 Commercial Inoculants

Commercial applications have been developed based on these engineered microbes for agricultural purposes by a number of companies. As an example, living cultures of genetically modified *Rhizobium* and *Azospirillum* biofertilizers are now being sold to farmers to enhance crops such as pulses and cereal grains through nitrogen fixation [102].

10.8.1.2 Integrated Pest Management

Recently, engineered microbes are being incorporated in the INTEGRATED PEST MANAGEMENT (IPM) approaches. For instance, a bacterium, *Bacillus thuringiensis*, that has been genetically modified is used for controlling a certain class of insect pests as a replacement to chemical insecticide. This application helps limit the environmental costs associated with pest management without compromising on the crop yields [103].

10.8.1.3 Activation and Remediation of Contaminated Soil

As evidenced in field trials, engineered microorganisms have inspired successful bioremediation measures. For example, transgenic fungi that can break down certain toxins have been used in polluted farmlands, with results showing an impressive reduction of the toxic materials, allowing healthy soil for further farming operations [104].

Table 10.3 Experimental research on genetically engineered microbes for agricultural applications.

Experimental research activity	Description	Key findings	References
Biocontrol of plant pathogens	Studies on *Bacillus* and *Pseudomonas* strains for controlling soil-borne pathogens	Engineered *Bacillus subtilis* reduced fungal infections (e.g., *Fusarium* wilt) in tomato crops, leading to healthier plants and higher yields compared to controls	[98]
Focus on improving nutrient efficiency	Research on genetically modified *Rhizobium* and *Azospirillum* for nitrogen fixation and nutrient assimilation	GMO *Rhizobium* in soybean fields resulted in increased nitrogen availability, biomass, and grain yield compared to non-GMO methods	[99]
Resistance to drought conditions	Development and testing of genetically modified microorganisms to improve drought tolerance in crops	Recombinant *Pseudomonas syringae* enhanced water retention and drought tolerance in wheat and maize by producing osmoprotectants, aiding growth in dry conditions	[100]
Enhancement of soil health	Use of modified *Trichoderma* spp. to improve soil quality and structure	Engineered *Trichoderma* increased microbial diversity and soil organic matter, enhancing grape production quality and quantity in vineyards	[101]

10.8.1.4 Environmental Initiative and Sustainable Agriculture

There are numerous agricultural sustainability initiatives that today include the use of engineered microbes. Farmers are taught how to make use of such microbial benedict solutions in as much as they will help increase productivity while reducing the use of chemicals for the sake of sustainable development [105].

The use of genetically modified beneficial microbes in agriculture through field trials and for practical applications has shown their ability to improve crop productivity, increase soil quality and contribute to environmentally sensitive agriculture. Figure 10.3 illustrates the journey of genetically engineered microbes from lab to land, highlighting their diverse agricultural applications such as biofertilizers, bioremediation, seed treatment, biocontrol, and encapsulation for sustainable agriculture. The scope of these technologies will be limited until progress in research and availability of more data are made to keep up with modern agricultural practices. Engineered microbes can help to make agriculture more economical and environmentally friendly provided that social, political, and economic challenges are surmounted. Figure 10.1 demonstrates the genetic engineering of microbes using CRISPR-Cas and RNAi, producing biofertilizers, bioremediation agents, seed treatments, biocontrol agents, and encapsulated products. These applications enhance crop quality and promote sustainable agriculture by improving nutrient availability and environmental resilience [41].

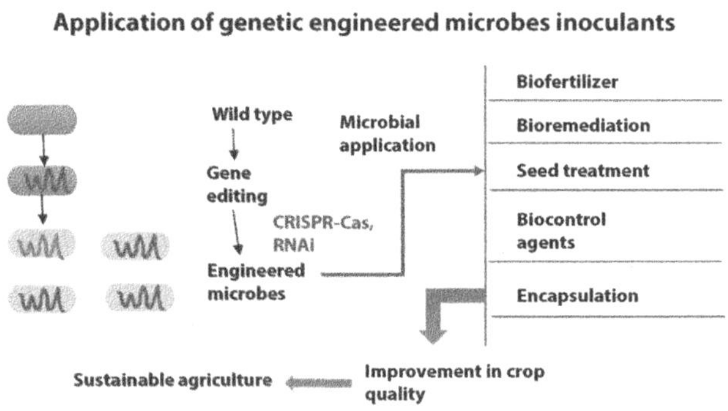

Figure 10.3 From lab to land: The journey of genetically engineered microbes. Application of genetically engineered microbes, developed through techniques like CRISPR-Cas9 and RNAi, for various agricultural purposes. These include biofertilizers, bioremediation, seed treatment, biocontrol agents, and encapsulation, ultimately contributing to sustainable agriculture and improved crop quality [106].

10.9 Challenges and Limitations

10.9.1 Biosafety and Regulatory Hurdles

Despite these strategies, the use of GEMs for improving siderophore production poses major biosafety/regulatory hurdles. The mechanisms of protecting the biologic safety of GEMs include the syncopation of risk control data and how to avoid HGT or any other explosive environmental effect. Overall, the regulatory systems are not uniform across the application markets, and the differences concerning them make the GEM acknowledgment more challenging [107]. For instance, engineered microbes should meet biosafety requirements that govern the release of potentially dangerous organisms within society. As can be observed from the previous discussions, extremely stringent evaluations can hamper the advancement of research as well as outcome commercialization. However, public regards created by genetically modified organisms (GMOs) make it even more complicated. Thus, the management of biosafety issues entails understanding, protection, and the establishment of common regulatory frameworks to address recurrent risks of inventions on biosphere sustainability [108].

10.9.2 Potential Ecological Impacts

The discharge of GEMs to environment may compete with native microbial assemblage thus having ecological implications. Although the advantage of these microbes is improved siderophore productivity, its effect on other microorganisms in the soil is unpredictable—interactions might result in the displacement of one strain by another or changes in nutrient cycling. For example, the overproduction of siderophore may have an impact on iron bioavailability to non-target organisms. However, the constant presence of GEMs in the environment gives a reason to worry about possible problems affecting ecosystems. Inoculation techniques and other related countermeasures are preventive ways of employing containment methods like engineered kill switches and locked containment release mechanisms. Long-term and multiperspective examinations of the effects of using GEMs to plants and different characteristics of environmentally sustainable agriculture have to be implemented [10].

10.9.3 Economic Viability and Scalability

The decision to use GEMs to produce siderophores sufficiently cost efficient methods of production and the feasibility of the scale. Although the experimental-scale results are positive, scientific attempts to scale up these findings and apply them in a large industrial scale are a totally different ball game [109]. This is about economic aspect or issues like cost of the genetic engineering tools and cost of fermentation and downstream applications that affect the economic feasibility. Moreover, utilizing GEM achievements for agriculture's need can equally face additional infrastructural and financial challenges. These challenges can be solved through technology implementation by integration of academia, industry, and policymakers that help in developing and implementing innovative funding models. For this reason, the use of GEMs in the agricultural sector will still not be enhanced unless these challenges are overcome [109, 110].

10.10 Future Directions in Genetic Engineering for Siderophore Production

10.10.1 Innovations in Gene Editing Tools

Newer gene editing tools including CRISPR-Cas and RNA interference (RNAi) are dramatically changing the microbial engineering world. These tools allow for accurate and rapid genetic manipulations of microbes so as to increase the siderophore synthesis rate [111]. For example, CRISPR-Cas systems are used today to control siderophore biosynthetic pathways to achieve enhanced productivity. Moreover, new techniques, such as base editors and prime editors, are more precise than CRISPR/Cas9, and they practically have no off-target effects [112]. It remains possible to improve engineered microbes by using synthetic biology strategies including metadata optimization involving metabolic pathways and genome-wide engineering. The advancements in gene editing provide a vast opportunity to create GEMs having enhanced siderophore synthesis rates for creating sustainable agricultural practices CRISPR-Cas Technology: Emerging Applications in Clinical Microbiology and Infectious Diseases [113].

10.10.2 Integrating Multi-Functional Traits

Engineering microbes having multiple traits will improve their functions in agriculture. For instance, linking siderophore producing traits with biocontrol characteristics or nutrient solubilization potential results in the production of multifunctional bioinoculants. Such microbes can perform multiple tasks in agriculture such as in nutrient availability, pest management, and disease management. Modern techniques in systems biology and in omics allow for the discovery of genes that underlie these phenotypes and their insertion to the genome of microbes. Future generation of advanced GEMs will need to be multi-functional, which calls for the consideration of its interactions with plants, soil and other microbial inhabitants of the biosphere. This integration of traits also enhances the quality of crops and yield as well as answers to the sustainable farming question by decreasing the use of chemicals [56].

10.10.3 Potential for Commercialization and Global Agricultural Impact

The use of GEMs for commercialization of siderophore production could help to overcome many of the challenges facing agriculture today. To make these microbes available to farmers across the world posit that sound and sustainable means of production can always be developed. siderophore-producing GEMs can potentially beneficially affect crop nutrition and its ability to withstand such stresses by increasing access of the crop to available iron in the soils of regions short of the metal [114, 115]. In this regard, great emphasis has been made toward the need to find and cultivate substantial partnerships and cooperation linkages between the academia and researchers, the industry players and policy makers to foster the necessary environments for commercialization. In the same respect, public awareness can help some myths about GMOs, and therefore promote the acceptance of the GEM-based solutions. Optimization of these microbes can revolutionize agri-business through increased sustainability, minimized effects on the environment and improved food security [116–118].

10.11 Conclusion

The engineering of microbes to produce siderophores is a revolutionary process in the advanced practice of agriculture. Siderophores are considered as small molecules which help bacteria to secrete iron particles and makes available for the plant uptakes. This is very important in enhancing availability of micronutrients to crops and at the same time maintaining of soil health. By further refining the top-notch methodologies like CRISPR-Cas and RNAi strategies, siderophore-production by those microbial strains can be strengthened, and the microbial structure can be tuned for better efficiency and adaptability for different environmental conditions. Alas, this kind of study has great problems. Biosafety and regulatory challenges together with the variable effects of GEMS on the environment call for careful and ethical application of GEMs. This continues to present some economic challenges in its accessibility due to the fact that it has to find its way in terms of cost and scalability thus time and capital intensive in terms of research. This paper envisages the future of GEMs in siderophore production depending on the improvement of the advanced gene editing that makes strains of higher utility to address several agricultural challenges. Attempt to combine multiple beneficial characteristics like nitrogen fixing, phosphate dissolving and suppression of pathogens into one microbial species can alter the very concept of biotechnology in agriculture. In addition, it is possible to spur collaborations of the researchers, policymakers, industries to facilitate the commercialization of such developments while feeding the farmers around the world. Lastly, genetic engineering of microbes involved in siderophores synthesis can be concluded as the environmentally friendly approach to increasing yield of crops and a fertility rate of the soil. As this paper has highlighted, with the elimination of current challenges and adoption of technology, GEMs have the potential to totally reclaim and modernize agriculture in the world boosting food security and conservation.

AI disclosure statement: The author confirms that no artificial intelligence (AI)-generated content or AI-assisted writing tools were used in the preparation of this chapter (including summaries and the abstract). All text, analysis, and conclusions are the result of the author's original work.

References

1. Giannelli, G., Del Vecchio, L., Cirlini, M., Gozzi, M., Gazza, L., Galaverna, G., *et al.*, Exploring the rhizosphere of perennial wheat: potential for plant growth promotion and biocontrol applications. *Sci. Rep.*, 14, 1, 22792, 2024 Dec 1.
2. Khasheii, B., Mahmoodi, P., Mohammadzadeh, A., Siderophores: Importance in bacterial pathogenesis and applications in medicine and industry. *Microbiol. Res.*, 250, 2021 Sep 1.
3. Almuhawish, M.A., Kotb, E., Alkhaldi, E., Ahmed, A.A., Production and Antibacterial Activity of Atypical Siderophore from Pseudomonas sp. QCS59 Recovered from Harpachene schimperi. *Pharmaceuticals (Basel)* [Internet], 17, 9, 1126, 2024 Aug 26 [cited 2024 Dec 17]. Available from: http://www.ncbi.nlm.nih.gov/pubmed/39338291.

4. Albelda-Berenguer, M., Monachon, M., Joseph, E., Siderophores: From natural roles to potential applications. *Adv. Appl. Microbiol.*, 106, 193–225, 2019 Jan 1.
5. Radhakrishnan, N. and Krishnasamy, C., Isolation and characterization of salt-stress-tolerant rhizosphere soil bacteria and their effects on plant growth-promoting properties. *Sci. Rep.*, 14, 1, 24909, 2024 Dec 1.
6. Cen, X., Li, H., Zhang, Y., Huang, L., Luo, Y., Isolation and Plant Growth Promotion Effect of Endophytic Siderophore-Producing Bacteria: A Study on Halophyte Sesuvium portulacastrum. *Plants (Basel Switzerland)* [Internet], 13, 19, 2024 Sep 27 [cited 2024 Dec 17]. Available from: http://www.ncbi.nlm.nih.gov/pubmed/39409573.
7. Miethke, M. and Marahiel, M.A., Siderophore-Based Iron Acquisition and Pathogen Control. *Microbiol. Mol. Biol. Rev.*, 71, 3, 413–51, 2007 Sep.
8. Johnstone, T.C. and Nolan, E.M., Beyond iron: Non-classical biological functions of bacterial siderophores. *Dalt. Trans.*, 44, 14, 6320–39, 2015 Apr 14.
9. Puja, H., Mislin, G.L.A., Rigouin, C., Engineering Siderophore Biosynthesis and Regulation Pathways to Increase Diversity and Availability. *Biomol.* [Internet], 13, 6, 959, 2023 Jun 1 [cited 2024 Dec 17]. Available from: https://pmc.ncbi.nlm.nih.gov/articles/PMC10296737/.
10. Ahmed, E. and Holmström, S.J.M., Siderophores in environmental research: roles and applications. *Microb. Biotechnol.* [Internet], 7, 3, 196, 2014 [cited 2024 Dec 18]. Available from: https://pmc.ncbi.nlm.nih.gov/articles/PMC3992016/.
11. Swayambhu, G., Bruno, M., Gulick, A.M., Pfeifer, B.A., Siderophore natural products as pharmaceutical agents. *Curr. Opin. Biotechnol.*, 69, 242–51, 2021 Jun 1.
12. Gupta, P., Rani, R., Usmani, Z., Chandra, A., Kumar, V., The role of plant-associated bacteria in phytoremediation of trace metals in contaminated soils. *New Futur. Dev. Microb. Biotechnol. Bioeng. Microb. Biotechnol. Agro-environmental Sustain*, 69–76, 2019 Jan 1.
13. Khan, A., Singh, P., Srivastava, A., Synthesis, nature and utility of universal iron chelator – Siderophore: A review. *Microbiol. Res.*, 212–213103–11, 2018 Jul 1.
14. Kurth, C., Kage, H., Nett, M., Siderophores as molecular tools in medical and environmental applications. *Org. Biomol. Chem.*, 14, 35, 8212–27, 2016.
15. Hider, R.C. and Kong, X., Chemistry and biology of siderophores. *Nat. Prod. Rep.*, 27, 5, 637–57, 2010.
16. Li, C., Zhu, L., Pan, D., Li, S., Xiao, H., Zhang, Z., *et al.*, Siderophore-Mediated Iron Acquisition Enhances Resistance to Oxidative and Aromatic Compound Stress in Cupriavidus necator JMP134. *Appl. Environ. Microbiol.* [Internet], 85, 1, e01938–18, 2018 Jan 1 [cited 2024 Dec 17]. Available from: https://pmc.ncbi.nlm.nih.gov/articles/PMC6293101/.
17. Tanabe, T., Naka, A., Aso, H., Nakao, H., Narimatsu, S., Inoue, Y., *et al.*, A novel aerobactin utilization cluster in Vibrio vulnificus with a gene involved

in the transcription regulation of the iutA homologue. *Microbiol. Immunol.*, 49, 9, 823–34, 2005 Sep 20.
18. Timofeeva, A.M., Galyamova, M.R., Sedykh, S.E., Bacterial Siderophores: Classification, Biosynthesis, Perspectives of Use in Agriculture. *Plants* [Internet], 11, 22, 3065, 2022 Nov 12 [cited 2024 Dec 17]. Available from: https://www.mdpi.com/2223-7747/11/22/3065/htm.
19. Briat, J.F., Dubos, C., Gaymard, F., Iron nutrition, biomass production, and plant product quality. *Trends Plant Sci.*, 20, 1, 33–40, 2015 Jan 1.
20. Singh, P., Chauhan, P.K., Upadhyay, S.K., Singh, R.K., Dwivedi, P., Wang, J., *et al.*, Mechanistic Insights and Potential Use of Siderophores Producing Microbes in Rhizosphere for Mitigation of Stress in Plants Grown in Degraded Land. *Front. Microbiol.*, 13, 2022 Jul 11.
21. Tripathi, D.K., Singh, S., Gaur, S., Singh, S., Yadav, V., Liu, S., *et al.*, Acquisition and homeostasis of Iron in higher plants and their probable role in abiotic stress tolerance. *Front. Environ. Sci.*, 5, FEB, 2018 Feb 9.
22. Ganz, T. and Nemeth, E., Iron homeostasis in host defence and inflammation. *Nat. Rev. Immunol.*, 15, 8, 500–10, 2015 Aug 25.
23. Sheng, Q., Li, L., Dong, W., He, L., Sheng, X., Impact of the Iron-Response Regulator Genes on the Release of Iron and Aluminum from Biotite by Rhizobium pusense S41. *Geomicrobiol. J.*, 41, 3, 298–307, 2024.
24. Leroux, J., Leroux, J., Charron-Lamoureux, V., Roy, S., Beauregard, P.B., Bellenger, J.P., Bacillus subtilis Modulates Its Usage of Biofilm-Bound Iron in Response to Environmental Iron Availability. *Appl. Environ. Microbiol.*, 86, 22, 2020 Jan 1.
25. Xie, B., Wei, X., Wan, C., Zhao, W., Song, R., Xin, S., *et al.*, Exploring the Biological Pathways of Siderophores and Their Multidisciplinary Applications: A Comprehensive Review. *Mol.* [Internet], 29, 10, 2318, 2024 May 15 [cited 2024 Dec 17]. Available from: https://www.mdpi.com/1420-3049/29/10/2318/htm.
26. Raymond, K.N., Dertz, E.A., Kim, S.S., Enterobactin: An archetype for microbial iron transport. *Proc. Natl. Acad. Sci. U.S.A.*, 100, 7, 3584–8, 2003 Apr 1.
27. Hasnain, A. and Kanwal, S., Biological Uses and Importance of Iron Regulation. *MARKHOR J. Zool*, 11–3, 2020 Jun 30.
28. Saha, R., Saha, N., Donofrio, R.S., Bestervelt, L.L., Microbial siderophores: A mini review. *J. Basic Microbiol.*, 53, 4, 303–17, 2013 Apr.
29. Liu, L., Wang, W., Wu, S., Gao, H., Recent Advances in the Siderophore Biology of Shewanella. *Front. Microbiol.*, 13, 2022 Feb 17.
30. Nandre, V., Kumbhar, N., Battu, S., Kale, Y., Bagade, A., Haram, S., *et al.*, Siderophore mediated mineralization of struvite: A novel greener route of sustainable phosphate management. *Water Res.*, 203, 2021 Sep 15.
31. Zhao, S., Wang, Z.P., Lin, Z., Wei, G., Wen, X., Li, S., *et al.*, Drug Repurposing by Siderophore Conjugation: Synthesis and Biological Evaluation of

Siderophore-Methotrexate Conjugates as Antibiotics. *Angew. Chem. Int. Ed*, 61, 36, 2022 Sep 5.

32. Smithies, O., Gregg, R.G., Boggs, S.S., Koralewski, M.A., Kucherlapati, R.S., Insertion of DNA sequences into the human chromosomal β-globin locus by homologous recombination. *Nature*, 317, 6034, 230–4, 1985.
33. Eid, A. and Mahfouz, M.M., Genome editing: The road of CRISPR/Cas9 from bench to clinic. *Exp. Mol. Med.*, 48, 10, 2016 Oct 7.
34. Jacinto, F.V., Link, W., Ferreira, B.I., CRISPR/Cas9-mediated genome editing: From basic research to translational medicine. *J. Cell. Mol. Med.* [Internet], 24, 7, 3766, 2020 Apr 1 [cited 2024 Dec 17]. Available from: https://pmc.ncbi.nlm.nih.gov/articles/PMC7171402/.
35. Komor, A.C., Badran, A.H., Liu, D.R., Erratum: CRISPR-Based Technologies for the Manipulation of Eukaryotic Genomes (Cell (2017) 168(2) (20–36) (S0092867416314659)(10.1016/j.cell.2016.10.044)). *Cell.*, 169, 3, 559, 2017 Apr 20.
36. Butterton, J.R. and Calderwood, S.B., Identification, cloning, and sequencing of a gene required for ferric vibriobactin utilization by Vibrio cholerae. *J. Bacteriol.*, 176, 18, 5631–8, 1994.
37. Crosa, J.H. and Walsh, C.T., Genetics and Assembly Line Enzymology of Siderophore Biosynthesis in Bacteria. *Microbiol. Mol. Biol. Rev.* [Internet], 66, 2, 223, 2002 Jun [cited 2024 Dec 17]. Available from: https://pmc.ncbi.nlm.nih.gov/articles/PMC120789/.
38. Zhao, J., Fang, H., Zhang, D., Expanding application of CRISPR-Cas9 system in microorganisms. *Synth. Syst. Biotechnol.* [Internet], 5, 4, 269, 2020 Dec 1 [cited 2024 Dec 17]. Available from: https://pmc.ncbi.nlm.nih.gov/articles/PMC7451738/.
39. Ding, X., Yu, L., Chen, L., Li, Y., Zhang, J., Sheng, H., *et al.*, Recent Progress and Future Prospect of CRISPR/Cas-Derived Transcription Activation (CRISPRa) System in Plants. *Cells* [Internet], 11, 19, 3045, 2022 Oct 1 [cited 2024 Dec 17]. Available from: https://pmc.ncbi.nlm.nih.gov/articles/PMC9564188/.
40. O'Brien, A.R., Wilson, L.O.W., Burgio, G., Bauer, D.C., Unlocking HDR-mediated nucleotide editing by identifying high-efficiency target sites using machine learning. *Sci. Rep.* [Internet], 9, 1, 2788, 2019 Dec 1 [cited 2024 Dec 17]. Available from: https://pmc.ncbi.nlm.nih.gov/articles/PMC6391469/.
41. Yang, P., Condrich, A., Lu, L., Scranton, S., Hebner, C., Sheykhhasan, M., *et al.*, Genetic Engineering in Bacteria, Fungi, and Oomycetes, Taking Advantage of CRISPR. *DNA* [Internet], 4, 4, 427–54, 2024 Nov 14 [cited 2024 Dec 17]. Available from: https://www.mdpi.com/2673-8856/4/4/30/htm.
42. Paul, N.C., Park, S.W., Liu, H., Choi, S., Ma, J., MacCready, J.S., *et al.*, Plant and Fungal Genome Editing to Enhance Plant Disease Resistance Using the CRISPR/Cas9 System. *Front. Plant Sci.*, 12, 2021 Aug 10.

43. Ansori, A.N.M., Antonius, Y., Susilo, R.J.K., Hayaza, S., Kharisma, V.D., Parikesit, A.A., *et al.*, Application of CRISPR-Cas9 genome editing technology in various fields: A review. *Narra J.* [Internet], 3, 2, e184, 2023 Aug 1 [cited 2024 Dec 18]. Available from: https://pmc.ncbi.nlm.nih.gov/articles/PMC10916045/.
44. Brillet, K., Ruffenach, F., Adams, H., Journet, L., Gasser, V., Hoegy, F., *et al.*, An ABC transporter with two periplasmic binding proteins involved in iron acquisition in pseudomonas aeruginosa. *ACS Chem. Biol.*, 7, 12, 2036–45, 2012 Dec 21.
45. Ghssein, G. and Ezzeddine, Z., A Review of Pseudomonas aeruginosa Metallophores: Pyoverdine, Pyochelin and Pseudopaline. *Biol. (Basel)*, 11, 12, 2022 Dec 1.
46. Barry, S.M. and Challis, G.L., Recent advances in siderophore biosynthesis. *Curr. Opin. Chem. Biol.*, 13, 2, 205–15, 2009 Apr.
47. Kang, Y., Wang, Y., Hou, B., Wang, R., Ye, J., Zhu, X., *et al.*, Adpalin, a pleiotropic transcriptional regulator, is involved in the cascade regulation of lincomycin biosynthesis in streptomyces lincolnensis. *Front. Microbiol.*, 10, OCT, 2019.
48. Li, L., Liu, X., Jiang, W., Lu, Y., Recent Advances in Synthetic Biology Approaches to Optimize Production of Bioactive Natural Products in Actinobacteria. *Front. Microbiol.*, 10, 495220, 2019 Nov 5.
49. Roell, G.W., Zha, J., Carr, R.R., Koffas, M.A., Fong, S.S., Tang, Y.J., Engineering microbial consortia by division of labor. *Microb. Cell Fact* [Internet], 18, 1, 1–11, 2019 Feb 8 [cited 2024 Dec 17]. Available from: https://microbialcellfactories.biomedcentral.com/articles/10.1186/s12934-019-1083-3.
50. Glick, B.R., Metabolic load and heterologous gene expression. *Biotechnol. Adv*, 13, 2, 247–61, 1995.
51. Crits-Christoph, A., Bhattacharya, N., Olm, M.R., Song, Y.S., Banfield, J.F., Transporter genes in biosynthetic gene clusters predict metabolite characteristics and siderophore activity. *Genome Res.* [Internet], 31, 2, 239–50, 2021 [cited 2024 Dec 17]. Available from: https://pmc.ncbi.nlm.nih.gov/articles/PMC7849407/.
52. Berntsson, R.P.A., Smits, S.H.J., Schmitt, L., Slotboom, D.J., Poolman, B., A structural classification of substrate-binding proteins. *FEBS Lett.*, 584, 12, 2606–17, 2010 Jun.
53. Leitão, A.L. and Enguita, F.J., Fungal extrolites as a new source for therapeutic compounds and as building blocks for applications in synthetic biology. *Microbiol. Res.*, 169, 9–10, 652–65, 2014 Sep 1.
54. Martin-Pascual, M., Batianis, C., Bruinsma, L., Asin-Garcia, E., Garcia-Morales, L., Weusthuis, R.A., *et al.*, A navigation guide of synthetic biology tools for Pseudomonas putida. *Biotechnol. Adv.*, 49, 107732, 2021 Jul 1.
55. LeBlanc, A.R. and Wuest, W.M., Siderophores: A Case Study in Translational Chemical Biology. *Biochem.* [Internet], 63, 15, 1877–91, 2024 Aug 6 [cited

2024 Dec 17]. Available from: https://pubs.acs.org/doi/full/10.1021/acs.biochem.4c00276.

56. Nunes, P.S.O., Lacerda-Junior, G.V., Mascarin, G.M., Guimarães, R.A., Medeiros, F.H.V., Arthurs, S., *et al.*, Microbial consortia of biological products: Do they have a future? *Biol. Control*, 188, 105439, 2024 Jan 1.
57. Basu, A., Prasad, P., Das, S.N., Kalam, S., Sayyed, R.Z., Reddy, M.S., *et al.*, Plant growth promoting rhizobacteria (Pgpr) as green bioinoculants: Recent developments, constraints, and prospects. *Sustain*, 13, 3, 1–20, 2021 Feb 1.
58. Busby, P.E., Soman, C., Wagner, M.R., Friesen, M.L., Kremer, J., Bennett, A., *et al.*, Research priorities for harnessing plant microbiomes in sustainable agriculture. *PLoS Biol.*, 15, 3, 2017 Mar 28.
59. Saeed, Q., Xiukang, W., Haider, F.U., Kučerik, J., Mumtaz, M.Z., Holatko, J., *et al.*, Rhizosphere bacteria in plant growth promotion, biocontrol, and bioremediation of contaminated sites: A comprehensive review of effects and mechanisms. *Int. J. Mol. Sci.*, 22, 19, 2021 Oct 1.
60. Igiehon, B.C., Babalola, O.O., Hassen, A.I., Rhizosphere competence and applications of plant growth-promoting rhizobacteria in food production – A review. *Sci. Afr.*, 23, e02081, 2024 Mar 1.
61. Tang, L., Shi, Y., Zhang, Y., Yang, D., Guo, C., Effects of Plant-Growth-Promoting Rhizobacteria on Soil Bacterial Community, Soil Physicochemical Properties, and Soil Enzyme Activities in the Rhizosphere of Alfalfa under Field Conditions. *Diversity*, 15, 4, 2023 Apr 1.
62. Ambrosi, C., Leoni, L., Putignani, L., Orsi, N., Visca, P., Pseudobactin Biogenesis in the Plant Growth-Promoting Rhizobacterium Pseudomonas Strain B10: Identification and Functional Analysis of the l-Ornithine N5-Oxygenase (psbA) Gene. *J. Bacteriol.* [Internet], 182, 21, 6233, 2000 [cited 2024 Dec 17]. Available from: https://pmc.ncbi.nlm.nih.gov/articles/PMC94761/.
63. Gaballa, A. and Helmann, J.D., Substrate induction of siderophore transport in Bacillus subtilis mediated by a novel one-component regulator. *Mol. Microbiol.* [Internet], 66, 1, 164–73, 2007 Oct [cited 2024 Dec 17]. Available from: https://pubmed.ncbi.nlm.nih.gov/17725565/.
64. Zou, Z., Robinson, J.I., Steinberg, L.K., Henderson, J.P., Uropathogenic Escherichia coli wield enterobactin-derived catabolites as siderophores. *J. Biol. Chem.* [Internet], 300, 1, 105554, 2023 Jan 1 [cited 2024 Dec 17]. Available from: https://pmc.ncbi.nlm.nih.gov/articles/PMC10788543/.
65. Sumbul, A., Ansari, R.A., Rizvi, R., Mahmood, I., Azotobacter: A potential bio-fertilizer for soil and plant health management. *Saudi J. Biol. Sci.* [Internet], 27, 12, 3634, 2020 Dec 1 [cited 2024 Dec 17]. Available from: https://pmc.ncbi.nlm.nih.gov/articles/PMC7714982/.
66. Zahran, H.H., Rhizobium-Legume Symbiosis and Nitrogen Fixation under Severe Conditions and in an Arid Climate. *Microbiol. Mol. Biol. Rev.* [Internet], 63, 4, 968, 1999 Dec [cited 2024 Dec 17]. Available from: https://pmc.ncbi.nlm.nih.gov/articles/PMC98982/.

67. Salwan, R., Sharma, V., Sharma, A., Singh, A., Molecular imprints of plant beneficial Streptomyces sp. AC30 and AC40 reveal differential capabilities and strategies to counter environmental stresses. *Microbiol. Res.*, 235, 126449, 2020 May 1.
68. Tatung, M. and Deb, C.R., Isolation, Characterization, and Investigation on Potential Multi-trait Plant Growth Promoting Rhizobacteria from Wild Banana (Musa itinerans) Rhizospheric Soil. *J. Pure Appl. Microbiol.*, 17, 3, 1578–90, 2023 Sep 1.
69. Serda, M., FG, B., Cleary M, Holtermann H, *et al.*, *Synteza i aktywność biologiczna nowych analogów tiosemikarbazonowych chelatorów żelaza*, vol. 7, G. Balint, B. Antala, C. Carty, J.-M.A. Mabieme, I.B. Amar, A. Kaplanova (Eds.), pp. 343–54, Uniwśląski [Internet], Katowice, Poland, 2013 [cited 2024 Dec 17], Available from: https://desytamara.blogspot.com/2017/11/sistem-pelayanan-perpustakaan-dan-jenis.html.
70. Tatung, M. and Deb, C.R., Screening and characterization of heavy metal tolerant rhizobacteria from wild Musa rhizosphere from coal mining area of Changki, Nagaland, India and assessment of their growth promoting potential under Cd/Cu contaminated conditions. *South Afr. J. Bot.*, 165, 217–27, 2024 Feb 1.
71. Deb, C.R. and Tatung, M., Siderophore producing bacteria as biocontrol agent against phytopathogens for a better environment: A review. *South Afr. J. Bot.*, 165, 153–62, 2024 Feb 1.
72. Aguiar, M., Orasch, T., Shadkchan, Y., Caballero, P., Pfister, J., Sastré-Velásquez, L.E., *et al.*, Uptake of the Siderophore Triacetylfusarinine C, but Not Fusarinine C, Is Crucial for Virulence of Aspergillus fumigatus. *MBio* [Internet], 13, 5, 2022 Sep 1 [cited 2024 Dec 17]. Available from: https://pubmed.ncbi.nlm.nih.gov/36125294/.
73. Tyśkiewicz, R., Nowak, A., Ozimek, E., Jaroszuk-ściseł, J., Trichoderma: The Current Status of Its Application in Agriculture for the Biocontrol of Fungal Phytopathogens and Stimulation of Plant Growth. *Int. J. Mol. Sci.* [Internet], 23, 4, 2329, 2022 Feb 1 [cited 2024 Dec 17]. Available from: https://pmc.ncbi.nlm.nih.gov/articles/PMC8875981/.
74. Chakraborty, N., Mitra, R., Dasgupta, D., Ganguly, R., Acharya, K., Minkina, T., *et al.*, Unraveling lipid peroxidation-mediated regulation of redox homeostasis for sustaining plant health. *Plant Physiol. Biochem.*, 206, 2024 Jan 1.
75. Zuriegat, Q., Zheng, Y., Liu, H., Wang, Z., Yun, Y., Current progress on pathogenicity-related transcription factors in Fusarium oxysporum. *Mol. Plant Pathol.* [Internet], 22, 7, 882, 2021 Jul 1 [cited 2024 Dec 17]. Available from: https://pmc.ncbi.nlm.nih.gov/articles/PMC8232035/.
76. Carroll, C.S., Grieve, C.L., Murugathasan, I., Bennet, A.J., Czekster, C.M., Lui, H., *et al.*, The rhizoferrin biosynthetic gene in the fungal pathogen Rhizopus delemar is a novel member of the NIS gene family. *Int. J. Biochem. Cell Biol.*, 89, 136–46, 2017 Aug 1.

77. Singh, H.B., Keswani, C., Ray, S., Yadav, S.K., Singh, S.P., Singh, S., *et al.*, Beauveria bassiana: Biocontrol Beyond Lepidopteran Pests. *Biocontrol Lepid Pests* [Internet], 43, 219, 2015 [cited 2024 Dec 17]. Available from: https://pmc.ncbi.nlm.nih.gov/articles/PMC7123603/.
78. Singh, A., Rajput, V.D., Varshney, A., Ghazaryan, K., Minkina, T., Small Tech, Big Impact: Agri-nanotechnology Journey to Optimize Crop Protection and Production for Sustainable Agriculture. *Plant Stress*, 10, 100253, 2023 Dec 1.
79. Guo, K., Yang, J., Yu, N., Luo, L., Wang, E., Biological nitrogen fixation in cereal crops: Progress, strategies, and perspectives. *Plant Commun.*, 4, 2, 100499, 2023 Mar 13.
80. Cheng, Y., Narayanan, M., Shi, X., Chen, X., Li, Z., Ma, Y., Phosphate-solubilizing bacteria: Their agroecological function and optimistic application for enhancing agro-productivity. *Sci. Total Environ.*, 901, 166468, 2023 Nov 25.
81. Ahmed, T., Noman, M., Qi, Y., Shahid, M., Hussain, S., Masood, H.A., *et al.*, Fertilization of Microbial Composts: A Technology for Improving Stress Resilience in Plants. *Plants* [Internet], 12, 20, 3550, 2023 Oct 1 [cited 2024 Dec 17]. Available from: https://pmc.ncbi.nlm.nih.gov/articles/PMC10609736/.
82. Kumar, S., Diksha, SS, S., Kumar, R., Biofertilizers: An ecofriendly technology for nutrient recycling and environmental sustainability. *Curr. Res. Microb. Sci.*, 3, 100094, 2022 Jan 1.
83. Aasfar, A., Bargaz, A., Yaakoubi, K., Hilali, A., Bennis, I., Zeroual, Y., *et al.*, Nitrogen Fixing Azotobacter Species as Potential Soil Biological Enhancers for Crop Nutrition and Yield Stability. *Front. Microbiol.*, 12, 2021 Feb 25.
84. Pirttilä, A.M., Tabas, H.M.P., Baruah, N., Koskimäki, J.J., Biofertilizers and biocontrol agents for agriculture: How to identify and develop new potent microbial strains and traits. *Microorganisms* [Internet], 9, 4, 817, 2021 Apr 1 [cited 2024 Dec 17]. Available from: https://pmc.ncbi.nlm.nih.gov/articles/PMC8069042/.
85. Ameen, M., Mahmood, A., Sahkoor, A., Zia, M.A., Ullah, M.S., The role of endophytes to combat abiotic stress in plants. *Plant Stress*, 12, 100435, 2024 Jun 1.
86. Ammar, E.E., Rady, H.A., Khattab, A.M., Amer, M.H., Mohamed, S.A., Elodamy, N.I., *et al.*, A comprehensive overview of eco-friendly bio-fertilizers extracted from living organisms. *Environ. Sci. Pollut. Res. Int.* [Internet], 30, 53, 113119, 2023 Nov 1 [cited 2024 Dec 17]. Available from: https://pmc.ncbi.nlm.nih.gov/articles/PMC10663222/.
87. Elbasiouny, H., El-Ramady, H., Elbehiry, F., Rajput, V.D., Minkina, T., Mandzhieva, S., Plant Nutrition under Climate Change and Soil Carbon Sequestration. *Sustain* [Internet], 14, 2, 914, 2022 Jan 14 [cited 2024 Dec 17]. Available from: https://www.mdpi.com/2071-1050/14/2/914/htm.
88. Tudi, M., Ruan, H.D., Wang, L., Lyu, J., Sadler, R., Connell, D., *et al.*, Agriculture Development, Pesticide Application and Its Impact on the Environment. *Int. J. Environ. Res. Public Health* [Internet], 18, 3, 1112, 2021

Feb 1 [cited 2024 Dec 17]. Available from: https://pmc.ncbi.nlm.nih.gov/articles/PMC7908628/.

89. Haq, I.U., Rahim, K., Yahya, G., Ijaz, B., Maryam, S., Paker, N.P., Eco-smart biocontrol strategies utilizing potent microbes for sustainable management of phytopathogenic diseases. *Biotechnol. Rep.*, 44, e00859, 2024 Dec 1.
90. Savary, S., Bregaglio, S., Willocquet, L., Gustafson, D., Mason D'Croz, D., Sparks, A., *et al.*, Crop health and its global impacts on the components of food security. *Food Secur.*, 9, 2, 311–27, 2017 Apr 1.
91. Mahdizade Ari, M., Dadgar, L., Elahi, Z., Ghanavati, R., Taheri, B., Genetically Engineered Microorganisms and Their Impact on Human Health. *Int. J. Clin. Pract.* [Internet], 2024, 6638269, 2024 [cited 2024 Dec 17]. Available from: https://pmc.ncbi.nlm.nih.gov/articles/PMC10944348/.
92. Kour, D., Negi, R., Khan, S.S., Kumar, S., Kaur, S., Kaur, T., *et al.*, Microbes mediated induced systemic response in plants: A review. *Plant Stress*, 11, 100334, 2024 Mar 1.
93. Zia, R., Nawaz, M.S., Siddique, M.J., Hakim, S., Imran, A., Plant survival under drought stress: Implications, adaptive responses, and integrated rhizosphere management strategy for stress mitigation. *Microbiol. Res.*, 242, 126626, 2021 Jan 1.
94. Arzanesh, M.H., Alikhani, H.A., Khavazi, K., Rahimian, H.A., Miransari, M., Wheat (Triticum aestivum L.) growth enhancement by Azospirillum sp. under drought stress. *World J. Microbiol. Biotechnol.*, 27, 2, 197–205, 2011 Feb 1.
95. Ayaz, M., Li, C.-H., Ali, Q., Zhao, W., Chi, Y.-K., Shafiq, M., *et al.*, Bacterial and Fungal Biocontrol Agents for Plant Disease Protection: Journey from Lab to Field, Current Status, Challenges, and Global Perspectives. *Mol.* [Internet], 28, 18, 6735, 2023 Sep 21 [cited 2024 Dec 18]. Available from: https://www.mdpi.com/1420-3049/28/18/6735/htm.
96. Farzand, A., Moosa, A., Zubair, M., Khan, A.R., Massawe, V.C., Tahir, H.A.S., *et al.*, Suppression of sclerotinia sclerotiorum by the induction of systemic resistance and regulation of antioxidant pathways in tomato using fengycin produced by Bacillus amyloliquefaciens FZB42. *Biomolecules*, 9, 10, 2019 Oct 1.
97. O'Callaghan, M., Ballard, R.A., Wright, D., Soil microbial inoculants for sustainable agriculture: Limitations and opportunities. *Soil Use Manag.* [Internet], 38, 3, 1340–69, 2022 Jul 1 [cited 2024 Dec 18]. Available from: https://onlinelibrary.wiley.com/doi/full/10.1111/sum.12811.
98. Etesami, H., Jeong, B.R., Glick, B.R., Biocontrol of plant diseases by Bacillus spp. *Physiol. Mol. Plant Pathol.*, 126, 102048, 2023 Jul 1.
99. Goyal, R.K., Schmidt, M.A., Hynes, M.F., Molecular Biology in the Improvement of Biological Nitrogen Fixation by Rhizobia and Extending the Scope to Cereals. *Microorganisms* [Internet], 9, 1, 125, 2021 Jan 1 [cited 2024 Dec 18]. Available from: https://pmc.ncbi.nlm.nih.gov/articles/PMC7825764/.

100. Khan, S., Anwar, S., Yu, S., Sun, M., Yang, Z., Gao, Z.Q., Development of Drought-Tolerant Transgenic Wheat: Achievements and Limitations. *Int. J. Mol. Sci.* [Internet], 20, 13, 3350, 2019 Jul 1 [cited 2024 Dec 18]. Available from: https://pmc.ncbi.nlm.nih.gov/articles/PMC6651533/.
101. Asghar, W., Craven, K.D., Kataoka, R., Mahmood, A., Asghar, N., Raza, T., *et al.*, The application of Trichoderma spp., an old but new useful fungus, in sustainable soil health intensification: A comprehensive strategy for addressing challenges. *Plant Stress*, 12, 2024 Jun 1.
102. Aloo, B.N., Tripathi, V., Makumba, B.A., Mbega, E.R., Plant growth-promoting rhizobacterial biofertilizers for crop production: The past, present, and future. *Front Plant Sci.* [Internet], 13, 1002448, 2022 Sep 16 [cited 2024 Dec 18]. Available from: https://pmc.ncbi.nlm.nih.gov/articles/PMC9523260/.
103. Abbas, M.S.T., Genetically engineered (Modified) crops (bacillus thuringiensis crops) and the world controversy on their safety. *Egypt J. Biol. Pest Control* [Internet], 28, 1, 1–12, 2018 Jun 19 [cited 2024 Dec 18]. Available from: https://ejbpc.springeropen.com/articles/10.1186/s41938-018-0051-2.
104. Pande, V., Pandey, S.C., Sati, D., Bhatt, P., Samant, M., Microbial Interventions in Bioremediation of Heavy Metal Contaminants in Agroecosystem. *Front Microbiol.* [Internet], 13, 824084, 2022 May 6 [cited 2024 Dec 18]. Available from: https://pmc.ncbi.nlm.nih.gov/articles/PMC9120775/.
105. Sudheer, S., Bai, R.G., Usmani, Z., Sharma, M., Insights on Engineered Microbes in Sustainable Agriculture: Biotechnological Developments and Future Prospects. *Curr. Genomics* [Internet], 21, 5, 321, 2020 Jun 4 [cited 2024 Dec 18]. Available from: https://pmc.ncbi.nlm.nih.gov/articles/PMC7536804/.
106. Batista, B.D. and Singh, B.K., Realities and hopes in the application of microbial tools in agriculture. *Microb. Biotechnol.* [Internet], 14, 4, 1258–68, 2021 Jul 1 [cited 2024 Dec 18]. Available from: https://onlinelibrary.wiley.com/doi/full/10.1111/1751-7915.13866.
107. Lamb, A.L., Breaking a pathogen's iron will: inhibiting siderophore production as an antimicrobial strategy. *Biochim. Biophys. Acta* [Internet], 1854, 8, 1054, 2015 May 30 [cited 2024 Dec 18]. Available from: https://pmc.ncbi.nlm.nih.gov/articles/PMC4457648/.
108. Passari, A.K., Ruiz-Villafán, B., Cruz-Bautista, R., Díaz-Domínguez, V., Rodríguez-Sanoja, R., Sanchez, S., Opportunities and challenges of microbial siderophores in the medical field. *Appl. Microbiol. Biotechnol.* [Internet], 107, 22, 6751, 2023 Nov 1 [cited 2024 Dec 18]. Available from: https://pmc.ncbi.nlm.nih.gov/articles/PMC10589192/.
109. Saha, M., Sarkar, S., Sarkar, B., Sharma, B.K., Bhattacharjee, S., Tribedi, P., Microbial siderophores and their potential applications: a review. *Environ. Sci. Pollut. Res.*, 23, 5, 3984–99, 2016 Mar 1.
110. Yoder, M.F. and Kisaalita, W.S., Iron specificity of a biosensor based on fluorescent pyoverdin immobilized in sol-gel glass. *J. Biol. Eng.* [Internet], 5, 4,

2011 May 10 [cited 2024 Dec 18]. Available from: http://www.ncbi.nlm.nih.gov/pubmed/21554740.

111. Doudna, J.A. and Charpentier, E., The new frontier of genome engineering with CRISPR-Cas9. *Science (80-)*, 346, 6213, 2014 Nov 28.
112. Luo, M.L., Leenay, R.T., Beisel, C.L., Current and future prospects for CRISPR-based tools in bacteria. *Biotechnol. Bioeng.*, 113, 5, 930–43, 2016 May 1.
113. Serajian, S., Ahmadpour, E., Rodrigues Oliveira, S.M., Pereira M de, L., Heidarzadeh, S., CRISPR-Cas Technology: Emerging Applications in Clinical Microbiology and Infectious Diseases. *Pharmaceuticals* [Internet], 14, 11, 1171, 2021 Nov 1 [cited 2024 Dec 18]. Available from: https://pmc.ncbi.nlm.nih.gov/articles/PMC8625472/.
114. Musani, T. and Das, M., Effect of Siderophore Producing Cadmium Tolerant Streptomyces pactum OR958669 on Plant Growth of Groundnut (Arachis hypogaea L.). *Curr. Agric. Res. J.*, 12, 1, 485–95, 2024 Apr 20.
115. Raymond, K.N., Müller, G., Matzanke, B.F., Complexation of Iron by Siderophores — A Review of Their Solution and Structural Chemistry and Biological Function. *Struct. Chem.*, 49–102, 2022 Feb 9.
116. Shirling, E.B. and Gottlieb, D., Methods for characterization of Streptomyces species. *Int. J. Syst. Bacteriol.*, 16, 3, 313–40, 1966 Jul 1.
117. Schwyn, B. and Neilands, J.B., Universal chemical assay for the detection and determination of siderophores. *Anal. Biochem.*, 160, 1, 47–56, 1987.
118. Meagher, R.B., Phytoremediation of toxic elemental and organic pollutants. *Curr. Opin. Plant Biol.*, 3, 2, 153–62, 2000 Apr 1.

11

Integrated Approaches: Combining Siderophores with Other Microbial Agents

Vedant Gautam[1]*, Vibhootee Garg[2], Jaywardhan Singh Khichi[3] and Shreyashi Singh[4]

[1]*Department of Mycology and Plant Pathology, Institute of Agricultural Sciences, Banaras Hindu University, UP, India*
[2]*Department of Horticulture, Jawaharlal Nehru Krishi Vishwavidyalaya, Jabalpur, MP, India*
[3]*Department of Genetics and Plant Breeding, Jawaharlal Nehru Krishi Vishwavidyalaya, Jabalpur, MP, India*
[4]*Department of Plant Physiology, Institute of Agricultural Sciences, Banaras Hindu University, UP, India*

Abstract

Siderophores are low-molecular-weight compounds produced by various microorganisms, primarily bacteria, that play a crucial role in iron acquisition and biocontrol of plant pathogens. This chapter explores the multifaceted roles of siderophores in suppressing phytopathogens and enhancing plant defense mechanisms. By chelating iron, siderophores inhibit the growth of harmful microbes in the rhizosphere, thereby promoting plant health. Additionally, they contribute to induced systemic resistance (ISR), empowering plants to mount a more robust defense against subsequent pathogen attacks. This chapter reviews the mechanisms through which siderophore-producing bacteria interact with plants and pathogens, evaluates their effectiveness as biocontrol agents, and discusses their potential applications in sustainable agriculture. Furthermore, it highlights recent advancements in research and technology aimed at harnessing these natural compounds for improved crop protection and yield.

***Keywords*:** Biofilm, induced systemic resistance, Trojan horse, siderophore-conjugated antibiotics, biosensors

**Corresponding author*: vedantgautam@bhu.ac.in

Debasis Mitra, Ayush Madan and Anju Rani (eds.) Microbial Siderophores for Agricultural Productivity: Perspectives of Use in Agriculture, (269–302) © 2026 Scrivener Publishing LLC

11.1 Introduction

Siderophores are low-molecular-weight secondary metabolites, ranging from 400 to 2,000 Da, produced by microorganisms like bacteria and fungi to sequester iron from the environment. These compounds preferentially bind to ferric iron (Fe^{3+}), forming stable hexadentate complexes that transport the metal into cells, crucial for microbial survival and growth under iron-limited conditions. The best siderophores typically feature three bidentate ligand groups, providing higher stability and efficient iron chelation. Their coordination geometry around iron is usually octahedral, and in some cases, siderophores can be chiral, influencing their binding and interactions. Siderophore research has broad applications in medicine,

Table 11.1 Types of siderophore.

Types of siderophores	Characteristic functional group	Characteristic	Siderophore-producing microorganism
Hydroxamic salt-type	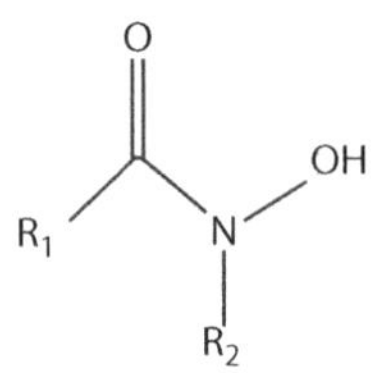	Most common in nature, the structure is more complex; it is more hydrophilic and prone to photooxidation	*Pseudomonas fluorescens* [27] *Aspergillus nidulans* [21]
Catechol salt-type	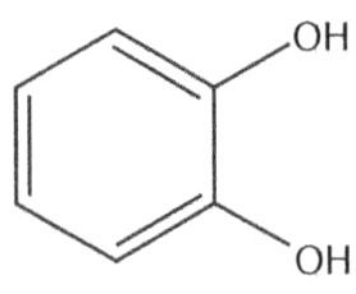	Strong lipophilicity and high affinity with Fe; strong resistance to environmental pH changes	*Escherichia coli*, *Klebsiella pneumoniae* [16, 39]
Carboxylate-type	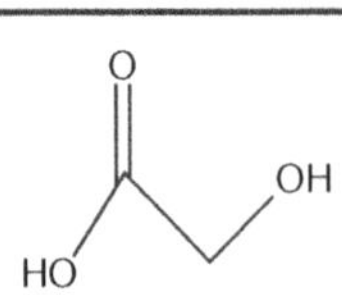	Potential ligands for the ocean iron cycle; some of them are photoactive	*Rhizobium meliloti*, *Staphylococcus aureus* [3, 11]

environmental sciences, and agriculture, highlighting their importance in various fields with Table 11.1 presenting the major types of siderophores.

Siderophores can be classified based on their chemical structure and the ligands used to chelate iron:

- **Catecholates**: These contain catechol groups and are among the strongest iron binders.
- **Hydroxamates**: These have hydroxamate groups and include compounds like ferrichrome.
- **Carboxylates**: These are less common but include derivatives of citric acid.

11.1.1 Principal Functions of Siderophores

1. **Iron chelation**: Siderophores exhibit a pronounced affinity for ferric iron (Fe^{3+}), enabling their capacity to bind and solubilize iron from the surrounding environment, thereby rendering it accessible for microbial assimilation.
2. **Transport mechanism**: Subsequent to the binding of iron, the siderophore–iron complex is translocated into the cellular interior *via* specific receptor-mediated mechanisms, thereby facilitating the internal acquisition of iron requisite for a variety of cellular activities.
3. **Enzymatic cofactor**: Upon entering the cellular milieu, iron is frequently reduced to its ferrous state (Fe^{2+}) and utilized as a cofactor in critical enzymatic processes that are integral to respiration, DNA synthesis, and energy metabolism.
4. **Microbial competition**: Siderophores function as competitive molecules among microorganisms. By sequestering available iron, they possess the capability to impede the proliferation of rival species, consequently augmenting the survival probability of the siderophore-producing organism.
5. **Plant growth promotion**: Siderophores synthesized by microorganisms can significantly enhance plant growth through the augmentation of iron availability within the soil. They facilitate the acquisition of iron under conditions of limited availability, which is essential for plant health and development.
6. **Pathogen virulence**: In pathogenic bacterial species, siderophores play a crucial role in virulence by promoting iron acquisition from the host organism. This function is especially pertinent for pathogens such as *Mycobacterium*

tuberculosis and uropathogenic *E. coli*, which necessitate substantial amounts of iron for their survival within host tissues.

7. **Bioremediation and environmental applications**: Siderophores can be employed in bioremediation initiatives aimed at mobilizing and extracting heavy metals from contaminated ecosystems, owing to their capacity to bind a range of metals in addition to iron.
8. **Oxidative stress tolerance**: Certain siderophores may confer resilience to oxidative stress in microorganisms by sequestering free metal ions that could otherwise catalyze detrimental reactions involving reactive oxygen species.

11.2 Synthesis and Secretion of Siderophores

11.2.1 Synthesis

There are different pathways to synthesize the siderophores. We discuss a few methods below:

1. NRPS pathway synthesis of siderophores

Siderophores are primarily produced *via* the non-ribosomal peptide synthetase (NRPS) pathway, involving multi-modular enzymes with key domains: adenylation (A), peptidyl carrier protein (PCP), and condensation (C). The A domain activates specific amino acids, forming an aminoacyl-AMP intermediate, which is transferred to the PCP domain and then to the C domain, where peptide bonds are formed, extending the peptide chain. Some siderophores, like liposiderin produced by *Pseudomonas putida*, also include fatty acid chains at the N-terminus. Fatty acids are activated into coenzyme A or acyl carrier proteins before being incorporated into the non-ribosomal lipopeptide structure. The schematic representation of siderophore synthesis *via* the nonribosomal peptide synthetase (NRPS) pathway is illustrated in Figure 11.1, highlighting the sequential involvement of adenylation (A), peptidyl carrier protein (PCP), and condensation (C) domains in peptide chain assembly and modification (Figure 11.1).

2. Synthesis of siderophores independent of the NRPS pathway (NIS)

The non-ribosomal-independent synthesis (NIS) pathway, involved in siderophore production like aerobactin in *E. coli*, complements the NRPS pathway. It utilizes enzymes and compounds such as dicarboxylic acids, diamines, and amino alcohols to form hydroxamate- or carboxylate-type

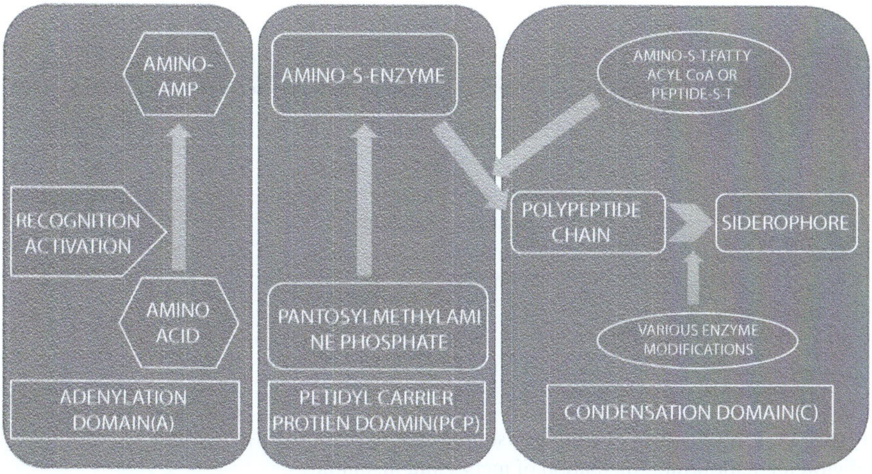

Figure 11.1 Process of synthesizing siderophores through the NRPS pathway.

siderophores. NIS synthases are classified into three types based on substrate specificity: class A (citric acid derivatives), class B (α-ketoglutaric acid derivatives), and class C (esterified/amidated carboxylic acids). A key example is petrobactin synthesis in *Bacillus anthracis*. NIS pathways, including those identified in *Streptomyces*, highlight their importance in microbial growth and host interactions.

11.2.2 Secretion

Siderophores are exported from cells by efflux systems, which include RND and ABC transporters in Gram-negative bacteria. *Pseudomonas aeruginosa* utilizes the ABC-type trimeric system PvdRT-OpmQ, whereas *Pseudomonas putida* KT2440 employs both PvdRT-OpmQ (ABC) and MdtABC-OpmB (RND). In the ABC system, the PvdRT-OpmQ complex recognizes siderophores and moves them to the periplasm by using energy from the inner membrane. Once inside the periplasm, the siderophore moves through the OpmQ channel out of the outer membrane. In RND systems, MdtA and MdtC proteins can recognize and transport siderophores across membranes.

11.3 Integration of Siderophores with Other Microbial Organisms

The integration of siderophores with other microbial organisms plays a significant role in enhancing plant health and suppressing pathogens.

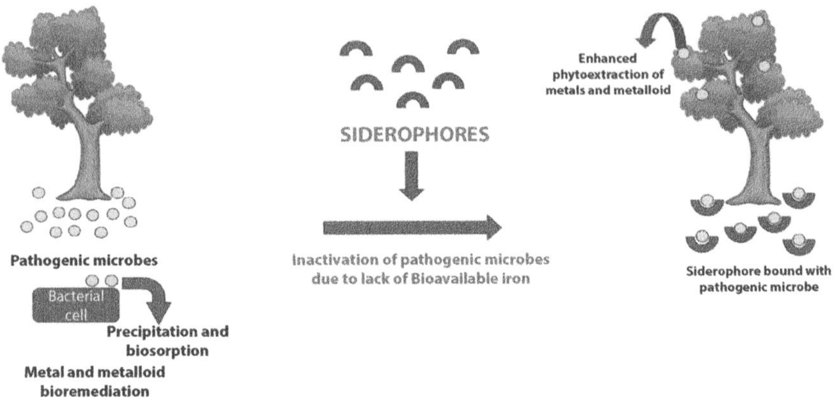

Figure 11.2 Mechanism of pathogen suppression in plant rhizosphere and phytoremediation/bioremediation of metals and metalloid.

Siderophores are iron-chelating molecules produced by various microorganisms, including bacteria and fungi, to compete for limited iron in the environment. When different microbial species coexist, their siderophore production can influence each other's growth and pathogenicity.

For instance, studies have shown that co-culturing *Pseudomonas* spp. with pathogens like *Ralstonia solanacearum* can enhance the pathogen suppression effects due to increased siderophore production. This competition for iron limits the availability of this essential nutrient to pathogens, thereby inhibiting their growth and virulence. Additionally, specific interactions among microbial communities can lead to variations in siderophore types produced, which may further enhance their effectiveness against different pathogens. Furthermore, the diversity of siderophores among microbial species allows for tailored interactions that can optimize biocontrol strategies. For example, certain *Pseudomonas* strains produce distinct siderophores that can selectively inhibit specific pathogens while promoting beneficial microbes. This strategic use of siderophores not only aids in nutrient acquisition but also fosters a balanced microbial ecosystem that supports plant health and resilience against diseases (Figure 11.2).

11.3.1 Siderophore-Producing Rhizobacteria

Siderophore-producing rhizobacteria play a pivotal role in enhancing plant growth and health by facilitating iron acquisition in nutrient-deficient soils. These beneficial bacteria, often classified as plant growth-promoting rhizobacteria (PGPR), secrete siderophores—high-affinity iron-chelating compounds that bind to ferric iron, making it more available to plants. This mechanism is particularly crucial in environments where iron is limited, as it not only supports plant nutrition but also helps suppress pathogenic organisms by depriving them of essential iron. For example, *Bacillus subtilis* has been shown to produce bacillibactin, a siderophore that inhibits fungal pathogens such as *Alternaria porri* and *Fusarium equiseti*, thereby reducing disease severity in crops like tomatoes and eggplants. The synergistic effects of these rhizobacteria with mycorrhizal fungi further amplify their benefits, improving nutrient uptake efficiency and overall plant vigor. As a result, the application of siderophore-producing rhizobacteria is gaining attention in sustainable agriculture practices aimed at enhancing crop resilience and productivity while minimizing chemical inputs, as summarized in Table 11.2, which lists key siderophore-producing rhizobacteria and their roles in mitigating plant pathogens.

Table 11.2 List of siderophore-producing rhizobacteria and its role in plant pathogen mitigation.

Isolate name	Siderophore type	Target pathogen	Plant host	References
Pseudomonas sp. *LHSP1*	Hydroxamate	*Fusarium equiseti, Fusarium solani*	*Vigna radiata*	[25]
Bacillus subtilis CAS15	Bacillibactin	*Alternaria porri, Fusarium equiseti*	*Solanum lycopersicum, Solanum melongena*	[28]
Pseudomonas monteilii	Hydroxamate	*Aspergillus, Fusarium*	Various crops	[52]
Pseudomonas fluorescens MPF47	Pyoverdine	*Rhizoctonia solani*	*Phaseolus vulgaris*	[51]
Enterobacter cloacae	Catechol	*Fusarium oxysporum*	*Brassica oleracea*	[59]
Bacillus cereus	Hydroxamate	*Rhizoctonia solani*	*Zea mays*	[18]
Pseudomonas aeruginosa KA19	Hydroxamate	*Pythium aphanidermatum*	*Cucumis sativus*	[14]
Serratia marcescens	Siderobactin	*Sclerotinia sclerotiorum*	*Glycine max*	[33]
Bacillus amyloliquefaciens	Bacillibactin	*Botrytis cinerea*	*Vitis vinifera*	[38]
Rhizobium leguminosarum	Hydroxamate	*Phytophthora infestans*	*Phaseolus vulgaris*	[8]

11.3.2 Development of Siderophore-Conjugated Antibiotics

The emergence of antimicrobial resistance (AMR) has prompted innovative strategies in antibiotic development, particularly the design of siderophore-conjugated antibiotics. This approach leverages the natural iron-uptake mechanisms of bacteria, utilizing siderophores—high-affinity iron-chelating compounds produced by microorganisms—to enhance the delivery and efficacy of antibiotics. The concept operates on the "Trojan horse" principle, wherein the antibiotic is disguised as a nutrient (iron) that bacteria actively transport into their cells. Siderophore–antibiotic conjugates are designed to exploit specific bacterial transport systems, allowing for increased uptake of the antibiotic and enhanced antibacterial activity against resistant strains. The initial research into these conjugates dates back to the 1980s, focusing on various classes of antibiotics, particularly β-lactams. However, many early attempts faced challenges such as poor solubility, inadequate membrane permeability, and ineffective release of the active antibiotic within the bacterial cell.

Recent advancements have led to more successful designs, with cefiderocol being a notable example. This cephalosporin–siderophore conjugate has demonstrated potent activity against multi-drug resistant Gram-negative bacteria, including those producing serine β-lactamases. The structural design of cefiderocol incorporates a chloro-catechol moiety that mimics natural siderophores, enhancing its ability to utilize bacterial iron transport systems effectively. The ability of cefiderocol to effectively bind PBPs and disrupt cell wall synthesis is what makes it particularly effective against Pseudomonas syringae and other Gram-negative pathogens (Figure 11.3) and additional siderophore–antibiotic conjugates are listed in Table 11.3.

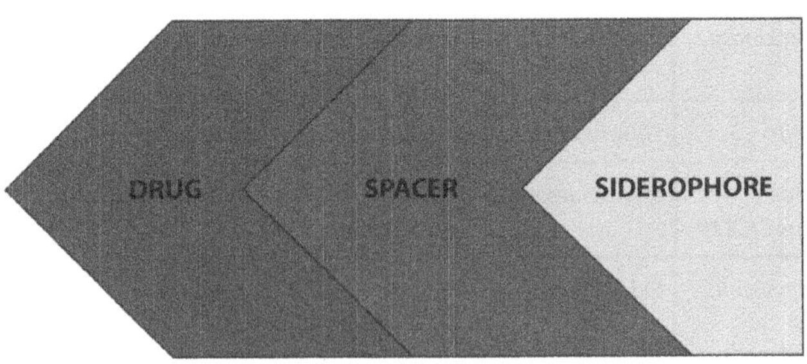

Figure 11.3 Siderophores inhibit the growth of drug-resistant microorganisms through the "Trojan horse strategy" by binding to the drug through the spacer.

Table 11.3 List of other siderophore-antibiotic conjugates to mitigate plant pathogens.

Conjugate name	Type	Target pathogen	Mechanism of action	References
Bacillibactin-enhanced *Bacillus subtilis*	Siderophore	*Fusarium oxysporum*	Inhibits fungal growth	[12]
Chlorodactyloferrin–pyridomycin	Natural siderophore	*Dactylosporangium fulvum*	Antibiotic delivery *via* iron chelation	[7]
Siderophore-conjugated daptomycin	Lipopeptide	*Acinetobacter baumannii*	Disrupts cell membrane	[43]
Siderophore-linked amikacin	Aminoglycoside	*Pseudomonas syringae*	Inhibits protein synthesis	[6]
Siderophore-linked vancomycin	Glycopeptide	*Staphylococcus aureus*	Inhibits cell wall synthesis	[43]
Siderophore-conjugated gentamicin	Aminoglycoside	*Escherichia coli*	Inhibits protein synthesis	[34]
Anachelin–vancomycin complex	Cyanobacterial siderophore	*Bacillus subtilis*	Enhances antibiotic stability	[15]
Pyoverdine–ciprofloxacin	Natural siderophore	*Pseudomonas aeruginosa*	Targets DNA gyrase	[55]
Fimsbactin–daptomycin	Mixed siderophore	*Acinetobacter baumannii*	Disrupts cell membrane	[20]
Enterobactin–ureidopenicillin	Synthetic	*Escherichia coli*	Inhibits cell wall synthesis	[35]
Hydroxamate–catechol conjugates	Synthetic	Various Gram-negative bacteria	Enhances antibiotic uptake	[19]
Siderophore-conjugated rifampicin	Rifamycin	*Mycobacterium tuberculosis*	Inhibits RNA synthesis	[30]

11.3.2.1 Cefiderocol and Its Mechanism in Pseudomonas Syringae Management

Cefiderocol is a novel cephalosporin antibiotic that has garnered attention for its unique mechanism of action, particularly in the context of managing infections caused by multidrug-resistant (MDR) Gram-negative bacteria, including *Pseudomonas syringae*. This pathogen is known to cause significant agricultural losses in various crops, making the development of effective treatments essential.

Cefiderocol (formerly known as S-649266) is characterized by a structure that combines features of traditional cephalosporins with those of siderophores. It contains a chlorocatechol moiety at the C-3 position, which allows it to mimic natural siderophores. This structural element is crucial for its ability to chelate iron, a vital nutrient for bacterial growth, particularly in iron-limited environments such as those encountered during plant infections.

The antibiotic's design enhances its stability against hydrolysis by various β-lactamases, including extended-spectrum β-lactamases (ESBLs) and carbapenemases. This stability is attributed to the presence of a carboxypropanoxyimino group at the C-7 position, similar to that found in ceftazidime. The combination of these features enables cefiderocol to maintain efficacy against resistant strains of bacteria.

Cefiderocol employs a "Trojan horse" strategy to penetrate bacterial cells. The process begins when cefiderocol binds to extracellular iron, forming a chelate complex. This complex is recognized and transported into the bacterial cell through specific iron transporters located in the outer membrane, such as CirA and Fiu. Once inside the periplasmic space, cefiderocol dissociates from the iron and binds to penicillin-binding proteins (PBPs), primarily PBP3. This binding inhibits peptidoglycan synthesis, leading to cell lysis and death. The ability of cefiderocol to effectively bind PBPs and disrupt cell wall synthesis is what makes it particularly effective against *Pseudomonas syringae* and other Gram-negative pathogens.

11.3.2.2 Efficacy against Pseudomonas Syringae

Cefiderocol's efficacy against *Pseudomonas syringae* can be attributed to several factors, namely:

High affinity for PBPs: Cefiderocol demonstrates a strong binding affinity for PBP3, which is critical for maintaining bacterial cell wall integrity. This high affinity translates into potent bactericidal activity against *Pseudomonas syringae*, even in strains resistant to other antibiotics.

Resistance overcoming: The unique structural features of cefiderocol allow it to evade many common resistance mechanisms employed by *Pseudomonas syringae*. Its stability against β-lactamases means that it can remain effective even when faced with enzymatic degradation that would typically render other β-lactam antibiotics ineffective.

Iron acquisition: By utilizing the bacterial iron transport systems, cefiderocol enhances its uptake into bacterial cells, ensuring that sufficient concentrations are achieved in the periplasmic space where it exerts its antibacterial effects.

Broad-spectrum activity: While cefiderocol is particularly noted for its activity against carbapenem-resistant Enterobacteriaceae, its spectrum also includes other Gram negative pathogens like *Pseudomonas aeruginosa*, which shares similar resistance profiles with *Pseudomonas syringae*.

11.3.3 Challenges and Resistance Mechanisms

The integration of siderophores with other microbial agents has opened new avenues for managing plant pathogens. However, this approach is not without challenges, particularly concerning the adaptive resistance mechanisms employed by target bacteria. Understanding these mechanisms is crucial for developing effective strategies to overcome resistance and enhance the efficacy of siderophore-based treatments.

11.3.3.1 Adaptive Resistance in Target Bacteria

Adaptive resistance refers to the ability of bacteria to modify their physiological state in response to environmental stresses, including the presence of antimicrobial agents. In the context of siderophore integration, several mechanisms have been identified that allow pathogens such as *Pseudomonas syringae* and other Gram-negative bacteria to survive despite the presence of iron-chelating compounds.

1. **Altered siderophore production**: Some pathogens can downregulate or modify their siderophore production in response to competition or the presence of exogenous siderophores. For instance, *Pseudomonas syringae* may alter its pyoverdine production when co-cultured with other

microorganisms that produce competing siderophores, thereby reducing its ability to sequester iron from the environment [48]. This adaptive response can lead to decreased effectiveness of siderophore-based treatments.

2. **Siderophore receptor modification**: Bacteria may also modify or lose receptors responsible for siderophore uptake. This alteration can prevent the effective internalization of siderophore–antibiotic conjugates, rendering them ineffective. For example, mutations in outer membrane receptors can hinder the transport of iron–siderophore complexes into bacterial cells [10].
3. **Biofilm formation**: Many plant pathogens, including *Pseudomonas syringae*, can form biofilms that provide a protective environment against antimicrobial agents. Biofilms are structured communities of bacteria encased in a self-produced matrix that can impede the penetration of siderophores and associated antibiotics [40]. The biofilm lifestyle not only protects bacteria from environmental stresses but also enhances their ability to persist in hostile conditions.
4. **Efflux pumps**: The presence of multidrug efflux pumps allows bacteria to actively expel antimicrobial agents, including those delivered *via* siderophores. These pumps can reduce intracellular concentrations of antibiotics and siderophores, leading to decreased susceptibility [5]. For instance, *Pseudomonas aeruginosa* employs efflux systems like MexAB-OprM to extrude various antibiotics and potentially disrupt the action of siderophore-conjugated drugs.
5. **Iron limitation responses**: Under iron-limited conditions, bacteria may enhance their virulence factors and alternative iron acquisition strategies, such as utilizing heme or other iron sources present in host tissues [49]. This adaptability complicates treatment strategies that rely on iron deprivation as a mechanism for controlling bacterial growth.

11.3.4 Future Directions for Overcoming Resistance

To effectively combat the challenges posed by adaptive resistance mechanisms in target bacteria, several future directions can be explored, namely:

1. **Combination therapies**: Developing combination therapies that integrate siderophore–antibiotic conjugates with other antimicrobial agents may enhance efficacy against resistant strains. For instance, pairing siderophores with efflux pump inhibitors could prevent the active expulsion of antimicrobial agents and increase their intracellular concentrations [31]. Such approaches could also target multiple pathways in bacterial metabolism and virulence.
2. **Targeting biofilm formation**: Research into biofilm-disrupting agents can enhance the effectiveness of siderophore-based treatments. Compounds that inhibit biofilm formation or promote biofilm dispersal may improve access to embedded pathogens and increase susceptibility to antimicrobial agents [29]. Developing formulations that combine siderophores with biofilm inhibitors could be a promising strategy.
3. **Novel siderophore structures**: The design of novel synthetic or modified siderophores that exhibit enhanced binding affinities for bacterial receptors or improved stability against degradation could improve their effectiveness in combating resistant strains [44]. Advances in genetic engineering and synthetic biology may facilitate the development of such compounds.
4. **Understanding microbial interactions**: Further research into microbial interactions and their effects on siderophore dynamics is essential for optimizing treatment strategies. Understanding how different microbial communities influence each other's iron acquisition strategies could lead to innovative applications in agriculture and biocontrol.
5. **Genomic and proteomic approaches**: Utilizing genomic and proteomic tools to study resistance mechanisms at a molecular level can provide insights into bacterial adaptations and inform the development of targeted therapies [2]. Identifying specific genes involved in resistance pathways may reveal potential targets for new antimicrobial agents or adjuvants.

6. **Field trials and practical applications**: Conducting field trials to assess the efficacy of integrated approaches combining siderophores with other microbial agents under real-world conditions will be crucial for validating laboratory findings [24]. Understanding how these treatments perform in agricultural settings will help refine application strategies and improve crop protection against pathogens like *Pseudomonas syringae.*

The integration of siderophores with other microbial agents presents a promising strategy for managing plant pathogens; however, adaptive resistance mechanisms pose significant challenges. By understanding these mechanisms and exploring innovative approaches to overcome them, researchers can enhance the efficacy of these integrated treatments. Future directions should focus on combination therapies, novel compound development, understanding microbial interactions, and practical applications in agricultural settings.

11.4 Biocontrol Applications in Agriculture

11.4.1 Enhancing Plant Growth with Siderophores

Siderophores are low-molecular-weight organic compounds produced by microorganisms, primarily to sequester iron from the environment. Given that iron is an essential micronutrient for plant growth and development, the role of siderophores in enhancing plant growth has garnered significant attention in agricultural research. Their ability to increase iron bioavailability, promote nutrient uptake, and protect against pathogens positions them as valuable tools in sustainable agriculture.

The interactions between plants and siderophore-producing microorganisms (SPM) are complex and multifaceted as further illustrated in Table 11.4, which summarizes synergistic interactions between siderophore-producing bacteria and mycorrhizal fungi across various plant species. These interactions can be categorized into several key mechanisms, namely:

1. **Iron acquisition**: Under conditions of low iron availability, plants often struggle to absorb sufficient amounts of this critical nutrient. Siderophores produced by soil bacteria such as *Pseudomonas fluorescens* and *Bacillus subtilis* can bind ferric iron (Fe^{3+}) with high affinity, forming stable complexes that

Table 11.4 Synergistic interactions between siderophore-producing bacteria and mycorrhizal fungi across various plant species, demonstrating their collective impact on enhancing plant growth, nutrient uptake, and resistance to stressors.

Combination	Plant species	Effects observed	Mycorrhizal fungi	Siderophore-producing bacteria	Citation
Glomus intraradices + *Acinetobacter* sp.	*Avena sativa*	Enhanced growth under hydrocarbon stress	*Glomus intraradices*	*Acinetobacter* sp.	[58]
Glomus mosseae + *Pseudomonas fluorescens*	*Zea mays*	Improved nutrient uptake and plant growth	*Glomus mosseae*	*Pseudomonas fluorescens*	[13]
Rhizophagus irregularis + *Bacillus subtilis*	*Cinnamomum camphora*	Increased iron nutrition and biomass	*Rhizophagus irregularis*	*Bacillus subtilis*	[57]
Funneliformis mosseae + *Enterobacter cloacae*	*Triticum aestivum*	Enhanced chlorophyll content and growth parameters	*Funneliformis mosseae*	*Enterobacter cloacae*	[45]
Glomus versiforme + *Pseudomonas putida*	*Phaseolus vulgaris*	Improved drought tolerance	*Glomus versiforme*	*Pseudomonas putida*	[50]

(Continued)

Table 11.4 Synergistic interactions between siderophore-producing bacteria and mycorrhizal fungi across various plant species, demonstrating their collective impact on enhancing plant growth, nutrient uptake, and resistance to stressors. (*Continued*)

Combination	Plant species	Effects observed	Mycorrhizal fungi	Siderophore-producing bacteria	Citation
Rhizophagus irregularis + Burkholderia cepacia	*Solanum lycopersicum*	Increased yield and resistance to pathogens	*Rhizophagus irregularis*	*Burkholderia cepacia*	[53]
Glomus intraradices + *Klebsiella* sp.	*Brassica oleracea*	Enhanced nutrient uptake efficiency	*Glomus intraradices*	*Klebsiella sp.*	[47]
Funneliformis mosseae + Staphylococcus sciuri	*Zea mays*	Increased root biomass and chlorophyll content	*Funneliformis mosseae*	*Staphylococcus sciuri*	[1]
Rhizophagus irregularis + Bacillus licheniformis	*Vigna radiata*	Improved seed germination and plant growth	*Rhizophagus irregularis*	*Bacillus licheniformis*	[54]
Glomus fasciculatum + Pseudomonas syringae	*Glycine max*	Enhanced plant growth and nutrient acquisition	*Glomus fasciculatum*	*Pseudomonas syringae*	[22]

(*Continued*)

Table 11.4 Synergistic interactions between siderophore-producing bacteria and mycorrhizal fungi across various plant species, demonstrating their collective impact on enhancing plant growth, nutrient uptake, and resistance to stressors. (*Continued*)

Combination	Plant species	Effects observed	Mycorrhizal fungi	Siderophore-producing bacteria	Citation
Rhizophagus irregularis + *Chryseobacterium gleum*	*Hordeum vulgare*	Increased yield and stress tolerance	*Rhizophagus irregularis*	*Chryseobacterium gleum*	[9]
Funneliformis mosseae + *Enterobacter aerogenes*	*Medicago sativa*	Enhanced nitrogen fixation and growth	*Funneliformis mosseae*	*Enterobacter aerogenes*	[45]
Glomus intraradices + *Bacillus aryabhattai*	*Canola*	Increased seedling vigor and biomass	*Glomus intraradices*	*Bacillus aryabhattai*	[23]
Rhizophagus irregularis + *Pseudomonas aeruginosa*	*Cucumis sativus*	Improved resistance to root pathogens	*Rhizophagus irregularis*	*Pseudomonas aeruginosa*	[42]

can be taken up by plant roots [36]. This process not only enhances the iron nutritional status of plants but also promotes overall growth and development.

2. **Nutrient mobilization**: Siderophores can also facilitate the mobilization of other essential nutrients in the rhizosphere. For instance, studies have shown that siderophore-producing bacteria can enhance the solubilization of phosphates and other minerals, making them more available for plant uptake [56]. This dual action—chelation of iron and mobilization of other nutrients—contributes to improved plant health.
3. **Reduction of heavy metal toxicity**: Siderophores can play a significant role in phytoremediation by chelating heavy metals present in contaminated soils. By binding to toxic metal ions, siderophores reduce their bioavailability and toxicity to plants while simultaneously enhancing the uptake of essential nutrients [46]. For example, certain strains of *Pseudomonas* have been shown to increase the absorption of chromium (Cr) and lead (Pb) in maize, promoting both growth and detoxification [26].
4. **Biocontrol of pathogens**: One of the most significant advantages of using siderophore-producing microorganisms is their ability to suppress plant pathogens through competitive exclusion. By sequestering iron in the rhizosphere, these microorganisms limit the availability of this nutrient to pathogenic fungi and bacteria, thereby reducing their growth and virulence [32]. For instance, *Bacillus subtilis*, known for its production of bacillibactin, has been effective against *Fusarium oxysporum*, a notorious pathogen causing wilt disease in various crops [4]. Melatonin upregulation caused by *Arthrobotrys oligospora* shows high affinity for iron and helps in regulating the population of nematode *Meloidogyne graminicola* [17].
5. **Plant growth promotion**: The application of siderophore-producing bacteria has been associated with enhanced seed germination rates and improved seedling vigor. Research indicates that inoculation with such bacteria can significantly increase germination rates in crops like tomatoes and cucumbers [37]. Furthermore, siderophores may stimulate root development and increase chlorophyll content, leading to better photosynthetic capacity.

6. **Synergistic effects with mycorrhizae**: The combination of siderophore-producing bacteria with mycorrhizal fungi can further enhance plant growth. Mycorrhizal associations improve nutrient uptake efficiency, while siderophores facilitate iron acquisition [41]. This synergistic interaction is particularly beneficial under nutrient-deficient conditions.

11.4.2 Suppression of Phytopathogens

Siderophores are crucial in suppressing plant pathogens by competing for iron, a vital nutrient often limited in the soil. Produced by beneficial microorganisms like *Pseudomonas* spp., these compounds bind iron with high affinity, thereby reducing its availability to pathogenic bacteria. This competitive exclusion not only diminishes pathogen growth but also enhances plant health by promoting beneficial microbial communities in the rhizosphere. Studies have shown that siderophore-producing strains significantly inhibit the growth of pathogens such as *Ralstonia solanacearum* and *Fusarium oxysporum*, demonstrating their potential as biocontrol agents in sustainable agriculture.

11.4.2.1 Mechanism of Pathogen Suppression

Siderophores exhibit a pronounced affinity for ferric iron owing to their intrinsic structural properties. Typically, they form robust hexadentate complexes with Fe^{3+}, thereby enhancing the solubility and bioavailability of iron in environments where it may be otherwise precipitated or inaccessible. For example, pyoverdine binds iron *via* multiple functional groups, including hydroxamates and catecholates. Siderophores are secreted into the extracellular environment, where they bind to iron ions. Transport proteins move the siderophores from inside the cell to the outside, forming soluble iron complexes that are recognized and absorbed by microbial uptake systems. Inside the cell, iron is released through two main mechanisms: the most common method involves siderophore reductases, which convert ferric ions to ferrous ions, crucial for plant iron uptake. Alternatively, certain siderophores are hydrolyzed by enzymes, weakening their iron-binding and releasing the iron. These mechanisms are essential for biological iron acquisition, providing iron for various biosynthetic processes (Figure 11.4).

Additionally, siderophores can directly inhibit pathogens by exhibiting antibacterial or antifungal properties, often in combination with other secondary metabolites. For instance, *Bacillus subtilis* produces siderophores that help biocontrol *Fusarium* wilt in various crops. The synthesis

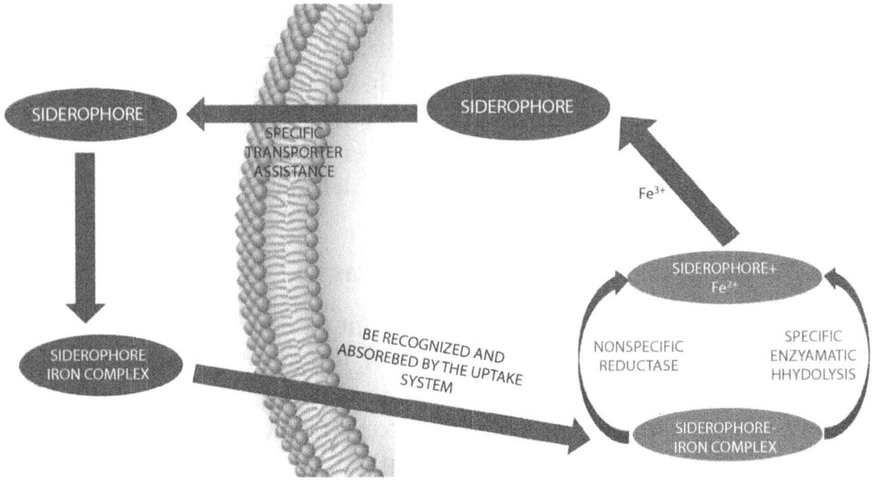

Figure 11.4 Schematic diagram of the action mechanism of siderophores.

of siderophores also fosters competition within microbial communities, as producing species limit access to iron for non-producing or incompatible species, significantly influencing the microbial structure of the rhizosphere.

Siderophores play a significant role in promoting plant health and resilience by inducing systemic resistance (ISR), modifying metal homeostasis, and improving plant growth. Siderophores produced by plant-growth-promoting rhizobacteria (PGPR) can trigger immunological responses in plants, priming their defenses against future pathogen attacks through the activation of systemic resistance mechanisms. Additionally, interactions between microbial siderophores and plant roots can perturb the plant's metal homeostasis, altering the distribution and absorption of essential metals like zinc and manganese. These changes can enhance stress responses and reduce the plant's vulnerability to pathogens. Moreover, bacteria that produce siderophores contribute to improved nutrient uptake, particularly iron, which indirectly enhances the plant's overall health and vigor. Healthier plants, with robust physiological states, are generally more resistant to biotic stresses, including infections, underscoring the multifaceted benefits of siderophore-producing microorganisms in agricultural systems.

11.4.2.2 Examples of Successful Biocontrol Agents

1. ***Pseudomonas fluorescens***: This well-studied biocontrol agent produces pyoverdine, a siderophore that effectively

sequesters iron, thereby inhibiting the growth of *Fusarium oxysporum*, a major pathogen affecting tomatoes. The ability of this bacterium to produce antifungal compounds further enhances its efficacy as a biocontrol agent.

2. ***Bacillus subtilis***: Known for its production of bacillibactin, *Bacillus subtilis* has demonstrated significant biocontrol potential against *Fusarium solani*, which causes root rot in peppers. Its dual action includes both siderophore-mediated iron acquisition and the secretion of antimicrobial substances that inhibit pathogen growth.
3. ***Pseudomonas monteilii***: This strain has been identified as a promising biocontrol agent due to its ability to produce hydroxamate-type siderophores. It has shown antagonistic activity against various fungal pathogens, including *Aspergillus calidoustus*, contributing to improved plant health in sissoo trees.
4. ***Enterobacter cloacae***: This bacterium produces catechol-type siderophores that enhance the growth of *Brassica* plants while suppressing pathogens like *Rhizoctonia solani*. Its ability to solubilize nutrients further supports plant development.
5. ***Bacillus amyloliquefaciens***: This species is effective against *Botrytis cinerea*, a common grapevine pathogen. The production of bacillibactin allows it to compete for iron, while its antifungal properties contribute to disease management.
6. ***Pseudomonas aeruginosa***: Known for its versatility, this bacterium produces pyoverdine and demonstrates strong competitive abilities against pathogens such as *Pythium aphanidermatum*. Its application in cucumber cultivation has shown promising results in disease suppression.
7. ***Serratia marcescens***: This bacterium utilizes siderobactin to acquire iron from the environment, effectively suppressing *Sclerotinia sclerotiorum*, which affects soybean crops. Its production of lytic enzymes further enhances its biocontrol capabilities.
8. ***Azospirillum brasilense***: This nitrogen-fixing bacterium not only promotes plant growth but also suppresses pathogens like *Fusarium culmorum*. Its ability to enhance nutrient uptake combined with pathogen inhibition makes it a valuable biocontrol agent in maize cultivation.

9. ***Rhizobium leguminosarum***: Known for its role in nitrogen fixation, this bacterium also produces hydroxamate-type siderophores that can induce systemic resistance in peas against pathogens like *Phytophthora infestans*. This dual functionality enhances both growth and disease resistance.
10. ***Bacillus cereus***: This strain has been shown to produce siderophores that effectively inhibit various fungal pathogens, contributing to improved crop yields and health through enhanced disease resistance mechanisms.

11.5 Bioremediation of Environmental Pollutants

Bioremediation is a sustainable process that utilizes living microorganisms, such as bacteria, fungi, and plants, to degrade and remove environmental pollutants. This technique effectively addresses various contaminants, including heavy metals, petroleum hydrocarbons, and pesticides, by transforming them into less harmful substances. Bioremediation can be categorized into three primary approaches: biostimulation, which enhances the growth of native microbes; bioaugmentation, which involves adding specific microorganisms to contaminated sites; and phytoremediation, where plants are used to absorb and detoxify pollutants. This eco-friendly method not only cleans up polluted environments but also promotes the restoration of natural ecosystems as summarized in Table 11.5, which lists various pollutants and the microorganisms that utilize siderophores for effective bioremediation across different environments.

Table 11.5 Various pollutants and the corresponding microorganisms that utilize siderophores for effective bioremediation strategies across different environments.

Pollutant	Microorganism	Bioremediation technique	Target environment
Heavy metals (lead, cadmium)	*Pseudomonas aeruginosa*	Siderophore-mediated uptake	Soil and water
Polychlorinated biphenyls	*Rhodococcus* sp.	Biodegradation	Soil

(*Continued*)

Table 11.5 Various pollutants and the corresponding microorganisms that utilize siderophores for effective bioremediation strategies across different environments. (*Continued*)

Pollutant	Microorganism	Bioremediation technique	Target environment
Petroleum hydrocarbons	*Mycobacterium* sp.	Bioaugmentation	Soil and water
Dyes (e.g., azo dyes)	*Klebsiella* sp.	Biofiltration	Water
Phenols	*Enterobacter cloacae*	Biodegradation	Soil
Heavy metals (arsenic)	*Bacillus subtilis*	Phytoremediation	Soil
Cyanide	*Nocardia* sp.	Biostimulation	Soil
Organic solvents	*Burkholderia cepacia*	*In situ* bioremediation	Soil
Mercury	*Desulfovibrio desulfuricans*	Anaerobic bioremediation	Water
Industrial effluents	*Acinetobacter* sp.	Bioaugmentation	Soil and groundwater
Chlorinated solvents	*Mycobacterium* sp.	Siderophore-assisted degradation	Soil
Polycyclic aromatic hydrocarbons (PAHs)	*Mycobacterium vanbaalenii*	Bioremediation	Soil and water
Agricultural pesticides	*Sphingomonas* sp.	Biodegradation	Soil
Landfill leachate	*Methanogenic archaea*	Anaerobic digestion	Landfill sites
Heavy metals (copper)	*Pseudomonas fluorescens*	Siderophore-mediated uptake	Soil and water

11.5.1 Metals

Although metals are essential to the development of human civilizations, metal pollution has been caused by mining operations, sludge applications, nuclear power plants, and the industrial industry. The solubility and mobility of a wide range of metals, such as Cd, Cu, Ni, Pb, Zn, and the actinides Th(IV), U(IV), and Pu(IV), are significantly improved by siderophores. Siderophores' effectiveness in this respect is mostly dependent on their ligand functions, which enable a strong affinity or selectivity for particular metals—apart from iron—in respect to the stability constants of the metal–siderophore complexes. As a result, siderophores become an important tool in bioremediation, offering an economical and environmentally friendly method.

In recent years, there has been a burgeoning interest in exploring the potential applications of siderophores in the context of metal bioremediation. Desferrioxamine B exhibited superior chelation of Co(III) compared to Fe(III) under alkaline pH conditions. *Azotobacter vinelandii*, known for its production of siderophores, namely, azotochelin and azotobactin, displayed the capability to utilize these siderophores for the acquisition of both Mo and V. Furthermore, the siderophore pyochelin synthesized by *Pseudomonas aeruginosa* demonstrated the capacity to chelate an extensive variety of metals, including Ag+, Al^{3+}, Cd^{2+}, Co^{2+}, Cr^{2+}, Cu^{2+}, Eu^{3+}, Ga^{3+}, Hg^{2+}, Mn^{2+}, Ni^{2+}, Pb^{2+}, Sn^{2+}, Tb^{3+}, Tl^{+}, and Zn^{2+}; however, the assimilation process did not appear to incorporate any metal apart from Fe^{3+}. Siderophores produced by *Fusarium solani* facilitated the solubilization of Cu and Zn *in vitro*. Siderophores also significantly contribute to the mobilization of metals from mining waste or metal-contaminated soils. A variety of metals (i.e., Fe, Ni, and Co) were mobilized from waste materials (acid-leached ore) of a former uranium mine in the presence of siderophores generated by *P. fluorescens*. It has been demonstrated that the siderophores synthesized by *Agrobacterium radiobacter* were capable of removing approximately 54% of the As content from metal-contaminated soil. Recently, it has been established that pyoverdines effectively mobilized U(VI), Np(V), and other metals from uranium mine waste.

Not solely do microbial siderophores play a crucial role in metal bioremediation, but numerous studies have also illustrated that phytosiderophores are adept at mobilizing metals within soil environments. Phytosiderophores exhibit a high affinity for complexation with several metals, arranged in the following order ($Cd^{2+} > Ni^{2+} > Pb^{2+} > Sn^{2+} > AsO_4^{-2} > AsO_2^{-1} > Mn^{2+} > Co^{2+} > Cu^{2+} > Fe^{+3}$) while demonstrating a significantly weak binding capacity for Al3+ and Cr3+. Investigations conducted in both uncontaminated and contaminated soils have revealed that phytosiderophores are markedly more effective in mobilizing Fe, Cu, Zn, Ni, and Cd from the soil when compared to synthetic chelators and microbial siderophores.

11.5.2 Petroleum Hydrocarbons

One of the main environmental issues in marine environments is the presence of petroleum hydrocarbons. The removal of petroleum hydrocarbons from the marine environment may benefit greatly from the use of microorganisms. Through an indirect method, microbial siderophores aid in the biodegradation of petroleum hydrocarbons by enabling the degraded microbes to acquire iron in the presence of Fe-limiting circumstances. Petrobactin was the first siderophore produced by the oil-degrading marine bacterium to be structurally described. A small number of research indicated that siderophores might be an effective method for cleaning up oil spills. Following the Deepwater Horizon oil leak in 2010, marine *Vibrio* species that were isolated from the Gulf of Mexico produced amphiphilic siderophores known as "ochrobactins," which were thought to effectively aid in the breakdown of petroleum hydrocarbons.

11.5.3 Nuclear Fuel Reprocessing

Anionic hydroxamate or catecholate functional groups found in siderophores create hard oxodonors that bind to Lewis acids with great force, producing complexes with unusually high stability constants. It has been proposed that siderophores could bind actinides with a complexation constant estimated to be K = 1,016 because actinides form strong complexes with hard oxygen anions.

The Purex technology has been used commercially to separate U and Pu for reuse from fission products like Ti and Np and reprocess irradiated nuclear fuel using solvent extraction techniques. U and Pu enter the solvent during this process and get polluted with Np. Since siderophores have been demonstrated to enable the selective extraction of Np from the solvent phase, they may be employed in the Purex process to make the actinide removal procedure simpler. With U(VI), desferrioxamine B forms a stable complex. Acetohydroxamic acid, a ligand suggested for actinide complexation, is identical to the hydroxamate functional group of desferrioxamine B.

According to recent reports, a small amount of siderophores could potentially affect the disintegration of spent nuclear fuel, and there did not appear to be much of a difference between using pyoverdine made by

P. fluorescens and synthetic desferrioxamine B. These investigations have led to the proposal of siderophores for nuclear fuel reprocessing and radioactive waste remediation.

11.5.4 Optical Biosensor

In order to enhance the signal to noise ratio and enable the detection of different kinds of reactions through specially designed systems, a biosensor is a biomolecule connected to an electrical device, such as a transducer, amplifier, or noise filter. Pyoverdines are fluorescent siderophores that are yellow-green and soluble in water. They have the following characteristics: (i) they form a strong complex with Fe(III) and have a weak or negligible affinity for Fe(II) and (ii) the stability constants of the Fe(III) complexes are very high (about K = 1,032). Pyoverdine is a promising drug for the development of optical biosensors because of these properties. For the molecular recognition component of the sensor that might be used to determine the bioavailability of iron in soils or oceanic water, using a siderophore with an extraordinary Fe(III)-binding constant would be the best option.

A siderophore has been used as a biosensor to measure the amount of iron in the water. The parabactin generated by *Paracoccus denitrificans* was employed as a biosensor in that study by encasing it in a sol-gel thin film on a quartz substrate. A flow cell installed in the fluorescence spectrometer's sample partition was used to analyze the seawater samples. Additionally, siderophores offer the possibility of a very sensitive and specific detection system that would replicate the biological uptake mechanism. For example, *A. vinelandii*'s azotobactin has been employed as an optical biosensor for Fe(III) in a modified design that relies on the azotobactin's encapsulation in sol-gel matrices without causing a noticeable reduction in its fluorescence signal.

Furthermore, by immobilizing the Fe biosensor pyoverdine in three porous sol-gel glass formulations (A, B, and C) with varying amounts of water supplied, the Fe(II and III) specificity has been adjusted. Pyoverdine immobilized in sol-gel C, which included more water than A and B, showed the highest specific and linear response for binding to Fe(II and III) in that investigation.

11.5.5 Bio-Bleaching of Pulps

Numerous environmental issues, such as photochemical oxidation, acidification, nitrification, global warming, human toxicity, ecotoxicity, and solid wastes, are mostly caused by the pulp and paper sector. The bleaching operations are the primary cause of pulp and paper manufacturing issues.

While some contaminants are released into wastewater, others are released into the atmosphere.

Since siderophores can cut down on 70% of the chemicals required to bleach Kraft pulp, they are regarded as efficient pulp treatment agents that are also environmentally benign. One of the most significant categories of wood-decaying microorganisms is thought to be brown-rot fungus. According to certain research, fungi that break down wood produce siderophores that are hydroxamate and catecholate. For example, hydroxamate siderophores that were separated from the fungus that causes brown rot (*Gloephyllum trabeum*) were able to mediate the reduction of iron in redox cycling processes. Following a reaction between hydrogen peroxide and the reduced iron, oxygen radical species are produced, which depolymerize cellulose, hemicellulose, and lignocellulose.

It was thought that siderophores' primary function in the bio-bleaching of pulps was this depolymerization process. Furthermore, they examined the ability of siderophores produced by various microbial species to bio-bleach pulps in terms of viscosity loss. They discovered that siderophores produced by *G. trabeum* could degrade pulp with a loss of viscosity of 10.8%, while siderophores produced by *Tinea versicolor* and *Perenniporia medula-panis* degraded pulps with losses of 13.6% and 14.4%, respectively. It has also been noted that the siderophores that *Coriolus versicolor* produces may change the structure of lignin, making it more prone to degradation and, as a result, aid in pulp bio-bleaching.

11.5.6 Enhancing Siderophore Production for Improved Biodegradation

To optimize the biodegradation efficiency of siderophores, several strategies can be implemented, namely:

1. Optimization of growth conditions
Studies have demonstrated that siderophore synthesis can be greatly increased by modifying environmental parameters such pH, nitrogen sources (such as ammonium nitrate), and iron concentration. One study, for example, optimized these parameters using a Box–Behnken design, which led to higher siderophore yields.

2. Genetic engineering
To increase the synthesis of siderophores in particular microbial strains, genetic changes can be used. One promising method for removing vital

nutrients from contaminated environments is to modify *Pseudomonas* organisms to produce more pyoverdine.

3. Synergistic effects with other compounds
Synergistic effects between siderophores and low molecular mass organic acids (LMMOAs) can raise the rates at which minerals dissolve more quickly than either could on its own. Together they can increase the biodegradation processes' overall effectiveness.

4. Application in bioremediation
In contaminated locations, using bacteria that produce siderophores as bio-inoculants has two advantages: it promotes plant development while also breaking down pollutants. By using less chemical pesticides and fertilizers, this method not only aids in the cleanup of damaged soils but also promotes sustainable farming methods. Growing consumer demand for sustainable agriculture methods is reflected in the expanding worldwide market for biopesticides and biofertilizers. Among the major participants in this industry are Novozymes, UPL, and Syngenta. In India alone, the biopesticide market is anticipated to expand at a compound annual growth rate (CAGR) of around 9.38% between 2022 and 2029. Growing public awareness of the environmental harm caused by chemical agriculture and government initiatives supporting organic farming are the main drivers of this expansion. Although there are many formulation, legal, and market acceptability issues with the commercial production of biofertilizers and biopesticides, there are also many chances for agricultural innovation and sustainability. In the upcoming years, this industry is expected to increase significantly as research progresses and customer attitudes change in favour of eco-friendly solutions.

11.5.7 Limitations

Molecular understanding: The current scientific understanding of the molecular mechanisms behind siderophore uptake systems is limited. The complexity of membrane-bound proteins and multicomponent systems makes it challenging to fully characterize these mechanisms, hindering the development of effective siderophore-antibiotic conjugates for enhanced therapeutic efficacy.
Transport mechanisms: Despite progress in understanding siderophore transport pathways, challenges remain in fully grasping their selectivity and efficiency. Biochemical interactions with host immune proteins, like siderocalin, add complexity, as certain siderophores may be recognized

and neutralized by the immune system, limiting their therapeutic effectiveness in clinical settings.

Resistance development: The rise of bacterial resistance mechanisms, such as mutations in outer membrane receptors, challenges the effectiveness of siderophore-antibiotic conjugates. Bacteria's ability to adapt underscores the need for ongoing research and innovation in developing these conjugates to combat the evolving landscape of microbial resistance.

AI disclosure statement: The author acknowledges the use of AI-based tools for minor language editing or grammatical correction. However, all intellectual contributions, analyses, and conclusions presented in this chapter are entirely the author's own.

References

1. Ain, Q.U., Hussain, H.A., Zhang, Q., Maqbool, F., Ahmad, M., Mateen, A., Imran, A., Coordinated influence of Funneliformis mosseae and different plant growth-promoting bacteria on growth, root functional traits, and nutrient acquisition by maize. *Mycorrhiza*, 34, 5, 477–488, 2024.
2. Aita, S.E., Ristori, M.V., Cristiano, A., Marfoli, T., De Cesaris, M., La Vaccara, V., Angeletti, S., Proteomic Insights into Bacterial Responses to Antibiotics: A Narrative Review. *Int. J. Mol. Sci.*, 26, 15, 7255, 2025.
3. Arora, N.K., Kang, S.C., Maheshwari, D.K., Isolation of siderophore-producing strains of Rhizobium meliloti and their biocontrol potential against Macrophomina phaseolina that causes charcoal rot of groundnut. *Curr. Sci.*, 673–677, 2001.
4. Boulahouat, S., Cherif-Silini, H., Silini, A., Bouket, A.C., Luptakova, L., Alenezi, F.N., Belbahri, L., Biocontrol efficiency of rhizospheric Bacillus against the plant pathogen Fusarium oxysporum: A promising approach for sustainable agriculture. *Microbiol. Res.*, 14, 3, 892–908, 2023.
5. Braun, V. and Braun, M., Active transport of iron and siderophore antibiotics. *Curr. Opin. Microbiol.*, 5, 2, 194–201, 2002.
6. Bremner, J.B., An update review of approaches to multiple action-based antibacterials. *Antibiotics*, 12, 5, 865, 2023.
7. Caradec, T., Anoz-Carbonell, E., Petrov, R., Billamboz, M., Antraygues, K., Cantrelle, F.X., Hartkoorn, R.C., A novel natural siderophore antibiotic conjugate reveals a chemical approach to macromolecule coupling. *ACS Cent. Sci.*, 9, 11, 2138–2149, 2023.
8. Carrillo-Castañeda, G. and Cano, M.E.A., Characterization of siderophore-mediated iron transport from Rhizobium leguminosarum bv. phaseoli. *J. Plant Nutr.*, 23, 11-12, 1669–1683, 2000.

9. Chhabra, R. and Bisht, A., Current Prospects in Enhancing Productivity and Nutritional Security of Crop Plants Exposed to Salt Stress, in: *Optimizing Plant Health under Abiotic-Stress Environments*, pp. 96–122, CRC Press, 2026.
10. Crosa, J.H., Genetics and molecular biology of siderophore-mediated iron transport in bacteria. *Microbiol. Rev.*, 53, 4, 517–530, 1989.
11. Dale, S.E., Doherty-Kirby, A., Lajoie, G., Heinrichs, D.E., Role of siderophore biosynthesis in virulence of Staphylococcus aureus: identification and characterization of genes involved in production of a siderophore. *Infect. Immun.*, 72, 1, 29–37, 2004.
12. Dimopoulou, A., Theologidis, I., Benaki, D., Koukounia, M., Zervakou, A., Tzima, A., Skandalis, N., Direct antibiotic activity of bacillibactin broadens the biocontrol range of Bacillus amyloliquefaciens MBI600. *Msphere*, 6, 4, 10–1128, 2021.
13. Džafić, E., Pongrac, P., Likar, M., Regvar, M., Vogel-Mikuš, K., The arbuscular mycorrhizal fungus Glomus mosseae alleviates autotoxic effects in maize (Zea mays L.). *Eur. J. Soil Biol.*, 58, 59–65, 2013.
14. Farda, B., Djebaili, R., Del Gallo, M., Pellegrini, M., Endophytic Bacteria: Plant Growth-promoting Activities and their Application in Plant Disease Control, in: *Biofungicides: Eco-Safety and Future Trends*, pp. 40–64, CRC Press, 2023.
15. Gademann, K., Kobylinska, J., Wach, J.Y., Woods, T.M., Surface modifications based on the cyanobacterial siderophore anachelin: From structure to functional biomaterials design. *Biometals*, 22, 4, 595–604, 2009.
16. Garénaux, A., Caza, M., Dozois, C.M., The Ins and Outs of siderophore mediated iron uptake by extra-intestinal pathogenic Escherichia coli. *Vet. Microbiol.*, 153, 1-2, 89–98, 2011.
17. Gautam, V., Garg, V., Nagar, R., Meena, N., Kumari, S., Dashrath, H.A., Singh, R.K., Characterizing Metabolic Changes in Rice Roots Induced by Meloidogyne graminicola and Modulated by Arthrobotrys oligospora: A Pathway-Based Approach. *Physiol. Mol. Plant Pathol.*, 102541, 2024, https://doi.org/10.1016/j.pmpp.2024.102541.
18. Ghazy, N. and El-Nahrawy, S., Siderophore production by Bacillus subtilis MF497446 and Pseudomonas koreensis MG209738 and their efficacy in controlling Cephalosporium maydis in maize plant. *Arch. Microbiol.*, 203, 3, 1195–1209, 2021.
19. Ghosh, M. and Miller, M.J., Synthesis and *in vitro* antibacterial activity of spermidine-based mixed catechol-and hydroxamate-containing siderophore—Vancomycin conjugates. *Bioorg. Med. Chem.*, 4, 1, 43–48, 1996.
20. Ghosh, M., Miller, P.A., Möllmann, U., Claypool, W.D., Schroeder, V.A., Wolter, W.R., Miller, M.J., Targeted antibiotic delivery: Selective siderophore conjugation with daptomycin confers potent activity against multidrug resistant Acinetobacter baumannii both *in vitro* and *in vivo*. *J. Med. Chem.*, 60, 11, 4577–4583, 2017.

21. Haas, H., Schoeser, M., Lesuisse, E., Ernst, J.F., Parson, W., Abt, B., Oberegger, H., Characterization of the Aspergillus nidulans transporters for the siderophores enterobactin and triacetylfusarinine C. *Biochem. J*, 371, 2, 505–513, 2003.
22. Ilbas, A.I., and Sahin, S., Glomus fasciculatum inoculation improves soybean production. *Acta Agric. Scand. Sect. B-Soil Plant Sci.*, 55, 4, 287–292, 2005.
23. Jalal-Ud-Din, S., Elahi, N.N., Mubeen, F., Significance of zinc-solubilizing plant growth-promoting rhizobacterial strains in nutrient acquisition, enhancement of growth, yield, and oil content of canola (Brassica napus L.). *Front. Microbiol.*, 15, 1446064, 2024.
24. Karmakar, B., Thakuria, D., Begum, R.H., Joga, R.J., Recent advances in experimental design of synthetic microbial communities for biocontrol application. *BioControl*, 70, 2, 229–244, 2025.
25. Kasana, R.C. and Singh, S.K., Isolation and Identification of a Siderophore Producing and Antagonistic Cold Tolerant (psychrotrophic) Pseudomonas sp. LHSP1: Identification of cold tolerant pseudomonas SP. *Ann. Arid Zone*, 63, 1, 23–31, 2024.
26. Kaur, H., Katyal, P., Chandel, S., Singh, D., Kumar, P., Choudhary, M., Microbes mediated alleviation of chromium (Cr VI) stress for improved phytoextraction in fodder maize (Zea mays L.) cultivar. *Heliyon*, 10, 23, 2024.
27. Kloepper, J.W., Leong, J., Teintze, M., Schroth, M.N., Pseudomonas siderophores: A mechanism explaining disease-suppressive soils. *Curr. Microbiol.*, 4, 5, 317–320, 1980.
28. Kumar, R., Singh, A., Shukla, E., Singh, P., Khan, A., Singh, N.K., Srivastava, A., Siderophore of plant growth promoting rhizobacterium origin reduces reactive oxygen species mediated injury in Solanum spp. caused by fungal pathogens. *J. Appl. Microbiol.*, 135, 2, lxae036, 2024.
29. Landini, P., Antoniani, D., Burgess, J.G., Nijland, R., Molecular mechanisms of compounds affecting bacterial biofilm formation and dispersal. *Appl. Microbiol. Biotechnol.*, 86, 3, 813–823, 2010.
30. Lysenko, V., Gao, M.L., Sterk, F.A., Innocenti, P., Slingerland, C.J., Martin, N.I., Design, Synthesis, and Antibacterial Evaluation of Rifampicin–Siderophore Conjugates. *ACS Infect. Dis.*, 2025.
31. Mahamoud, A., Chevalier, J., Alibert-Franco, S., Kern, W.V., Pagès, J.M., Antibiotic efflux pumps in Gram-negative bacteria: The inhibitor response strategy. *J. Antimicrob. Chemother.*, 59, 6, 1223–1229, 2007.
32. Maheshwari, D.K., Kumar, S., Maheshwari, N.K., Patel, D., Saraf, M., Nutrient availability and management in the rhizosphere by microorganisms, in: *Bacteria in agrobiology: Stress management*, pp. 301–326, Springer, Berlin, Heidelberg, 2012.
33. Mason, C.J., Peiffer, M., St Clair, A., Hoover, K., Felton, G.W., Concerted impacts of antiherbivore defenses and opportunistic Serratia pathogens on the fall armyworm (Spodoptera frugiperda). *Oecologia*, 198, 1, 167–178, 2022.

34. Negash, K.H., Norris, J.K., Hodgkinson, J.T., Siderophore–antibiotic conjugate design: New drugs for bad bugs? *Molecules*, 24, 18, 3314, 2019.
35. Ohi, N., Aoki, B., Moro, K., Kuroki, T., Sugimura, N., Noto, T., Nehashi, T., Matsumoto, M., Okazaki, H., Matsunaga, I., Semisynthetic P-lactam antibiotics. II. Effect on antibacterial activity of ureido N-substituents in the 6-[(R)-2-[3-(3,4-dihy droxybenzoyl)-l-ureido]-2-phenylacetamido]penicillanic acids. *J. Antibiot.*, 39, 242–250, 1986.
36. Pahari, A., Pradhan, A., Nayak, S.K., Mishra, B.B., Bacterial siderophore as a plant growth promoter, in: *Microbial biotechnology: Volume 1. applications in agriculture and environment*, pp. 163–180, Springer Singapore, Singapore, 2018.
37. Pérez-García, L.A., Sáenz-Mata, J., Fortis-Hernández, M., Navarro-Muñoz, C.E., Palacio-Rodríguez, R., Preciado-Rangel, P., Plant-growth-promoting rhizobacteria improve germination and bioactive compounds in cucumber seedlings. *Agronomy*, 13, 2, 315, 2023.
38. Pinto, C., Sousa, S., Froufe, H., Egas, C., Clément, C., Fontaine, F., Gomes, A.C., Draft genome sequence of Bacillus amyloliquefaciens subsp. plantarum strain Fito_F321, an endophyte microorganism from Vitis vinifera with biocontrol potential. *Stand. Genomic Sci.*, 13, 1, 30, 2018.
39. Podschun, R., Fischer, A., Ullmann, U., Siderophore production of Klebsiella species isolated from different sources. *Zentralbl. Bakteriol.*, 276, 4, 481–486, 1992.
40. Price, S.L. and Skaar, E.P., Forging biofilms: metal-induced microbial responses in biofilm formation. *J. Bacteriol.*, e00247–25, 2025.
41. Rajapitamahuni, S., Kang, B.R., Lee, T.K., Exploring the roles of arbuscular mycorrhizal fungi in plant–iron homeostasis. *Agriculture*, 13, 10, 1918, 2023.
42. Ravnskov, S., Cabral, C., Larsen, J., Mycorrhiza induced tolerance in Cucumis sativus against root rot caused by Pythium ultimum depends on fungal species in the arbuscular mycorrhizal symbiosis. *Biol. Control*, 141, 104133, 2020.
43. Rayner, B., Verderosa, A.D., Ferro, V., Blaskovich, M.A., Siderophore conjugates to combat antibiotic-resistant bacteria. *RSC Med. Chem.*, 14, 5, 800–822, 2023.
44. Rodríguez, D. and González-Bello, C., Siderophores: chemical tools for precise antibiotic delivery. *Bioorg. Med. Chem. Lett.*, 87, 129282, 2023.
45. Roy, S., Chakraborty, A.P., Chakraborty, R., Understanding the potential of root microbiome influencing salt-tolerance in plants and mechanisms involved at the transcriptional and translational level. *Physiol. Plant.*, 173, 4, 1657–1681, 2021.
46. Scavino, A.F. and Pedraza, R.O., The role of siderophores in plant growth-promoting bacteria, in: *Bacteria in agrobiology: Crop productivity*, pp. 265–285, Springer Berlin Heidelberg, Berlin, Heidelberg, 2013.

47. Sellappan, R. and Thangavel, K., Role of Arbuscular Mycorrhizal Fungi (AMF) in Organic Vegetables Production, in: *Organic Production of Vegetable Crops*, pp. 245–266, Apple Academic Press, 2025.
48. Shao, Z., Gu, S., Zhang, X., Xue, J., Yan, T., Guo, S., Wei, Z., Siderophore interactions drive the ability of Pseudomonas spp. consortia to protect tomato against Ralstonia solanacearum. *Hortic. Res.*, 11, 9, uhae186, 2024.
49. Sheldon, J.R., Laakso, H.A., Heinrichs, D.E., Iron acquisition strategies of bacterial pathogens, in: *Virulence Mechanisms of Bacterial Pathogens*, pp. 43–85, Wiley, 2016.
50. Singh, K., Organic amendments to soil inoculated arbuscular mycorrhizal fungi and Pseudomonas fluorescens treatments reduce the development of root-rot disease and enhance the yield of Phaseolus vulgaris L. *Eur. J. Soil Biol.*, 47, 5, 288–295, 2011.
51. Solanki, M.K., Singh, R.K., Srivastava, S., Kumar, S., Kashyap, P.L., Srivastava, A.K., Arora, D.K., Isolation and characterization of siderophore producing antagonistic rhizobacteria against Rhizoctonia solani. *J. Basic Microbiol.*, 54, 6, 585–597, 2014.
52. Srivastava, P., Sahgal, M., Sharma, K., Enshasy, H.A.E., Gafur, A., Alfarraj, S., Sayyed, R.Z., Optimization and identification of siderophores produced by Pseudomonas monteilii strain MN759447 and its antagonism toward fungi associated with mortality in Dalbergia sissoo plantation forests. *Front. Plant Sci.*, 13, 984522, 2022.
53. Tallapragada, P., Dikshit, R., Seshagiri, S., Influence of Rhizophagus spp. and Burkholderia seminalis on the growth of tomato (Lycopersicon esculatum) and bell pepper (Capsicum annuum) under drought stress. *Commun. Soil Sci. Plant Anal.*, 47, 17, 1975–1984, 2016.
54. Tarigan, S.I., Toth, S., Szalai, M., Kiss, J., Turoczi, G., Toepfer, S., Biological control properties of microbial plant biostimulants. A review. *Biocontrol Sci. Technol.*, 32, 12, 1351–1371, 2022.
55. Tewes, F., Bahamondez-Canas, T.F., Smyth, H.D., Efficacy of ciprofloxacin and its copper complex against Pseudomonas aeruginosa biofilms. *AAPS PharmSciTech*, 20, 5, 205, 2019.
56. Timofeeva, A.M., Galyamova, M.R., Sedykh, S.E., Plant growth-promoting soil bacteria: Nitrogen fixation, phosphate solubilization, siderophore production, and other biological activities. *Plants*, 12, 24, 4074, 2023.
57. Wang, Y., Li, T., Wu, A., Li, Y., Zhang, N., Mycorrhizal benefits of salt-stressed Cinnamomum camphora (L.) Presl. may be related to P and Mn2+ contents in shoots, biomass allocation, and K+/Na+ in roots and shoots. *Forests*, 13, 11, 1882, 2022.

58. Xun, F., Xie, B., Liu, S., Guo, C., Effect of plant growth-promoting bacteria (PGPR) and arbuscular mycorrhizal fungi (AMF) inoculation on oats in saline-alkali soil contaminated by petroleum to enhance phytoremediation. *Environ. Sci. Pollut. Res.*, 22, 1, 598–608, 2015.
59. Yuan, Y., Liu, D., Huang, X., Wang, S., Qiu, R., Zhang, Z., Ming, J., Effect of Enterobacter sp. EG16 on Selenium biofortification and speciation in pak choi (Brassica rapa ssp. chinensis). *Sci. Hortic.*, 310, 111723, 2023.

12

Challenges and Opportunities in the Commercialization of Siderophore-Based Products

Siya Sharma[1], Prerna Sutar[1], Bipin Sati[2], Durgesh Pant[2] and Subhajit Basu[1]*

[1]*School of Health Science and Technology (SoHST), UPES University, Dehradun, Uttarakhand, India*
[2]*Uttarakhand Council of Science and Technology (UCOST), Vigyna Dham, Jhajra, Premnagar, Dehradun, Uttarakhand, India*

Abstract

In response to the limitation of bioavailable iron (Fe), many microbes secrete diverse low-molecular Fe chelating organic molecules known as siderophores as part of high-affinity uptake systems. Siderophores hold promising potentials for applications in various sectors of agriculture, medicine, and metal bioremediation. Of particular importance are the well-known roles of various siderophores for plant growth promotion and bio-control by increasing the bioavailability of Fe in soil and improving the antagonistic activity against plant pathogens, thereby reducing the dependency on chemical fertilizers and pesticides in agricultural as well as medicinal sectors. However, the commercialization of products based on siderophores still faces multiple challenges for large-scale production, such as the high cost of product purification, issues in stability and viability in field applications, and other regulatory-related issues. Despite all of these barriers, opportunities for the utilization of siderophores as a product are emerging through advancements in synthetic biology, formulation technology, and advanced bioengineering. Additionally, growing consumer demands for sustainable and more environment-friendly agriculture offers a viable market for bio-fertilizer and bio-pesticide formulations containing siderophores. This paper discusses the applications of siderophores and challenges faced during the commercialization of siderophores and explores the strategies for improving applications in the field of agriculture and medicine.

**Corresponding author*: subhajit.basu@ddn.upes.ac.in

Debasis Mitra, Ayush Madan and Anju Rani (eds.) Microbial Siderophores for Agricultural Productivity: Perspectives of Use in Agriculture, (303–318) © 2026 Scrivener Publishing LLC

Keywords: Siderophores, commercialization, antagonistic activity, bioengineering

12.1 Introduction

Iron (Fe) is an essential trace nutrient required for diverse cellular functions, like photosynthesis, respiration, chlorophyll production, nitrogen fixation, processing of reactive oxygen species, nutrient acquisition, and synthesis of various biomolecules [24]. pH plays an important role in governing the redox speciation and bioavailability of Fe [43]. At a slightly alkaline pH, the more soluble Fe(II) rapidly oxidizes within minutes to the thermodynamically stable Fe(III) state and precipitates as Fe-oxyhydroxide minerals. This limits the bioavailability of Fe as cells can only uptake Fe in soluble or dissolved forms. One of the most prevalent strategies evolved by microorganisms like bacteria, fungi, and also plants is the synthesis of siderophores. Siderophores are diverse low molecular weight organic chelators ranging from 400 to 2,000 Da with a high affinity for Fe (III). Bacterial cells sense Fe limitation *via* the Fur (ferric uptake regulator) and upregulate the synthesis of diverse siderophores—employing either NRS (non-ribosomal peptidal synthetase) or NRPS (NRS-independent non-ribosomal peptidal synthetases) pathways [18]. The production of siderophores has been reported to be controlled by quorum sensing [27]. The production of siderophores relies on the action of repressor proteins, which, in the presence of adequate cytoplasmic Fe, are responsible for inhibiting the gene expression involved in the biosynthesis of siderophores. In Gram-positive and Gram-negative bacteria with inadequate amounts of guanine (G) and cytosine (C), Fur is a primary Fe-dependent regulator. The Fur uses Fe (II) as a cofactor linking to the specific DNA operator and inhibits the transcription of the target gene when Fe levels are elevated. Siderophores are secreted by cells to scavenge Fe from environments when the organism recognizes an insufficient amount of intracellular Fe. Microorganisms can produce diverse siderophores bearing hydroxamates, catecholates, carboxylates, and mixed functional groups [1]. Many siderophores can also be photoactive, which leads to the dissociation of the Fe–siderophore complex when exposed to visible light [2, 8]. Of about 500 siderophores characterized to date, majority have been identified from terrestrial systems and only handfuls have been characterized from marine systems [42]. In general, the Fe(III) binding groups of siderophores act as lone electron pair donating Lewis bases (containing oxygen and/or nitrogen) to coordinate Fe(III) and can form strong octahedral hexadentate, tetradentate, and tridentate to even weak bidentate Fe–siderophore complexes. The commonly detected functional groups of strong octahedral Fe–siderophore complexes are the tris-catecholates (enterobactins, vibriobactins, corynebactins, etc.), tris-hydroxamates (e.g., desferrioxamines, amphibactins, ferrichromes, etc.), tris-α-hydroxycarboxylates (e.g., rhizobactins, achromobactins, etc.), and mixed functional groups containing hydroxamates or catecholates along with most hydroxycarboxylates [38].

Siderophores can be used to detoxify heavy-metal-contaminated materials, which explains their function in the aspect of bioremediation. Siderophores can identify the amount of Fe present in various settings, which presents their capability to be incorporated in a biosensor. In the medical industry, siderophores help deliver antibiotics to bacteria that are resistant to them by forming complexes with them using the "Trojan horse strategy" [15]. Siderophores have also been used in medicine for their antimalarial and anticancer properties, and apart from their pharmaceutical uses, they are also used in the food industry [1]. This chapter, therefore, aims at highlighting how various siderophores are used in sectors of agriculture, pharmaceutical, and bioremediation sites and identifying the challenges faced in the commercialization of the potential application of siderophores as shown in Figure 12.1.

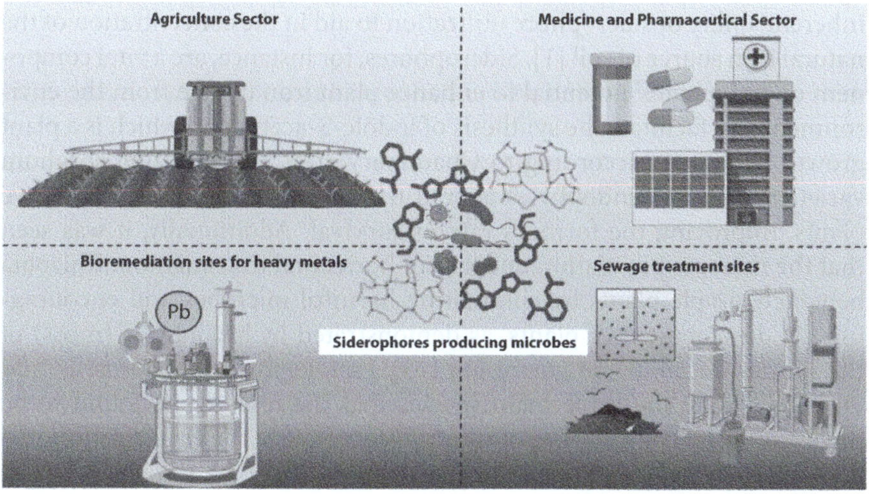

Figure 12.1 Graphical representation of the potential uses of siderophores in different fields like agriculture sector, medicine and pharmaceuticals, bioremediation purposes, and for sewage treatment.

12.2 Siderophores: Opportunities and Challenges in Commercialization

12.2.1 Role and Applications of Siderophores

12.2.1.1 Agriculture

In the agricultural sector, siderophores are showing their potential by promoting the growth and development in plants. They do this by improving their ability to uptake Fe from the environment and reduce its reliance on chemical-based fertilizers. Along with this, these small molecules help in the cultivation of disease-resilient crops through antagonistic activities [36]. Several microorganisms are releasing these siderophores that are helping in the synthesis of phytohormones and organic acids and solubilization of phosphates from the environment [29]. This multi-component system is defining the adequate functioning of siderophores in cultivation and yield of plants. Numerous research are investigating the interaction between microbes and plants along with the understanding that siderophores interact with plants in iron-deficient conditions [1]. These molecules function as natural biofertilizers and biopesticides, promoting sustainable farming practices [5]. By facilitating nutrient availability, siderophores improve crop yields and soil health, addressing the challenges posed by conventional agricultural inputs. Various facets of siderophore and iron usage for agriculture have been studied with the primary hypothesis focusing on the inherent ability of siderophore utilization to aid in the concentration of the natural iron source of soil [1]. Siderophores, for instance, are a vital component of endophytes' potential to enhance plant iron uptake from the environment and facilitate the synthesis of indole-3-acetic acid which is a plant growth hormone. According to Chaudhary *et al.* [9], invading sorghum varieties' bacterial endophytes altered the rhizobia community of native plants, improving the former's rate of survival. Additionally, it was seen that the rhizome population underwent horizontal transmission. Rhizobia benefited simple plants by eliminating harmful microbes and encouraging the development of plants, as demonstrated by bryophytes (moss) of the species *Racomitrium japonicum* [13], *Pseudomonas, Rhodococcus,* and *Duganella* were the three main species that the researchers found to be involved in the composition of moss rhizobia [40]. The investigation validated the formation of rhizobia siderophores. Rhizobia have been investigated for their possible use as biofertilizers, offering a viable means of enhancing moss growth and establishing a green roof industry [5]. Certain siderophores may be harmful to the host in bacterial infections in a clinical environment, but they may also be helpful in other situations. The creation of the siderophore pyoverdine [26] by *Pseudomonas fluorescens*, which is linked to the microbe's clinical virulence and pathogenicity, serves as an illustration of this claim. By contrasting Fe(II) iron with pyoverdine administration, researchers observed a substantial accumulation of Fe(III) iron in the leaves and roots [20]. Additionally, plants treated with the siderophore plus Fe(III) iron grew more than those supplemented with Fe(III) ion alone, as seen in the increase in plant weight and leaves. Reduced expression was found in molecular biology testing for iron-regulated transporter genes and Fe(III)-chelate reductase [16]. The production of pigment (cartenoids, anthocyanins, and chlorophyll) returned to the levels of plants treated with Fe(II) iron. Pyoverdine, which is produced by *Pseudomonas*, may therefore be utilized as a component of biofertilizer [39]. In iron-deficient environments, pyoverdine, which shows high affinity for attaching Fe(III) ions, is remarkably essential for scavenging iron and promoting bacterial cell growth [12]. As pyoverdine promotes bacterial growth within a biofilm, this increases the production of pyocyanin which is involved in stable biofilm formation. Additionally, the oxidative damage caused by pyocyanin to adjacent bacterial and fungal cells raises the Fe(III) ion availability, making it easier for pyoverdine to absorb iron. However, when pyocyanin is reduced, it converts Fe(III) ions to Fe(II) ions, which prevent pyoverdine from absorbing iron [23].

12.2.1.2 Food Industry

Catechols, one of the categories into which siderophores can be categorized, are produced by enteric bacteria such as *Escherichia coli*, *Aerobacter*, and *Salmonella*, where *Salmonella* is synthesized predominantly by yeast and fungi [11]. To produce foodstuffs like cheese and other products, aerobic and facultative anaerobic microorganisms are used. As mentioned in Table 12.1, the fungi *Penicillium roqueforti* and *Penicillium camemberti* are used for the ripening of blue and camembert-type cheese [31]. Different kinds of condiments and beverages are also produced by taking microbial siderophores into account. Studies show that the iron content in cheese ranges between 2 to 10 ppm, and there is also mention of lactoferrin, which is an iron-binding protein present in milk and milk products that denies bacteria iron for their survival [22]. The beverage sake has been found to

Table 12.1 Applications of diverse siderophores in various fields of medicine, agriculture, and bioremediation.

Application	Siderophore	Target	Use	References
Medical and pharmaceutical	Enterobactin analog—ciprofloxacin derivative, trimethoprim	*Pseudomonas aeruginosa, Escherichia coli*	Trojan horse	[20]
	Enterobactin and salmochelin conjugated to cholera toxin subunit B	Adherent-invasive *E. coli*	Immune stimulation	[45]
	Desferrioxamine B	MCF-7 (human breast), K 562 (human leukemia)	Reduction in proliferation and tumour growth	[41]
	Desferrioxamine B	*Plasmodium falciparum*	Antimalarial	[7]
Agriculture	Alginate preparation	*Pseudomonas fluorescens, Azospirillum brasilense*	Biocontrol, bio-fertilizing and bio-stimulating agent	[46]
	Consortium consisting of *Trichoderma harzianum, Epicoccum* spp., *Bacillus megatherium* and *B. amyloliquefaciens*	*Cochliobolus sativus complex, Alternaria alternata, and Fusarium graminearum*	Control of black spot in the caryopsis of wheat	[47]
	Combination of *Trichoderma harzianum* and *Pseudomonas fluorescens*	*Magnaporthe oryzae*	Biocontrol of rice blast	[48]
Food industry	Incorporated by *Penicillium roqueforti, Penicillium camemberti*	-	Ripen the blue and camembert-type cheese	[31]

(*Continued*)

Table 12.1 Applications of diverse siderophores in various fields of medicine, agriculture, and bioremediation. (*Continued*)

Application	Siderophore	Target	Use	References
	Incorporated by *Aspergillus oryzae, Rhizopus*	-	Oriental beverages, condiments, and sauces prepared from beans and rice	[49]
Bioremediation	Incorporated in Nitratireductor kimnyeongensis	-	Reduction in the pollutants [chromium (85%), sulphate (63%), chemical oxygen demand (COD) (69%), phosphate (76%)]	[50]
	Azotochelin and azotobactin produced by *Azotobacter vinelandii*	-	Used for Mo and V acquisition	[4]
	Pyochelin produced by *Pseudomonas aeruginosa*	-	Could chelate a wide range of metals like Ag+, Al_3^+, Cd2$^+$, Co_2^+, Cr_2^+, Cu^{2+}, Eu^{3+}, Ga_3^+, Hg_2^+, Mn_2^+, Ni_2^+, Pb_2^+, Sn_2^+, Tb^{3+}, Tl^+, and Zn_2^+	[51]

constitute several members of the Fe(III)chrome family [10]. In addition to this, *P. camemberti* is known to produce the siderophore coprogen when grown in a medium which is iron-deficient [11].

12.2.1.3 Bioremediation

Environmental contamination has been the biggest outcome of the present-day scenario of industrialized economy that includes the synthesis of products in industries, sludge management, radiology, nuclear power stations, and mining areas which are causing heavy metal pollution [3].

Heavy metals and metalloids (Cd, Cu, Ni, Co, Zn, Pb, and As), petroleum hydrocarbons, pesticides, halogenated hydrocarbons, and solvents are among the many pollutants [44]. Because metals and metalloids persist in the environment, it makes the remediation of these substances still a difficult task. Traditional *in situ* physicochemical remediation methods typically have impacted negatively on the subsurface environment and are quite costly [21]. In the last few years, biological techniques have been developed to efficiently and economically recover contaminated habitats. Metals like Cu, Co, Zn, or Cr are necessary nutrients for several physiological and biochemical processes, yet they can be harmful to bacteria at differing degrees. However, other metals, like Cd, Pd, and Hg, are generally harmful [17]. Microorganisms (MOs) have evolved defense mechanisms against metal toxicity, including metal reduction, decreased cell permeability, extracellular sequestration (metal precipitation), biosorption, and complexation, in response to prolonged exposure to metal pollution. It has been demonstrated that siderophores—organic molecules made by plants and MOs—are crucial to the detoxification of metals [35]. While increasing Fe (III) bioavailability is the main function of siderophores, their varied structure makes it evident that they are not selective to a single metal ion.

Bacteria synthesizing these small iron chelating molecules are showing potentials in the phytoremediation of heavy metal toxicity in soils [33]. Along with this, siderophores are serving as redox-cycling compounds that are mediating the production of reactive oxygen species and consecutively oxidizing organic compounds. Exploring the possibilities of engaging siderophores in the bioremediation of metals has gained attention. In highly acidic conditions, it was proven that siderophore molecules are capable of chelating Co (III) more effectively than Fe (III) by desferrioxamine B [1]. Siderophores like azotochelin and azotobactin were secreted by *Azotobacter vinelandii* and could be used for both Mo and V acquisition [4]. Along with this, a variety of heavy metals can be chelated by the siderophore released by *Pseudomonas aeruginosa* known as pyochelin, although no metal other than Fe(III) was shown to be assimilated during the absorption process. Additionally, it was observed that siderophores that are formed from *Fusarium solani* aided in the solubilization of copper and zinc under experimental circumstances. Siderophores were also essential for facilitating the extraction of metals from mining debris or metal-contaminated soils [6]. Numerous metals (which incorporates nickel, cobalt, and other metals plus iron) were released from the acid-leached ore waste of a former uranium mine where *P. fluorescens*-produced siderophores were present [1]. It has been demonstrated that *Agrobacterium radiobacter*'s siderophores eliminated almost 54% of As from a soil that was contaminated with metals. Pyoverdines have recently been discovered to mobilize U(VI), Np(V), and other metals from the waste of uranium mines [1]. Siderophores improve the microbe remediation of hazardous locations in an environmental bioremediation process by contributing to detoxify heavy metals. They work well as agents to restore the quality of the environment because of how well they can form complexes with detrimental metals.

12.2.1.4 Medicine and Pharmacology

Beyond agriculture, siderophores hold promise in medicine for their role in combating infections by depriving pathogenic microbes of essential iron. According to Miethke and Marahiel [28], siderophores are an essential component of microbial pathogenicity and are produced by Gram-positive as well as Gram-negative bacteria in anticipation of Fe scarcity. A significant element of the Trojan horse strategy is the way siderophores are utilized to cure illnesses like tumors and give pharmaceuticals to bacteria that are immune to antibiotics. In this technique, siderophores are usually useful for avoiding membrane-associated drug resistance [15]. The antimicrobial monomer utilizes the Fe–siderophore harvesting approach to interact with the siderophore through a "Trojan horse" transportation and go through the microbial cell. A linker typically appears in conjugated pharmaceuticals to attach the drug to the siderophore and allow for regulated pharmacological or biochemical release of the drug into an intended bacterial cell. It has been discovered that certain siderophores are effective in treating *Plasmodium falciparum* which causes malaria. *Streptomyces pilosus* produces desferrioxamine B, which is now also made by chemical synthesis both *in vitro* and *in vivo*, against *P. falciparum* [7]. Intracellular iron depletion is brought on by the agent's entry into the parasite. The activity of this siderophores is further enhanced by the employment of nalidixic acid as a conjugative agent against *P. falciparum* that is resistant to multiple drugs [28]. This conjugate's mechanism of action is oxidative DNA damage catalyzed by a metal. Under the trade name Fetroja®, the first siderophore cephalosporin was an authorized drug that was approved by the Food and Drug Administration (FDA) for use in people according to Bellotti *et al.* [41]. According to Niu *et al.* [30], cefiderocol utilizes the iron transport system to successfully combat bacteria by absorbing the catechol component with extracellular iron. Cefiderocol has been effective against clinically important strains producing a wide range of carbapenemases. The beneficial effects of cefiderocol, ceftazimide–avibactam, and meropenem–vaborbactam against severe enough infrequent infections caused by microorganisms that achieve carbapenemases have been identified [30].

They also serve as drug delivery agents and diagnostics tools due to their high specificity and affinity for iron.

12.3 Challenges and Strategies for Siderophore Commercialization

The commercialization of siderophores offer potential applications in medicine, agriculture, and environmental bio-remediation. However, there are several challenges that impose the widespread use of industrial adoption [32]. The synthesis of siderophores often depends on bio-fermentation, ranging from complex and resource-intensive research. The high yield and purity of isolates require optimization, selection, adequate growth conditions, and selective growth media for extensive research and commercialization [19]. The misuse of chemical-based pesticides and fertilizers in intensive agriculture has rendered the soil contaminated. The environment and human health is also at risk by the synthesis of chemicals and handling and disposal of bio-wastes from the agricultural sector. Therefore, finding substitutes for chemical fertilizers and managing agricultural waste is imperative given the current state of rising food demand. A viable choice is the valorization of plant residues to create natural bio-fertilizers through microbiological treatments. This makes the conversion of such residues into simple molecules that can be utilized by the plants and promotes plant growth [34]. The regulatory approvals and documentation process required for the commercialization of siderophore-based products is stringent especially in sectors like agriculture and pharmaceuticals that are associated with safety, efficacy, and stability which require field trials as well as pre-clinical and clinical trials [14]. In this case, agriculture-associated products need to go through proper field trials as there can be an impending concern over potential environmental impacts, toxicity to non-target organisms, and long-term ecological effects. Apart from these, there is the economic-wise competition that siderophore-based solutions face from already marketed synthetic metal chelators [25]. All of these alternatives can have low production costs and are already established in the market.

To overcome these challenges, there is a need for the utilization of synthetic biology and metabolic engineering to design microbial strains that can synthesize siderophores at high production yields and lesser cost compared to already marketed products [37]. Microbes that are genetically modified can also be used to enhance the uptake of iron through different pathways or overexpress siderophore-related genes. The exploration of non-fermentative synthetic routes could also be engaged. The production of optimized protocol for safety, efficacy, and impact on the environment can be streamlined through different experiments and trials and further going through regulatory approvals. The enhancements of the stability of siderophore-based products can be done by their synthesis through formulation by encapsulating or co-formulations with stabilizers to protect the siderophores from environmental damage.

12.4 Future Opportunities and Conclusion

The high demand of consumers for sustainable and organic solutions presents a significant market for siderophore-based commercialized products. The products associated with agriculture as biofertilizers and biopesticides will reduce the reliance on chemical-based fertilizers and their toxic impact on the environment [5]. Similarly, their applications in medicines and bioremediation are offering a high potential for further innovation and development. Emerging new technologies and development like AI-based optimized microbial understanding systems and precise agricultural tools can further promote the scope of siderophore utilization through a better understanding of microbial siderophore synthesizing pathways, help in precise farming, and enhance the utilization in medical industries [25]. Collaborative research between different industries from academic institutes to industries and policymakers are essential to unlock their full potential.

Siderophores represent a promising potential for sustainable development in different sectors of agriculture, medicine, and environment management. While issues are related with production, stability, and regulatory affairs and advancements in artificially synthesized synthetic products, bioengineering and formulation technologies are creating different pathways for their better commercialization. The growth in awareness and demands of organic and eco-friendly solutions will open wide the pathway for siderophores to become a sustainable practice in the near future.

AI disclosure statement: The author acknowledges the use of AI-based tools for minor language editing or grammatical correction. However, all intellectual contributions, analyses, and conclusions presented in this chapter are entirely the author's own.

References

1. Ahmed, E. and Holmström, S.J.M., Siderophores in environmental research: roles and applications. *Microb. Biotechnol.*, *7*, 3, 196–208, 2014, https://doi.org/10.1111/1751-7915.12117.
2. Amin, S.A., Green, D.H., Gärdes, A., Romano, A., Trimble, L., Carrano, C.J., Siderophore-mediated iron uptake in two clades of Marinobacter spp. associated with phytoplankton: the role of light. *BioMetals*, *25*, 1, 181–192, 2012, https://doi.org/10.1007/s10534-011-9495-5.
3. Arjoon, A., Olaniran, A.O., Pillay, B., Co-contamination of water with chlorinated hydrocarbons and heavy metals: challenges and current bioremediation strategies. *Int. J. Environ. Sci. Technol.*, *10*, 2, 395–412, 2013, https://doi.org/10.1007/s13762-012-0122-y.
4. Baars, O., Zhang, X., Morel, F.M.M., Seyedsayamdost, M.R., The Siderophore Metabolome of Azotobacter vinelandii. *Appl. Environ. Microbiol.*, *82*, 1, 27–39, 2016, https://doi.org/10.1128/AEM.03160-15.
5. Bhardwaj, D., Ansari, M.W., Sahoo, R.K., Tuteja, N., Biofertilizers function as key player in sustainable agriculture by improving soil fertility, plant tolerance and crop productivity. *Microb. Cell Fact.*, *13*, 1, 66, 2014, https://doi.org/10.1186/1475-2859-13-66.
6. Braud, A., Hannauer, M., Mislin, G.L.A., Schalk, I.J., The Pseudomonas aeruginosa pyochelin-iron uptake pathway and its metal specificity. *J. Bacteriol.*, *191*, 11, 3517–3525, 2009, https://doi.org/10.1128/JB.00010-09.
7. Bunnag, D., Poltera, A.A., Viravan, C., Looareesuwan, S., Harinasuta, K.T., Schindléry, C., Plasmodicidal effect of desferrioxamine B in human vivax or falciparum malaria from Thailand. *Acta Trop.*, 52, 1, 59–67, 1992, https://doi.org/10.1016/0001-706X(92)90007-K.
8. Butler, A., Harder, T., Ostrowski, A.D., Carrano, C.J., Photoactive siderophores: Structure, function and biology. *J. Inorg. Biochem.*, *221*, 111457, 2021, https://doi.org/10.1016/j.jinorgbio.2021.111457.
9. Chaudhary, P., Agri, U., Chaudhary, A., Kumar, A., Kumar, G., Endophytes and their potential in biotic stress management and crop production. *Front. Microbiol.*, *13*, 933017, 2022, https://doi.org/10.3389/fmicb.2022.933017.
10. Collins, J.H., Kunyeit, L., Weintraub, S., Sharma, N., White, C., Haq, N., Anu-Appaiah, K.A., Rao, R.P., Young, E.M., Genetic basis for probiotic yeast phenotypes revealed by nanopore sequencing. *G3 Genes|Genomes|Genetics*, *13*, 8, jkad093, 2023, https://doi.org/10.1093/g3journal/jkad093.

11. Prabhakar P.K., Bacterial Siderophores and Their Potential Applications: A Review. *Current molecular pharmacology*, 13, 4, 295–305, 2020. https://doi.org/10.2174/1874467213666200518094445.
12. Durán, O., Ramos, C., Chen, O., Castillo, J., de Mayorga, B., de Chial, M., Pyoverdine as an important virulence factor in Pseudomonas aeruginosa antibiotic resistance, in: *The Global Antimicrobial Resistance Epidemic—Innovative Approaches and Cutting-Edge Solutions*, 2022.
13. Fahde, S., Boughribil, S., Sijilmassi, B., Amri, A., Rhizobia: A Promising Source of Plant Growth-Promoting Molecules and Their Non-Legume Interactions: Examining Applications and Mechanisms. *Agriculture, 13*, 7, 2023, https://doi.org/10.3390/agriculture13071279.
14. Gomes, A.F.R., Almeida, M.C., Sousa, E., Resende, D.I.S.P., Siderophores and metallophores: Metal complexation weapons to fight environmental pollution. *Sci. Total Environ.*, 932, 173044, 2024, https://doi.org/10.1016/j.scitotenv.2024.173044.
15. Górska, A., Sloderbach, A., Marszałł, M.P., Siderophore-drug complexes: potential medicinal applications of the "Trojan horse" strategy. *Trends Pharmacol. Sci.*, *35*, 9, 442–449, 2014, https://doi.org/10.1016/j.tips.2014.06.007.
16. Guan, J., Zhang, Z., Shi, G., Genome-Wide Identification of the Ferric Chelate Reductase (FRO) Gene Family in Peanut and Its Diploid Progenitors: Structure, Evolution, and Expression Profiles. *Plants*, *13*, 3, 2024, https://doi.org/10.3390/plants13030418.
17. Igiri, B.E., Okoduwa, S.I.R., Idoko, G.O., Akabuogu, E.P., Adeyi, A.O., Ejiogu, I.K., Toxicity and Bioremediation of Heavy Metals Contaminated Ecosystem from Tannery Wastewater: A Review. *J. Toxicol.*, *2018*, 2568038, 2018, https://doi.org/10.1155/2018/2568038.
18. Izoré, T., Candace Ho, Y.T., Kaczmarski, J.A., Gavriilidou, A., Chow, K.H., Steer, D.L., Goode, R.J.A., Schittenhelm, R.B., Tailhades, J., Tosin, M., Challis, G.L., Krenske, E.H., Ziemert, N., Jackson, C.J., Cryle, M.J., Structures of a non-ribosomal peptide synthetase condensation domain suggest the basis of substrate selectivity. *Nat. Commun.*, *12*, 1, 2511, 2021, https://doi.org/10.1038/s41467-021-22623-0.
19. Kamińska, K., Mular, A., Olshvang, E., Nolte, N.M., Kozłowski, H., Wojaczyńska, E., Gumienna-Kontecka, E., The diversity and utility of arylthiazoline and aryloxazoline siderophores: challenges of total synthesis. *RSC Adv.*, *12*, 39, 25284–25322, 2022, https://doi.org/10.1039/d2ra03841b.
20. Kang, D., Kirienko, D.R., Webster, P., Fisher, A.L., Kirienko, N.V., Pyoverdine, a siderophore from Pseudomonas aeruginosa, translocates into C. elegans, removes iron, and activates a distinct host response. *Virulence*, *9*, 1, 804–817, 2018, https://doi.org/10.1080/21505594.2018.1449508.
21. Kapahi, M. and Sachdeva, S., Bioremediation Options for Heavy Metal Pollution. *J. Health Pollut.*, *9*, 24, 191203, 2019, https://doi.org/10.5696/2156-9614-9.24.191203.

22. Kell, D.B., Heyden, E.L., Pretorius, E., The Biology of Lactoferrin, an Iron-Binding Protein That Can Help Defend Against Viruses and Bacteria. *Front. Immunol.*, *11*, 1221, 2020, https://doi.org/10.3389/fimmu.2020.01221.
23. Koley, D., Ramsey, M.M., Bard, A.J., Whiteley, M., Discovery of a biofilm electrocline using real-time 3D metabolite analysis. *Proc. Natl. Acad. Sci.*, *108*, 50, 19996–20001, 2011.
24. Kraemer, S.M., Duckworth, O.W., Harrington, J.M., Schenkeveld, W.D.C., Metallophores and Trace Metal Biogeochemistry. *Aquat. Geochem.*, *21*, 2, 159–195, 2015, https://doi.org/10.1007/s10498-014-9246-7.
25. Kurth, C., Kage, H., Nett, M., Siderophores as molecular tools in medical and environmental applications. *Org. Biomol. Chem.*, *14*, 35, 8212–8227, 2016, https://doi.org/10.1039/C6OB01400C.
26. Lozano-González, J.M., Zuluaga, M.Y.A., Lucena, J.J., López-Rayo, S., Monterisi, S., Cesco, S., Pii, Y., Cultivating resilience: Harnessing pyoverdine-producing Pseudomonas to contrast iron deficiency in cucumber plants. *Plant Stress*, *14*, 100565, 2024, https://doi.org/https://doi.org/10.1016/j.stress.2024.100565.
27. McRose, D.L., Baars, O., Seyedsayamdost, M.R., Morel, F.M.M., Quorum sensing and iron regulate a two-for-one siderophore gene cluster *in Vibrio* harveyi. *Proc. Natl. Acad. Sci.*, *115*, 29, 7581–7586, 2018, https://doi.org/10.1073/pnas.1805791115.
28. Miethke, M. and Marahiel, M.A., Siderophore-based iron acquisition and pathogen control. *Microbiol. Mol. Biol. Rev. MMBR*, *71*, 3, 413–451, 2007, https://doi.org/10.1128/MMBR.00012-07.
29. Molnár, Z., Solomon, W., Mutum, L., Janda, T., Understanding the Mechanisms of Fe Deficiency in the Rhizosphere to Promote Plant Resilience. *Plants*, *12*, 10, 2023, https://doi.org/10.3390/plants12101945.
30. Niu, Z.-X., Wang, Y.-T., Zhang, S.-N., Li, Y., Chen, X.-B., Wang, S.-Q., Liu, H.-M., Application and synthesis of thiazole ring in clinically approved drugs. *Eur. J. Med. Chem.*, *250*, 115172, 2023, https://doi.org/10.1016/j.ejmech.2023.115172.
31. Ong, S.A. and Neilands, J.B., Siderophores in microbially processed cheese. *J. Agric. Food. Chem.*, *27*, 5, 990–995, 1979, https://doi.org/10.1021/jf60225a033.
32. Passari, A.K., Ruiz-Villafán, B., Cruz-Bautista, R., Díaz-Domínguez, V., Rodríguez-Sanoja, R., Sanchez, S., Opportunities and challenges of microbial siderophores in the medical field. *Appl. Microbiol. Biotechnol.*, *107*, 22, 6751–6759, 2023, https://doi.org/10.1007/s00253-023-12742-7.
33. Rajkumar, M., Ae, N., Prasad, M.N.V., Freitas, H., Potential of siderophore-producing bacteria for improving heavy metal phytoextraction. *Trends Biotechnol.*, *28*, 3, 142–149, 2010, https://doi.org/10.1016/j.tibtech.2009.12.002.
34. Robas Mora, M., Fernández Pastrana, V.M., Probanza Lobo, A., Jiménez Gómez, P.A., Valorization as a biofertilizer of an agricultural residue leachate: Metagenomic characterization and growth promotion test by PGPB in

the forage plant Medicago sativa (alfalfa). *Front. Microbiol.*, *13*, December, 1–14, 2022, https://doi.org/10.3389/fmicb.2022.1048154.

35. Roskova, Z., Skarohlid, R., McGachy, L., Siderophores: an alternative bioremediation strategy? *Sci. Total Environ.*, *819*, 153144, 2022, https://doi.org/https://doi.org/10.1016/j.scitotenv.2022.153144.
36. Saeed, Q., Xiukang, W., Haider, F.U., Kučerik, J., Mumtaz, M.Z., Holatko, J., Naseem, M., Kintl, A., Ejaz, M., Naveed, M., Brtnicky, M., Mustafa, A., Rhizosphere Bacteria in Plant Growth Promotion, Biocontrol, and Bioremediation of Contaminated Sites: A Comprehensive Review of Effects and Mechanisms. *Int. J. Mol. Sci.*, *22*, 19, 2021, https://doi.org/10.3390/ijms221910529.
37. Saha, M., Sarkar, S., Sarkar, B., Sharma, B.K., Bhattacharjee, S., Tribedi, P., Microbial siderophores and their potential applications: a review. *Environ. Sci. Pollut. Res.*, *23*, 3984–3999, 2016.
38. Sandy, M. and Butler, A., Microbial iron acquisition: marine and terrestrial siderophores. *Chem. Rev.*, *109*, 10, 4580–4595, 2009, https://doi.org/10.1021/cr9002787.
39. Santos, C.S., Ozgur, R., Uzilday, B., Turkan, I., Roriz, M., Rangel, A.O.S.S., Carvalho, S.M.P., Vasconcelos, M.W., Understanding the Role of the Antioxidant System and the Tetrapyrrole Cycle in Iron Deficiency Chlorosis. *Plants*, *8*, 9, 2019, https://doi.org/10.3390/plants8090348.
40. De Serrano, L.O., Biotechnology of siderophores in high-impact scientific fields. *Biomol. Concepts*, 8, 3–4, 169–178, 2017. https://doi.org/10.1515/bmc-2017-0016
41. Bellotti, D. and Remelli, M., Deferoxamine B: A natural, excellent and versatile metal chelator. *Molecules*, 26, 11, 2021, https://doi.org/10.3390/molecules26113255.
42. Vraspir, J.M. and Butler, A., Chemistry of marine ligands and siderophores. *Annu. Rev. Mar. Science*, *1*, 43–63, 2009, https://doi.org/10.1146/annurev.marine.010908.163712.
43. Xie, B., Wei, X., Wan, C., Zhao, W., Song, R., Xin, S., Song, K., Exploring the Biological Pathways of Siderophores and Their Multidisciplinary Applications: A Comprehensive Review. *Molecules*, *29*, 10, 2024, https://doi.org/10.3390/molecules29102318.
44. Zhang, T., Lowry, G.V., Capiro, N.L., Chen, J., Chen, W., Chen, Y., Dionysiou, D.D., Elliott, D.W., Ghoshal, S., Hofmann, T., Hsu-Kim, H., Hughes, J., Jiang, C., Jiang, G., Jing, C., Kavanaugh, M., Li, Q., Liu, S., Ma, J., Alvarez, P.J.J., *In situ* remediation of subsurface contamination: opportunities and challenges for nanotechnology and advanced materials. *Environ. Sci.: Nano*, *6*, 5, 1283–1302, 2019, https://doi.org/10.1039/C9EN00143C.
45. Wang, H., Zeng, X., Mo, Y., He, B., Lin, H., Lin, J., Enterobactin-Specific Antibodies Induced by a Novel Enterobactin Conjugate Vaccine. *Appl. Environ. Microbiol.*, 85, 10, e00358–19, 2019, https://journals.asm.org/doi/abs/10.1128/aem.00358-19.

46. Prasad, A.A. and Babu, S., Compatibility of Azospirillum brasilense and Pseudomonas fluorescens in growth promotion of groundnut (Arachis hypogea L.). *An. Acad. Bras. Cienc.*, 89, 02, 1027–40, 2017.
47. Seenivasagan, R. and Babalola, O.O., Utilization of Microbial Consortia as Biofertilizers and Biopesticides for the Production of Feasible Agricultural Product. *Biology (Basel)*, 10, 11, 1111, 2021, https://www.mdpi.com/2079-7737/10/11/1111.
48. Prisa, D., Fresco, R., Spagnuolo, D., Microbial Biofertilisers in Plant Production and Resistance: A Review. *Agriculture*, 13, 9, 1666, 2023, https://www.mdpi.com/2077-0472/13/9/1666.
49. Abd Razak, D.L., Abd Rashid, N.Y., Jamaluddin, A., Sharifudin, S.A., Abd Kahar, A., Long, K., Cosmeceutical potentials and bioactive compounds of rice bran fermented with single and mix culture of Aspergillus oryzae and Rhizopus oryzae. *J. Saudi Soc. Agric. Sci.* [Internet]. 16, 2, 27–34, 2017. Available from: https://www.sciencedirect.com/science/article/pii/S1658077X15000168
50. Vijayaraj, A.S., Mohandass, C., Joshi, D., Microremediation of Tannery Wastewater by Siderophore Producing Marine Bacteria. *Environ. Technol.*, 41, 27, 3619–3632, 2020, https://doi.org/10.1080/09593330.2019.1615995.
51. Ghssein, G. and Ezzeddine, Z., A Review of Pseudomonas aeruginosa Metallophores: Pyoverdine, Pyochelin and Pseudopaline. *Biology (Basel)*, 11, 12, 1711, 2022, https://www.mdpi.com/2079-7737/11/12/1711.

13

Field Trials and Case Studies of Siderophore Applications in Crops

Priya Chugh[1]*, Rahul Kumar[2], Divya Gunsola[2] and Apaarna[2]

[1]*School of Agriculture and Technology, Maya Devi University, Dehradun, Uttarakhand, India*
[2]*Department of Microbiology, Graphic Era Deemed to be University, Dehradun, Uttarakhand, India*

Abstract

Siderophores are produced by microorganisms, chelate iron, and are known to aid plants in iron uptake and improving their growth. Iron deficiency leads modern agriculture to constraints, especially in alkaline and calcareous soils, affecting the productivity of the crop. Siderophore-producing bacteria (SPB) and fungi are certainly excellent alternatives from synthesized chemical fertilizers and provide a sustainable approach. These SPB and fungi solubilize iron and make it available to the plants. This study, which is on the application of microbial siderophores in plant nutrition, disease suppression, and sustainable agriculture, compiles various field trials indicating that siderophore-producing microorganisms could increase iron uptake, biomass production, and crop yield. In addition, it states that SPB promote plant health by competing with soil-borne pathogens for iron, thereby reducing disease incidence. Recent findings have emphasized their role in the alleviation of abiotic stresses such as drought and salinity through enhanced nutrient uptake and osmotic adjustment. Siderophore-based biofertilizers have been integrated into agricultural practices, enhancing soil fertility while reducing synthetic input utilization. The review accepts the mechanisms, case studies, and limitations in using siderophores in agriculture. The findings indicate that microbial siderophores are a promising avenue for sustainable agriculture, enhancing iron uptake, plant growth, and ecological farming approaches. Future studies should enhance agronomic application through strain selection, formulation stability, and large-scale field implementation.

**Corresponding author*: priyachugh056@gmail.com

Debasis Mitra, Ayush Madan and Anju Rani (eds.) Microbial Siderophores for Agricultural Productivity: Perspectives of Use in Agriculture, (319–342) © 2026 Scrivener Publishing LLC

***Keywords*:** Siderophore, crop improvement, field studies, sustainable agriculture

13.1 Introduction

The application of fertilizers is the basic requirement in traditional agriculture aimed at feeding the fast-increasing population of the world. Chemical fertilizers boost agricultural productivity; the excess has hardened the soil, reduced fertility, polluted air and water, and generated greenhouse gases, all of which jeopardize the environment and public health [1]. Hence, due to the adverse impact of synthetic fertilizers in combination with the incidence of plant diseases, there is a need for biological platforms to enhance plant productivity apart from chemical-based strategies [2]. Siderophore-producing bacterial and fungal strains have excellent potential to confer benefits regarding plant growth. These are the organic iron chelators of low molecular weight produced by microorganisms and function together with plants and microorganisms in improving plant growth by inducing the uptake of most important iron molecules [3]. Siderophores are low-molecular-weight substances that help microorganisms and plants extract iron from the environment to make it available to plants. Iron is an important micronutrient for chlorophyll synthesis, respiration, and enzymatic processes but is usually found in soils in insoluble forms that further lead to deficiencies obstructive to plant growth and yield [4]. Siderophore-producing bacteria (SPB) and fungi, on the one hand, can enhance iron and other essential minerals for plant nutrition and health given nutrient-limited conditions [5].

The agricultural application of siderophores got great attention due to its role in improving the acquisition of nutrient and growth of plants. Several field tests revealed that microorganisms producing siderophores were capable of improving the productivity of crops while reducing their dependence on chemical fertilizers and minimizing exposure to environmental stresses such as drought and salinity [6]. Studies further showed that the application of siderophore-producing bacteria enhances iron uptake and also triggers induced systemic resistance in plants, offering protection against pathogens from the soil [7].

Recent studies are emphasizing on new advances regarding the incorporation of siderophore-based biofertilizers in sustainable agricultural practices. These include field trials on cereals, legumes, and horticultural crops, which conclude that siderophore-producing microbes can highly improve the efficiency of iron uptake in plants and increase their biomass significantly [8]. Nevertheless, the difference in efficacy of microbes with different factors such as soil condition, crop species, and environment necessitates further research studies to optimize their application in the field.

The chapter examines the multiple field reports and case studies on the performance of siderophore-producing organisms in various cropping systems. This chapter thus intends to fill the field gap between laboratory research findings and real-world constraints in the application of sustainable agriculture by discussing different field trials and challenges.

13.2 Mechanism of Siderophore in Plant Nutrition

Iron, an essential micronutrient in plants, is important for chlorophyll synthesis, photosynthesis, respiration, and enzyme activation, but its availability to plant is challenging because it is insoluble in the soils, especially in alkaline and calcareous soils [3]. Siderophore increases the bioavailability of ferric iron (Fe^{3+}) forms through chelation. The process of iron acquisition from siderophores in plants involves a variety of steps including the production and secretion of siderophores, Fe^{3+} chelation, and the internal release of iron from the iron–siderophore complexes into the plant system [9].

13.2.1 Siderophore Synthesis and Secretion and Fe^{3+} Chelation

The first step involves the considerable synthesis and secretion of siderophores by microorganism or plant roots themselves. Among the various microorganism, *Pseudomonas*, *Bacillus*, and *Azospirillum* species produce siderophores that are released into the rhizosphere to bind with Fe^{3+} ions; few plant species and especially some graminaceous members like maize, rice, wheat, and barley can produce other types of siderophores known as phytosiderophores that are members of the mugineic acid family [10]. Such molecules are secreted to the soil to help in the mobilization of iron from insoluble complexes. When the siderophores chelate Fe^{3+}, the newly established ferric–siderophore complex becomes soluble and is then shuttled back to the root surface for uptake [11].

13.2.2 Uptake of Iron–Siderophore Complexes

Iron uptake mechanisms differ largely from plant species to species and with types of siderophores involved. There are mainly two strategies of iron acquisition in plants: strategy I, which means reduction-based uptake, and strategy II, which means chelation-based uptake. Furthermore, strategy I is utilized by non-graminaceous plants such as dicots and non-graminaceous monocots. It involves the acidification of the rhizosphere through proton extrusion, ferric iron reduction by means of a membrane-bound ferric reductase, and eventual transport of ferrous iron (Fe^{2+}) into root cells through an iron-regulated transporter (IRT1) [4]. Furthermore, siderophores act indirectly in the mechanism, as they make iron more soluble in the soil, thus increasing Fe^{3+} availability for reduction and uptake. On the contrary, strategy II is common in graminaceous species, wherein the phytosiderophores chelated Fe^{3+} to form ferric–phytosiderophore complexes taken up directly by root cells through specialized transporters such as Yellow Stripe1 (YS1) and its homologs [10].

13.2.3 Iron Release

Once inside the plant, iron must be released from the siderophore complex and distributed to various plant tissues. In strategy I plants, ferric iron is reduced to Fe^{2+} at the plasma membrane before transport, ensuring that iron enters metabolic pathways efficiently. Strategy II plants transport the entire ferric–phytosiderophore complex into root cells, where iron is released by enzymatic activity or pH change [12]. The released iron is then translocated *via* xylem transport to the leaves, where it contributes to necessary functions such as chloroplast development, nitrogen metabolism, and oxidative stress defense. Furthermore, iron gained from siderophores can further enhance plant growth, yield, and stress tolerance by alleviating the iron deficiency symptoms of interveinal chlorosis, stunted growth, and reduced activity of enzymatic functions [13].

Aside from the uptake of iron, the importance of siderophores in plant nutrition goes further. These molecules have the potential to increase the bioavailability of other trace nutrients like zinc, manganese, and copper, which are often deficient in the soil. Furthermore, rhizobacteria that produce siderophores can promote the growth of plants by competing with soil pathogens for iron; therefore, they are expected to decrease the proliferation of harmful microbes and, ultimately, strengthen plant health indirectly [14]. This interaction between the plant and microbes has particular relevance to sustainable agriculture, where such non-chemical measures, including bioinoculants and PGPR, improve soil fertility while reducing reliance on synthetic fertilizers. Data support that inoculation with siderophore-producing bacteria leads to enhanced performance of crops in iron-deficient soils, thus heralding a possible avenue for sustainable practice in agriculture [15].

13.3 Application of Siderophore: Field Trials

In many different agricultural environments, field trials have demonstrated the effectiveness of siderophore-producing bacteria for improved crop growth and yield. These trials provide ethnographical data supporting the application of such bacteria as biofertilizers and biocontrol agents. Below are the details of field trial and siderophore mediated responses.

13.3.1 Wheat

A study conducted under no-till-agriculture soil investigated five potential bacterial strains. These bacterial strains have been used as seed treatments for increasing growth and yield under low-nitrogen fertilizer levels in two wheat cultivars. The field trials were set up as split-plot designs with two nitrogen levels (0% and 35%) and bacterial inoculation levels compared to controls of 100% nitrogen, 0% nitrogen (uninoculated), and 35% nitrogen without bacterial treatment. The differences manifested as marked improvements in yield, ranging from 51% to 96.5% with respect to the uninoculated control and 24.7%-43.1% with respect to the full fertilization controls. Dry biomass production of shoot showed differences due to bacterial strain, wheat cultivar, and growing season. All bacterial strains were able to produce ethylene, siderophores, and indoleacetic acid (IAA) in the range of ethylene production from 28.4 to 51.3 nmol L^{-1} h^{-} and IAA levels from 2.1 to 14.6 mg mL^{-}. The regular yield gains under reduced nitrogen showed promise for using these diazotrophic bacteria as additional biofertilizer options in optimizing wheat production with reduced synthetic nitrogen inputs [16]. Iron (Fe) is an important micronutrient that promotes plant growth, yield, and quality. Nevertheless, staple crops such as wheat (*Triticum aestivum*) and soybean (*Glycine max*) often have low concentrations of Fe in their grains. Foliar applications of Fe-enriched bacteriosiderophores are proposed to enhance both Fe accumulation and plant growth. In a preliminary study with soybean inoculated with 20 bacterial strains, *Arthrobacter* sp. (low siderophore producer) and *Lysinibacillus fusiformis* (high siderophore producer) were selected according to their capacity for siderophore production and induced plant growth response. Field validation experiments in wheat and soybean involved a foliar application of bacteriosiderophores with or without Fe at the onset of flowering, with observations taken at 6 days post-application. Further studies revealed that *L. fusiformis* siderophores with Fe superiorly improved various plant parameters in biomass accumulation, leaf area, and plant tissue

Fe concentration over *Arthrobacter* sp. Bacterial siderophores fortified with Fe increased the grain yield in wheat and soybean by 45% and 28%, respectively, while the nature of Fe concentration was increased by 1.7-fold in soybean and 2.0-fold in wheat. Following the foliar application of Fe, partitioning was more efficient in wheat than in soybean. This study highlights for the first time that Fe-enriched foliar application of bacterio-siderophores proves to be an effective and economically viable agronomic method for the biofortification of Fe in crop plants [17]. Pyoverdine-type siderophore producers also act as iron-chelating agents that could have applications in agriculture. A study screened the bacterial isolates from different sites of Faisalabad for showing fluorescence, producing siderophores, and producing indole-acetic acid equivalents. The most efficient strains were then evaluated under field conditions with wheat and maize. Out of 15 isolates, seven showed significant activities for plant growth promotion. Efficient strains produced increases in grain yield by 24.2% and 20.2%, increases in plant height by 30.9% and 23.7%, and total grain weight increases of 25.3% and 13.4% for wheat and maize, respectively, over the control. It also showed increased numbers of grains per spike or cob. The analysis of grain iron content gave an increase of 67% in maize over the control. These results indicate that biofortification of cereal crops using fluorescent siderophore-producing microbes is an effective means of enhancing growth and development in plants through better nutrient solubilization and mobilization [18].

13.3.2 Potato

Dietary iron bioavailability is generally considered very deficient among those depending on a monotonous plant-based diet. Iron deficiency is a very common occurrence in developing countries like Pakistan, India, and Sri Lanka, inflicting huge health and economic consequences on the population, especially poor pregnancy outcomes, low school performance, and loss of productivity among students. The present study was designed to assess the consortium of iron-solubilizing and siderophore-producing rhizobacteria and the plant growth regulator l-tryptophan in enhancing plant growth and iron uptake in potato tubers. The results show that the application of the plant growth-promoting rhizobacteria (PGPR) *Acinetobacter calcoaceticus* and *Bacillus simplex* along with l-tryptophan increased the plant height, aboveground biomass, tuber characteristics, physiological attributes, and most importantly, iron uptake into potato tubers due to inoculation. PGPR and PGR treatments have increased potato yield by 65% and iron uptake in tubers by up to onefold compared with control plants. The findings have proven that PGPR act as a bio-fortifying agent helping plants to absorb very high amounts of iron under deficient environments. These findings support the view that microbial-assisted biofortification of potato tubers might relieve micronutrient deficiency in people, particularly in resource-constrained regions [19]. Bacillibactin, an iron-scavenging compound secreted by the plant growth-promoting rhizobacterium (*Bacillus subtilis*), exhibits *in vitro* inhibitory antifungal activity against *Alternaria porri* and *Fusarium equiseti*, which were isolated from diseased *Solanum lycopersicum* (tomato) and *Solanum melongena* (eggplant) plants. The study evaluated the antifungal activity of bacillibactin by means of plate assay, dry fungal biomass evaluation, cell and scanning electron microscopy for cellular deformities, and antioxidative enzyme assays as well as membrane damage through lipid peroxidation analysis. Quantification of the intracellular iron was done using inductively coupled plasma atomic emission spectroscopy (ICP-AES). The extent of the role played by bacillibactin in reducing pathogenesis was tested on edema caused by pathogens in detached leaves treated with the siderophore followed by an assay for the increased levels of reactive oxygen species (ROS), necrosis, and cell death. The results show great damage to hyphae for *A. porri* and *F. equiseti* because of bacillibactin treatment, lowered biomass accumulation by the fungi, increased levels of antioxidative enzymes, and damage to membranes, all indicative of inhibitory effects. The ICP-AES analysis indicated the increased uptake of iron by bacillibactin. The results revealed that the application of bacillibactin at 1,500 μg/mL in infected leaves essentially inhibited ROS accumulation, necrosis, and cell death, indicating the potential of bacillibactin as a biocontrol agent to control fungal phytopathogens in crops [20]. The study was performed in order to test bacterial siderophore effectiveness in the alleviation of iron deficiency symptoms in tomato plants by quantifying siderophore production from two bacterial strains and evaluating their effectiveness in diminishing chlorotic symptoms. The measurement of siderophore produced in MM9 was performed, followed by the selection of the best strain for biological assays. Iron-containing bacterial media or with bacterium (BS) and without (S) were treated on 12-week-old hydroponically grown tomato plants for iron deficiency. The controls represented are full Hoagland solution, iron-free Hoagland solution, and water. The double treatment was performed, 1 week apart, after which plant yield, chlorophyll content, and leaf nutrient status were evaluated at harvest. On the other hand, both treatments with siderophores significantly enhanced plant biomass, chlorophyll content, and iron uptake in comparison to the positive control, such that with full Hoagland solution, indicating that the process of iron acquisition by plants was effectively mediated by the activity of siderophores, with or without the bacteria involved. Significantly, Chryseobacterium C138-derived siderophores have proven highly effective in supplying iron to tomato plants; thus, they can be developed for the production of eco-friendly and cost-effective organic iron chelators [21]. The organism *Pseudomonas fluorescens* produces pyoverdine, a high-affinity siderophore for ferric ions, while certain plants such as *Solanum lycopersicum* are unable to synthesize or directly utilize it. The primary objective of this study was to assess the effects of pyoverdine-chelated ferric iron on tomato iron nutrition. The results indicated bioassimilators in which plant biomass and photosynthesis pigment concentrations for pyoverdine treatments were restored to levels comparable with ferrous iron treatments. Also, more iron accumulation occurred in leaves and roots with pyoverdine-chelated ferric iron than iron alone. When an expression analysis was performed, the results indicated that the SlFRO1 and SlIRT1 levels showed downregulation of 20% and 50% in leaves and roots, respectively, in comparison to treatment with ferric iron only, which was most probably due to a kind of negative feedback regulation. These results indicate that pyoverdine-assisted ferric iron uptake could improve iron availability in tomato plants, providing a viable strategy for enhancing iron acquisition in iron-deficient environments [22]. This work examined the antagonistic actions of two strains of *Pseudomonas fluorescens*, SPs9 and SPs20, against *Fusarium oxysporum* f.sp. *lycopersici*, the wilting agent on tomato. Both strains produced significant amounts of siderophore, with SPs9 and SPs20 producing 59.30 and 52.74 $\mu g\ mL^{-1}$, respectively. Their antifungal properties were confirmed by inhibition zone assays performed on deferrated siderophore succinate medium (SSM), in which the zone of inhibition was 3.4 cm for SPs9 and 3.8 cm for SPs20. The increased concentration of iron ions in SSM resulted in reduced inhibition zones in a dosage-dependent fashion, thus corroborating the view of the antagonism being *via* siderophore action. In *in vivo* experiments against wilt, SPs9 protected 100% of tomato seedlings from infection, while SPs20 protected 75%. SPs20 also significantly increased the fresh and dry plant weight, respectively. *Pseudomonas fluorescens* thus presents as a good option as a biocontrol agent by way of iron competition to control tomato wilting [23].

13.3.3 Oil Seed Crops

Pseudomonas aeruginosa GRC1 is an isolate of potato rhizosphere and can show several plant growth-promoting activities, including producing phytohormones and antibiotics. Hydroxamate-type siderophore production in an iron-deficient environment was found to be prolific in this isolate. The siderophore was purified and characterized for GRC1. The analysis indicated that this purified siderophore was most likely pyoverdine, based on the typical amino acid composition. In field trials, *P. aeruginosa* GRC1 increased the growth of *Brassica campestris* var Pusa Gold (Indian mustard). There were highly significant increases in root and shoot weight, length, grains per plant, and total grain yield [24]. Plant growth-promoting bacteria (PGPB) are a sustainable alternative to chemical fertilizers. However, the performance of the PGPB often suffers from the controlled environment to field conditions. Bacterial endophytes shielded from rhizospheric competition and environmental stressors are, therefore, promising candidates for improving crop productivity in field conditions. The screening of bacterial endophytes from rapeseed for identifying an effective inoculant for biofertilization was carried out *in vivo*, *in vitro*, and *in silico* to select *Pseudomonas brassicacearum* CDVBN10 as the final candidate. The isolate demonstrated multilayer plant-beneficial traits such as siderophore production, phosphorus solubilization, and cellulose synthesis and was found to promote early-stage seedling growth. Under field trial conditions without fertilizer application, pod numbers, pod dry weight, and shoot dry weight were significantly enhanced. Notably, the metagenomic analysis revealed that CDVBN10 inoculation did not have any disruptive effects on the root bacterial microbiome, thus maintaining the plant–microbiome homeostasis, which is vital for plant health. Thus, this study establishes that *P. brassicacearum* CDVBN10 is a potential biofertilizer for rapeseed yield enhancement under field conditions without chemical input, which is a great leap in the advancement of crop-specific PGP inoculants [25]. An investigation of the potential of *Pseudomonas fluorescens* strains and an endophytic microbial consortium as biofertilizers for *Brassica napus* began in greenhouse and in field trials. The bacterial strains selected, including the well-characterized rhizospheric *P. fluorescens* F113, have been identified in the past for plant growth-promoting (PGP) traits like siderophore production, ACC deaminase activity, and inorganic phosphate solubilization. The Kmr- and GFP-tagged strains used as a seed coat were shown to effectively colonize the rhizosphere and roots of *B. napus* under greenhouse conditions, significantly increasing plant biomass over the non-inoculated controls. The field trials were under standard North-Western European agronomic conditions by spraying the inoculations on winter oilseed rape (*B. napus* var. Compass) with either individual strains or the consortium. With this treatment, significant increases in crop height accompanied increased stem/leaf biomass and pod biomass, particularly in the consortium-treated plants. Trends were upward toward seed and oil yield because of inoculation; though given the conditions, these increases were not statistically significant. Additional field trials should be conducted to assess the consistency of biofertilizers in varying agronomic conditions for the optimization of sustainable crop enhancement [26]. Drought stress is a global problem affecting crop productivity; therefore, sustainable strategies such as microbial biostimulants arising to enhance tolerance to drought are of key importance. This study focuses on PGPRs in ameliorating drought stress in oilseed crops, namely, *Acinetobacter calcoaceticus* AC06 and *Bacillus amyloliquefaciens* BA01. The microbial isolates were probed for drought tolerance and several PGP traits, such as phosphate solubilization, indole-3-acetic acid, siderophores, hydrogen cyanide, ammonia, and exopolysaccharides. Under osmotic stress conditions, both isolates produced considerable quantities of osmolytes: proline with the amounts of 2.21 and 1.75 μg/mL; salicylic acid, 18.59 and 14.21 μg/mL; trehalose, 28.35 and 22.74 μg/mg FW; and glycine betaine, 11.35 and 7.74 mg/g, respectively. The effectiveness of both isolates was confirmed in groundnut (*Arachis hypogaea* L.) in mild (60% field capacity) and severe (40% field capacity) drought stresses. Improvement in osmotic adjustment due to the inoculation with these strains resulted in an upsurge in plant growth, biomass, photosynthetic pigments, relative water content, proline, and soluble sugars while downregulating the parameters indicating plant susceptibility, namely, electrolyte leakage and malondialdehyde (MDA) content. Stress-responsive antioxidant enzymes, namely, catalase (CAT), ascorbate peroxidase (APX), and superoxide dismutase (SOD), were upregulated, leading to elevated drought tolerance. Therefore, there is evidence that *Acinetobacter* sp. AC06 and *Bacillus* sp. BA01 are osmolyte-producing microbial biostimulants which exert metabolic and physiological alterations in groundnuts and represent promising candidates for drought resilience enhancement in oilseed crops [27].

13.3.4 Intercropping

Recent research into peanut/maize intercropping has demonstrated the root-associated microbiota's effects on iron nutrition, with increased iron uptake in peanut but not statistically significant in maize. Interestingly, intercropping reconfigures the microbiome, such that both plant species show convergence. A primary finding is pyoverdine—siderophore secreted by *Pseudomonas*, enhancing nutritional benefits toward iron under both the glasshouse and field conditions—the critical role of siderophore-producing *Pseudomonas* with respect to iron availability for intercropping systems, and this valuable information would be an important and significant contribution for optimizing future intercropping strategies toward better nutrition for plants [28]. Crop performance is determined by interactions between plants and is also dependent on the microbiome composed of pathogens and mutualists. Intercropping is famous for many interactions favorably affecting yield, whereas very little is known on the microbiome's role in mediating pathogen suppression. Here we study the microbe interaction in a maize/faba bean intercropping system and focus on *Fusarium oxysporum* inhibition in faba beans. Long-term field observations indicated that intercropping promoted the yield of the crops while reducing *F. oxysporum* gene copies within a range of 30%–84% and showed increased bacterial wealth and diversity as compared to monocropping. Changes observed in the microbial communities were correlated with lineage distributions pertaining to siderophore-producing *rhizobacteria* found in maize and Faba bean pathogenicity. More specifically, 61% of maize *rhizobacteria* were found to be capable of producing siderophores compared to 30% for faba bean. Maize was 71% higher in numbers than faba with respect to siderophore-producing rhizobacteria. Moreover, it is also discovered to have 33% higher relative abundance of genes for nonribosomal peptide synthetases (especially pytochelin), further corroborated by enhanced populations of siderophore-producing bacteria and concomitantly reduced *F. oxysporum* gene copies from the intercropped faba bean rhizosphere. Out of 95 isolates, *Pseudomonas* spp. (B004, B021) and *Bacillus* spp. (B005, B208) exhibited antagonistic activity toward *F. oxysporum*, where B005, identified as a key hub in the network, demonstrated prolific siderophore production. These results together indicate that intercropping is enhancing the bacterial diversity and network complexity in the system, resulting in improved pathogen suppression and faba bean resistance to *F. oxysporum*. This study has great agronomic implications, indicating that microbiome-centric modalities may be deployed to fine-tune intercropping practices for the sustainable production of crops [29]. The cyanobacteria that produce siderophores are believed to be important for alleviating plant stress and stimulating physiological and biochemical processes, especially in iron-deficient conditions. The present study explored the capacity of producing siderophores from *Arthrospira platensis*, *Pseudanabaena limnetica*, *Nostoc carneum*, and *Synechococcus mundulus* to find the best strain to facilitate the germination of seedling of *Zea mays* in an environment lacking iron. Among the above-studied strains, one, *Synechococcus mundulus*, showed the maximum percentage of siderophore production (78% ± 2%), whereas the other, *Nostoc carneum*, showed the minimum (24.67% ± 0.58%). Thus, *Synechococcus mundulus* was selected for further experimentation. Siderophore characterization further proved that the compound from *S. mundulus* is hydroxamate, with optimum biomass (387.11 mg/L) and maximum siderophore production (91.84%) having been attained in the course of response surface optimization. Structural analyses of the samples carried out by proton nuclear magnetic resonance (^{1}H NMR) and Fourier transform infrared (FT-IR) spectroscopy further confirmed the hydroxamate nature of the siderophore. Under those conditions, *S. mundulus* siderophore addition improves growth compared with untreated controls. Increased levels of chlorophyll a, total carbohydrates, and protein content in plants treated with siderophores indicate that they are taking up more iron, thus suggesting a possible use of *Synechococcus mundulus* siderophores as a bioinoculant for enhancing plant resistance to iron-deficient environments [30].

13.3.5 Legumes

The recent investigations have reported an interactive effect of seed inoculation with *Rhizobium* and foliar application of bacteriosiderophore on the growth and productivity of legumes including chickpea and soybean. *Rhizobium* sp. isolated from chickpea nodules were used for seed inoculation, whereas the foliar application was done using *Bacillus megaterium*, a good siderophore producer which was supplemented with either sulphur or inorganic iron during flowering stages, with or without additional inorganic iron supplementation. Experimental field studies were conducted to establish agronomic parameters for plant height, number of nodules formed, pod numbers formed, grain yield and nutrient uptake, and iron content at harvest in grain. The cCombined application of *Rhizobium* and bacteriosiderophore resulted in an increased iron accumulation with values of 1.44 for soybean and 4.07 for chickpea concerning their respective control and treatments. The combined application of siderophore increased iron with *Rhizobium* showed maximum plant growth, nodule formation, and grain yield, giving 2.16-fold increase in yield for chickpea and 1.6-fold for soybean. A cost-effective strategy to enhance iron biofortification in legume crops grown on alkaline calcareous soils is through foliar application of those phytosiderophore-inducing bacteria particularly with iron supplementation [31]. Nodules of legumes contain a wide pool of rhizobia and endophytic bacteria with potential for plant growth promotion, making them good candidates for biostimulant applications. A recent work performed a genome-wide analysis of *Phyllobacterium endophyticum* PEPV15 and Rhizobium *laguerreae* PEPV16, both isolated from *Phaseolus vulgaris* nodules, and revealed the presence of genes involved in quorum sensing (N-acyl-homoserine lactone biosynthesis) and biofilm formation (cellulose biosynthesis), processes highly relevant for plant colonization. Genes potentially responsible for various mechanisms of plant growth promotion, including phosphate solubilization, indole acetic acid, siderophore biosynthesis, and nitrogen fixation, were also present in both genomes. Field trials indicated that PEPV15 and PEPV16 inoculation can stimulate an increase in more than 20% in lettuce shoot and 10% in carrot root yields. These results underscore the capacity of genome mining to forecast the potential of the studied bacterial strains as crop inoculants and highlight its entire role in evolving new-generation and effective bacteria-based biostimulants for plants [32]. The plant growth-promoting rhizobacteria, *Pseudomonas fluorescens*, enhance plant development by producing a yellow-green, fluorescent siderophore that aids high-affinity iron transport. Siderophore production has a direct correlation with the ability of an isolate of *P. fluorescens* to grow in media deficient in iron. Thus, from different locations in the region of Chhattisgarh, *P. fluorescens* isolates were collected from rhizosphere and non-rhizosphere soils of caves, forests, fallow lands, and agricultural fields using King's B media under 260-nm wavelengths. Screening in media deficient in iron succinate confirmed that most of them produced a high quantity of siderophore. Among them, isolate "P3" from Pakhanjore produced the highest and was further studied for plant growth-promoting effects on rice, urd, mung, and arhar crops. The results yielded improved root length, with rice showing the most improvement of 68.45%, compared to urd, which registered 40.1%, mung at 33.71%, and arhar with 29.53%. Improvement was also observed in shoot length, with rice having the highest improvement at 68.93%, followed by urd at 43.42%, mung at 39.96%, and arhar at 19.45%. These findings indicate that *P. fluorescens* has the ability to colonize plant roots efficiently and promote plant growth while probably reducing disease incidence, thus offering great potential for sustainable agriculture under greenhouse conditions [33]. *Sclerotium rolfsii* is a fungal pathogen which has caused extensive damage to common beans and several other crops. Potential biocontrol agents have been revealed among the *Rhizobium* strains with reference to its fungal pathogenicity. The study analyzed the efficacy of 78 rhizobial isolates from common bean through dual culture assay, volatile compound analysis, and determination of IAA and siderophore production and by 16S rRNA sequencing. It was found that out of these, 33 strains exhibited antagonistic activity, whereby at least 16 strains inhibited fungal mycelial growth by at least 84%. A postulated observation showed a positive correlation between IAA production (up to 36.5 $\mu g\ mL^{-1}$) and inhibition of fungal mycelium. The strain SEMIA 460 produced mycelial inhibition of 45% by virtue of volatile compound production. In pot experiments, disease symptoms were effectively suppressed in such soils infested with *S. rolfsii*, following inoculation with the strains SEMIA 4032, 4077, 4088, 4080, 4085, and 439. Under field conditions, SEMIA 439 and 4088 decreased the disease incidence by 18.3% and 14.5%, respectively. These results highlight that *Rhizobium* strains would also be very useful as biological control agents for collar rot in common beans, of which the best were SEMIA 4032, 4077, 4088, 4080, 4085, and 439 [34]. Cobalt (Co) is one of such important elements in agriculture as it contributes as vitamin B_{12}; however, when accumulated above optimum limits due to natural and anthropogenic sources, it becomes toxic to the crop plants, causing reduced yield and biomass. The oxidative stress induced by Co alters the iron (Fe) homeostasis, resulting in reduced chlorophyll content and causing Fe deficiency. Co toxicity is an important issue, and this study investigates the promise of siderophore-producing microbes for mitigating Co stress while activating the mobilization of Co and Fe. A total of 250 bacterial isolates from agricultural and non-agricultural soils were screened out, and 28.8% of them were siderophore producers—the isolation of which is by tolerance to 20,000 ppm Co and other plant-beneficial traits like production of ACC deaminase, phosphate solubilization (39.33–142.67 μg/mL), siderophore production (96.82%–99.67%), and indole-3-acetic acid synthesis (15.15–70.55 μg/mL). A plate assay experiment under 200 mM Co stress pointed toward some of the promising four isolates (KSBTS12, SBTS12, CWTS5, and CWTS10), which further augmented the growth of black gram (*Vigna mungo* L.). When co-stressed at 2,000 ppm in pot experiments, further confirmation showed maximum increases in root (60.69%–174.24%) and shoot length (3.27%–143.96%) when compared to the controls. The induced plants exhibited an improved uptake of nitrogen (37.33–42.36 mg/g) and phosphorus (3.12–3.92 mg/g), higher chlorophyll content (7.60–22.97 mg/g), and increased soil siderophore levels (282.41%–331.53 %). Enhanced soil respiration activity was also seen by fluorescein diacetate hydrolysis (11.33–24.88 μg/g), dehydrogenase enzyme activity (109.76–197.26 μg/g), and alkaline phosphatase activity (631.53–918.20 μg/g). Therefore, *Bacillus subtilis* (CWTS 5) and *Bacillus albus* (CWTS 10) are expected to serve as efficient bioinoculants in alleviating Co stress as well as mobilizing Fe and Co within plants, thereby providing an agronomic solution toward crop stress resilience [35]. Chlorosis caused by deficiency in iron is one of those abiotic stresses which counteract the production of groundnuts specifically on calcareous and alkaline soils with a pH value ranging from 7.5 to 8.5. The study attempts to assess the efficacy of siderophore-producing bacterial (SPB) strains for improving iron uptake and thus the growth of plants. Three SPB superior strains, *Bacillus subtilis*

SID 25, *Bacillus licheniformis* SID 30, and *Ochrobactrum grignonense* SID 33, were isolated from calcareous groundnut rhizosphere and evaluated by multifunctional experiments at various plant growth stages. The salient feature of these SPB is that they harbor plant growth-promoting traits like phosphate and zinc solubilization, thus making the nutrient available. Results from greenhouse trials indicated that SPB inoculation significantly increased chlorophyll content and iron levels in roots and shoots, active iron content, and total biomass while alleviating chlorotic symptoms. The effectiveness of the various siderophores in iron acquisition depended on both the genotype of the plants and the strain of the bacterium. Results from field trials further confirmed these observations: SPB-inoculated plants outperformed controls in terms of pod yield by 12.6%–17.6% (*B. subtilis* SID 25), 12.3%–16.0% (*B. licheniformis* SID 30), and 7.9%–18.2% (*O. grignonense* SID 33). In addition, SPB inoculation enhanced the pod number (59.5%), kernel yield (10.9%), and kernel iron content (17.3%) compared to the control. Among the genotypes tested, VRI 8 and CO 7 were identified as being Fe deficiency tolerant, further signifying the capability of SPB strains in the alleviation of iron stress while promoting yield and nutritional quality in groundnut under calcareous stress conditions. This suggests that SPB could act as one of the environmentally feasible biofertilizer strategies directed at enhancing fewer-yielding crops in iron-deficient soils [36]. A study aimed at isolating effective siderophore-producing bacterial strains and testing their plant growth-promoting abilities under iron deficient environments. During observation with the two strains, *Brucella* sp. E7 and *Pseudomonas brassicae* W7, isolated from rhizosphere soil, both strains produced an optimum amount of siderophore under certain conditions. A series of experiments carried out on inoculation conditions revealed that the respective bacteria E7 and W7, when both are applied together, could also significantly enhance the number of growth parameters of *Vigna radiata* seedlings from the two strains. Furthermore, it was noticed that in the inoculated status, irrespective of an iron deficiency and high iron stress, both strains managed to enhance growth in *Vigna radiata* and *Lolium multiflorum* seedlings. *Brucella* sp. E7 and *Pseudomonas brassicae* W7 are promising plant growth-promoting agents for improving agricultural productivity under iron-limited soils [37].

13.3.6 Effect on Plant Disease Reduction

The production of siderophores is considered one of the best methods of minimizing diseases in plants [38]. The PGPR secretes siderophores that can compete with ambient trace iron in order to limit the access of plant inter-root pathogens to iron. Rhizosphere-PGPR-mediated production of siderophores warded off fungal infections such as *Phytophthora*, *Pythium*, and *Sclerotinia* plant diseases [39]. Siderophores have been reported in some studies to impact fungi. The effect of the siderophore produced by *Pseudomonas aeruginosa* FP6 may greatly reduce *Colletotrichum gloeosporioides* growth in pepper [40]. Seven of the siderophore-producing strains introduced by Vijay Karuppiah *et al.* exhibited *in vitro* antifungal activity against *Fusarium oxysporum* f. sp. *lycopersici* MTCC10270 (Fol), *Fusarium equiseti* MFol, and *Sarocladium* sp. SWL [41]. Siderophore-producing rhizobacteria (SPR) have emerged as a prospective candidate for sustainable plant direct defense strategies in crop health management. In the present study, 220 bacterial isolates were screened from the tomato rhizosphere for the *in vitro* antagonism against *Rhizoctonia solani* AG-4. Among these, nine potent antagonistic strains were identified on the basis of physiological characterization and 16S rDNA sequencing, comprising *Alcaligenes* sp. (MUN1, MB21, MPF37), *Enterobacter* sp. (MPM1), *Pseudomonas* sp. (M10A, MB65), *P. aeruginosa* (MPF14, MB123), and *P. fluorescens* (MPF47). These strains exhibited some plant-growth-promoting traits, such as production of hydrolytic enzymes, hydrogen cyanide, and indole-acetic acid, whereas only limited phosphate solubilization activity was found. In glasshouse experiments, both strains MB123 and MPF47 considerably reduced disease severity and were further examined in field trials with diverse application methods. Significantly, soil mixing with *P. fluorescens* MPF47 and root dipping of seedlings before transplantation led to the greatest suppression of disease. Moreover, plants treated with MPF47 had much higher total chlorophyll and levels of available iron than infected controls. Thus, these findings affirm the strong biocontrol potential of *P. fluorescens* MPF47 and suggest that it may serve as a bio-management agent against *R. solani* using integrated soil and root application methods [42].

13.3.7 *Solanum* Species

Zinc-solubilizing bacteria (Zn-SB) are critical to ensuring soil fertility and plant health by allowing the availability of Zn in the rhizosphere. However, still, their dynamics in various cultivation systems remain uncertain since patterns of agricultural land use control microbial activities and plant growth efficiencies. Thus, the explorations studied the prospects of zinc solubilization by microbial antagonists in various agricultural systems with *Solanum lycopersicum*, *Solanum melongena*, and *Capsicum annuum* grown in polyhouse soil (PS) as well as in open fields (OF). Of 80 rhizobacterial isolates, 20 isolates from PS and only two from OF were able to solubilize zinc at a very high range as observed through atomic absorption spectroscopy. The bacterial strain PS4 solubilized 253.06 ppm of ZnO with high production of lactic acid (168.62 g/mL) and acetic acid (470.5 g/mL), whereas strain OF1 solubilized 16.02 ppm of ZnO together with glycolic acid (42.89 g/mL), lactic acid (22.30 g/mL), formic acid (106.03 g/mL), and acetic acid (48.5 μg/mL). *In vitro* studies indicated that PS1 produced enhanced levels of auxins, gibberellic acid, and siderophores as compared to OF1 [43].

13.3.8 Rice

Rice is a staple crop for large sectors of the world, including Vietnam, which is the second-largest exporter after Thailand. In Vietnam, rice cultivation is beset by certain adversities, one of which is the overstretching of agrochemical usage. This causes several harms to the environment as well as human health. Endophytic bacteria (EB) can be envisaged as a sustainable effort to meet such interests owing to their plant growth-promoting (PGP) and antagonistic activities against important plant pathogens. A total of 150 endophytic bacterial strains were isolated from the roots of upland rice grown in Tung village, Nam Co commune, Mu Cang Chai district, Yen Bai province, Vietnam. Siderophore production was manifested by the 119 isolates and categorized as low (1–10 mm halo zone), moderate (11–30 mm), and high (>30 mm) activity levels. Along with this, all of the isolates were screened for antifungal activity against *Fusarium oxysporum* and antibacterial activity against *Xanthomonas oryzae* pv. *oryzae*. Thus, out of 36 isolates with antimicrobial properties, five were antifungal, 15 were antibacterial, and 10 were both antifungal and antibacterial. These findings, therefore, can be a basis for further evaluation under net house and field conditions for identifying promising strains for bioinoculant development, thereby providing an environmentally sustainable approach to combat rice diseases caused by *F. oxysporum* and *X. oryzae* [44]. The sustainable agricultural alternative proposes the use of cyanobacterial siderophores to augment plant growth and yield in search of environmentally safe alternatives to chemical fertilizers [45]. In the case of *Bacillus aryabhattai* siderophore application in rice cultivation, it was found to enhance crop production by 60% in non-saline conditions and 43% in saline conditions [46].

13.3.9 As Herbicides

The increasing need toward sustainability in agriculture has brought search toward eco-friendly alternatives to synthetic herbicides, of which siderophores emerge as candidate bioherbicides. The research on the herbicidal principles of siderophores by *Amycolatopsis lurida* strain 407 on ryegrass and redroot weeds were conducted and it was found that under different environmental conditions, strain 407 exhibited two colony morphologies—red and white; both significantly inhibited weed growth. The quantification of siderophore production in iron-deficient medium was done at 613.4 ppm for red variant and 388.5 ppm for white, demonstrating strong iron-chelating activity. There has been a direct correlation—the more the siderophore in the plant culture medium, the more the inhibition in growth. Column chromatography analysis on the aqueous fractions further revealed that the fractions from the 407 red variant reduced ryegrass shoot length by 45% and root length by 83%–86%; redroot seedling lengths decreased by 36%. Similarly, those fractions from 407 white completely inhibited germination or reduced ryegrass root length by 94% and redroot seedling length by 52%. These fractions show iron chelation activity and thus could be considered the most effective metabolite—multiple synergistic with the action of the siderophores or otherwise may contribute to herbicidal effect. The results also confirm the potential of bioherbicides based on siderophores to be a sustainable choice and alternative of chemical weed control [47].

13.4 Challenges and Future Perspective

In the sustainable agricultural market, though siderophore-producing microorganisms display promise, their application is very much in a formative phase and challenges exist that hamper widespread application. One of the major limitations is that microbial efficacy often varies with different soil types, climatic conditions, and crop species. Environmental variables affecting the field performance of siderophore-producing bacteria (SPB) include soil pH, organic matter content, and competition between the organisms for ecological niches. Furthermore, it is technically and economically challenging to manufacture and formulate microbial inoculants in bulk, and it has been persistently difficult to maintain these biofertilizers' viability and effectiveness in commercial applications. Regulatory approvals and awareness among the farmers impede the wider use of microbial inoculants; extensive field validation is needed to convince farmers of their efficacy.

Future research should concentrate on optimizing microbial consortia that encompass multiple plant-beneficial traits, such as siderophore production, phosphate solubilization, and tolerance against stress. Leveraging the current advances in microbial genomics and synthetic biology would greatly help to develop genetically modified strains with enhanced siderophore production and resilience in varying environments. Furthermore, the stability and controlled release of these siderophores could be improved by utilizing nanotechnology-based formulations, thus ensuring their effective application in the field. Precision agriculture may also entail fittingly adopting siderophore-producing microbes through setting up sensor-based monitoring of soil health and nutrient status. On the other hand, actual cross-disciplinary collaboration amidst microbiologists, agronomists, and policymakers would enable the full translation of lab findings into immediately feasible and scalable solutions for application by the farmers. Solving those puzzles will finally provide ample platforms for the adoption of microbial siderophores as sustainable alternatives to conventional fertilizers to increase crop productivity and save the environment.

13.5 Conclusion

Rewarding the greener way to provide sustainable and organic iron nutrition for plants, mitigation of iron deficiencies and plant health improvement are but a few of the benefits of the application of siderophores. It highlights the importance of contemporary agriculture: improving solubilization of iron, reducing plant diseases, and improving stress tolerance. All field trials have shown that these microorganisms produce siderophores, invariably increasing crop yield and nutrient uptake but at the same time decreasing the requirement for synthetic fertilizers. Moreover, they induce systemic resistance in plants against excitor-borne pathogens, thus acting as excellent biocontrol agents.

Although siderophores have potentials, a few challenges need to be solved to maximize benefits from siderophore-based applications. Key concerns are microbial performance variability with environmental conditions, formulation stability, and scale-up issues. Development should pursue the optimization of microbial consortia, co-application of siderophore-producing microbes with other biofertilizers, and genetic and biotechnological manipulation to enhance the efficacy of the microorganisms. Regulatory aspects, as well as on-farm validation and farmer awareness programs for such technologies, must exist for their effective implementation. In conclusion, microbial siderophores hold immense promise for revolutionizing sustainable agriculture by enhancing soil fertility, improving nutrient bioavailability, and reducing environmental impact. With continued research and technological advancement, global food security and environmentally friendly farming may be supported, wherein coupling precision agriculture and biofortification strategies could also enhance the effects of these technologies, making them a viable option for sustainable crop production since climate change and soil degradation increasingly threaten present practices.

AI disclosure statement: The author confirms that no artificial intelligence (AI)-generated content or AI-assisted writing tools were used in the preparation of this chapter (including summaries and the abstract). All text, analysis, and conclusions are the result of the author's original work.

References

1. Wang, M. and Li, H., Structure, Function, and Biosynthesis of Siderophores Produced by Streptomyces Species. *J. Agric. Food. Chem.*, 2025, https://doi.org/10.1021/acs.jafc.4c08231.
2. Rai, A.K., Basak, N., Rai, A.K., Priyank, *Rhizospheric Interactions for Abiotic Stress Mitigation*, 2024.
3. Garg, G., Kumar, S., Bhati, S., Siderophore in Plant Nutritional Management: Role of Endophytic Bacteria, in: *Endophytes: Mineral Nutrient Management*, D.K. Maheshwari and S. Dheeman (Eds.), vol. 3, pp. 139–183, Springer International Publishing, Cambridge Scholars Publishing, 2021, https://doi.org/10.1007/978-3-030-65447-4_14.
4. Ahmed, E. and Holmström, S.J.M., Siderophores in environmental research: Roles and applications. *Microb. Biotechnol.*, *7*, 3, 196–208, 2014, https://doi.org/10.1111/1751-7915.12117.
5. Hider, R.C. and Kong, X., Chemistry and biology of siderophores. *Nat. Prod. Rep.*, *27*, 5, 637, 2010, https://doi.org/10.1039/b906679a.
6. Thakur, M., Khushboo, Shah, S., Kumari, P., Kumar, M., Vibhuti, R.K., Pramanik, A., Yadav, V., Raina, M., Negi, N.P., Gautam, V., Rustagi, A., Verma, S.K., Kumar, D., Unlocking the Secrets of Rhizosphere Microbes: A New Dimension for Agriculture. *Symbiosis*, *92*, 3, 305–322, 2024, https://doi.org/10.1007/s13199-024-00980-w.
7. Ghosh, S.K., Bera, T., Chakrabarty, A.M., Microbial siderophore – A boon to agricultural sciences. *Biol. Control*, *144*, 104214, 2020, https://doi.org/10.1016/j.biocontrol.2020.104214.
8. Negi, R., Sharma, B., Kumar, S., Chaubey, K.K., Kaur, T., Devi, R., Yadav, A., Kour, D., Yadav, A.N., Plant endophytes: Unveiling hidden applications toward agro-environment sustainability. *Folia Microbiol.*, *69*, 1, 181–206, 2024, https://doi.org/10.1007/s12223-023-01092-6.
9. Singh, P., Chauhan, P.K., Upadhyay, S.K., Singh, R.K., Dwivedi, P., Wang, J., Jain, D., Jiang, M., Mechanistic Insights and Potential Use of Siderophores

Producing Microbes in Rhizosphere for Mitigation of Stress in Plants Grown in Degraded Land. *Front. Microbiol.*, *13*, 2022, https://doi.org/10.3389/fmicb.2022.898979.

10. Srivastava, N., Siderophore Production in Iron Uptake and Plant Biofortification, in: *Plant Microbiome for Plant Productivity and Sustainable Agriculture*, S. Chhabra, R. Prasad, N.R. Maddela, N. Tuteja (Eds.), pp. 313–329, Springer Nature, 2023, https://doi.org/10.1007/978-981-19-5029-2_13.
11. Saha, M., Sarkar, S., Sarkar, B., Sharma, B.K., Bhattacharjee, S., Tribedi, P., Microbial siderophores and their potential applications: A review. *Environ. Sci. Pollut. Res.*, *23*, 5, 3984–3999, 2016, https://doi.org/10.1007/s11356-015-4294-0.
12. Miethke, M. and Marahiel, M.A., Siderophore-Based Iron Acquisition and Pathogen Control. *Microbiol. Mol. Biol. Rev.*, *71*, 3, 413–451, 2007, https://doi.org/10.1128/MMBR.00012-07.
13. Timofeeva, A.M., Galyamova, M.R., Sedykh, S.E., Bacterial Siderophores: Classification, Biosynthesis, Perspectives of Use in Agriculture. *Plants*, *11*, 22, 3065, 2022, https://doi.org/10.3390/plants11223065.
14. De Andrade, L.A., Santos, C.H.B., Frezarin, E.T., Sales, L.R., Rigobelo, E.C., Plant Growth-Promoting Rhizobacteria for Sustainable Agricultural Production. *Microorganisms*, *11*, 4, 1088, 2023, https://doi.org/10.3390/microorganisms11041088.
15. Xie, B., Wei, X., Wan, C., Zhao, W., Song, R., Xin, S., Song, K., Exploring the Biological Pathways of Siderophores and their Multidisciplinary Applications: A Comprehensive Review. *Molecules*, *29*, 10, 2024, Article 10, https://doi.org/10.3390/molecules29102318.
16. Otanga, R.R., Yobo, K.S., Laing, M.D., Response of two wheat cultivars to inoculation of diazotrophic bacteria in combination with reduced nitrogen fertilisation under field conditions. *S. Afr. J. Plant Soil*, *35*, 5, 343–350, 2018, https://doi.org/10.1080/02571862.2017.1419387.
17. Sharma, S., Chandra, S., Kumar, A., Bindraban, P., Saxena, A.K., Pande, V., Pandey, R., Foliar Application of Iron Fortified Bacteriosiderophore Improves Growth and Grain Fe Concentration in Wheat and Soybean. *Indian J. Microbiol.*, *59*, 3, 344–350, 2019, https://doi.org/10.1007/s12088-019-00810-4.
18. Ehsan, S., Qureshi, A., Chaudhary, N., Ali, A., Niaz, A., Javed, H., Ijaz, F., Anwar, S.A., Effect of Fluorescent-Producing Rhizobacteria on Cereal Growth Through Siderophore Exertion. *J. Appl. Res. Plant Sci.*, *4*, 02, 601–611, 2023, https://doi.org/10.38211/joarps.2023.04.02.168.
19. Mushtaq, Z., Nazir, A., Asghar, H.N., Zahir, Z.A., Interactive Effect of Siderophore-Producing Bacteria and l-Tryptophan on Physiology, Tuber Characteristics, Yield, and Iron Concentration of Potato. *Potato Res.*, *65*, 4, 1015–1027, 2022, https://doi.org/10.1007/s11540-022-09565-w.
20. Kumar, R., Singh, A., Shukla, E., Singh, P., Khan, A., Singh, N.K., Srivastava, A., Siderophore of plant growth promoting *rhizobacterium* origin reduces

reactive oxygen species mediated injury in *Solanum* spp. Caused by fungal pathogens. *J. Appl. Microbiol.*, *135*, 2, lxae036, 2024, https://doi.org/10.1093/jambio/lxae036.
21. Radzki, W., Gutierrez Mañero, F.J., Algar, E., Lucas García, J.A., García-Villaraco, A., Ramos Solano, B., Bacterial siderophores efficiently provide iron to iron-starved tomato plants in hydroponics culture. *Antonie van Leeuwenhoek*, *104*, 3, 321–330, 2013, https://doi.org/10.1007/s10482-013-9954-9.
22. Nagata, T., Oobo, T., Aozasa, O., Efficacy of a bacterial siderophore, pyoverdine, to supply iron to *Solanum lycopersicum* plants. *J. Biosci. Bioeng.*, *115*, 6, 686–690, 2013, https://doi.org/10.1016/j.jbiosc.2012.12.018.
23. Arya, N., Rana, A., Rajwar, A., Sahgal, M., Sharma, A.K., Biocontrol Efficacy of Siderophore Producing Indigenous *Pseudomonas* Strains Against *Fusarium Wilt* in Tomato. *Natl. Acad. Sci. Lett.*, *41*, 3, 133–136, 2018, https://doi.org/10.1007/s40009-018-0630-5.
24. Pandey, P., Kang, S.C., Gupta, C.P., Maheshwari, D.K., Rhizosphere Competent *Pseudomonas aeruginosa* GRC1 Produces Characteristic Siderophore and Enhances Growth of Indian Mustard (*Brassica campestris*). *Curr. Microbiol.*, *51*, 5, 303–309, 2005, https://doi.org/10.1007/s00284-005-0014-1.
25. Jiménez-Gómez, A., Saati-Santamaría, Z., Kostovcik, M., Rivas, R., Velázquez, E., Mateos, P.F., Menéndez, E., García-Fraile, P., Selection of the Root Endophyte *Pseudomonas brassicacearum* CDVBN10 as Plant Growth Promoter for *Brassica napus* L. Crops. *Agronomy*, *10*, 11, 1788, 2020, https://doi.org/10.3390/agronomy10111788.
26. Lally, R.D., Galbally, P., Moreira, A.S., Spink, J., Ryan, D., Germaine, K.J., Dowling, D.N., Application of Endophytic *Pseudomonas fluorescens* and a Bacterial Consortium to *Brassica napus* Can Increase Plant Height and Biomass under Greenhouse and Field Conditions. *Front. Plant Sci.*, *8*, 2193, 2017, https://doi.org/10.3389/fpls.2017.02193.
27. Eswaran, S.U.D., Sundaram, L., Perveen, K., Bukhari, N.A., Sayyed, R.Z., Osmolyte-producing microbial biostimulants regulate the growth of *Arachis hypogaea* L. under drought stress. *BMC Microbiol.*, *24*, 1, 165, 2024, https://doi.org/10.1186/s12866-024-03320-6.
28. Wang, N., Wang, T., Chen, Y., Wang, M., Lu, Q., Wang, K., Dou, Z., Chi, Z., Qiu, W., Dai, J., Niu, L., Cui, J., Wei, Z., Zhang, F., Kümmerli, R., Zuo, Y., Microbiome convergence enables siderophore-secreting-rhizobacteria to improve iron nutrition and yield of peanut intercropped with maize. *Nat. Commun.*, *15*, 1, 839, 2024, https://doi.org/10.1038/s41467-024-45207-0.
29. Sun, X., Zhang, C., Bei, S., Wang, G., Geisen, S., Bedoussac, L., Christie, P., Zhang, J., High bacterial diversity and siderophore-producing bacteria collectively suppress *Fusarium oxysporum* in maize/faba bean intercropping. *Front. Microbiol.*, *13*, 2022, https://doi.org/10.3389/fmicb.2022.972587.
30. Brick, M.B., Hussein, M.H., Mowafy, A.M., Hamouda, R.A., Ayyad, A.M., Refaay, D.A., Significance of siderophore-producing cyanobacteria on

enhancing iron uptake potentiality of maize plants grown under iron-deficiency. *Microb. Cell Fact.*, *24*, 1, 3, 2025, https://doi.org/10.1186/s12934-024-02618-4.

31. Ehsan, S., Waheed, S., Sarwar, A., Chaudhary, N., Ashraf, W., Affan, Q., Anwar, F., Zafar, H., Saqib, A.I., Foliar Application of Iron Fortified Bacterio-siderophore And Rhizobium Seed Inoculation Promote Growth and Grain Fe Contents in Soybean and Chickpea. *J. Microbiol. Sci.*, *3*, 01, 20–29, 2024, https://doi.org/10.38211/jms.2024.01.70.
32. Flores-Félix, J.D., Velázquez, E., Martínez-Molina, E., González-Andrés, F., Squartini, A., Rivas, R., Connecting the Lab and the Field: Genome Analysis of Phyllobacterium and Rhizobium Strains and Field Performance on Two Vegetable Crops. *Agronomy*, *11*, 6, 1124, 2021, https://doi.org/10.3390/agronomy11061124.
33. Mandal, L. and Kotasthane, A.S., Isolation and Assessment of Plant Growth Promoting Activity of Siderophore Producing *Pseudomonas fluorescens* in Crops. *Int. J. Agric. Environ. Biotechnol.*, *7*, 1, 63, 2014, https://doi.org/10.5958/j.2230-732X.7.1.009.
34. Volpiano, C.G., Lisboa, B.B., São José, J.F.B., De Oliveira, A.M.R., Beneduzi, A., Passaglia, L.M.P., Vargas, L.K., Rhizobium strains in the biological control of the phytopathogenic fungi *Sclerotium* (Athelia) *rolfsii* on the common bean. *Plant Soil*, *432*, 1–2, 229–243, 2018, https://doi.org/10.1007/s11104-018-3799-y.
35. Chandwani, S. and Amaresan, N., Siderophore-producing bacteria mitigate cobalt stress in black gram (*Vigna mungo* L.), and the mitigation strategies are associated with iron concentration. *Environ. Sci. Pollut. Res.*, *30*, 59, 123556–123569, 2023, https://doi.org/10.1007/s11356-023-31106-2.
36. Sreeramulu, R.K.K.V., Suresh, M., Subburamu, K., Durairaj, J., Siderophore producing *Bacillus* spp. and *Ochrobactrum grignonense* enhance the iron content and yield of groundnut genotypes (*Arachis hypogaea* L.) in calcareous soils. *Arabian J. Geosci.*, *16*, 11, 624, 2023, https://doi.org/10.1007/s12517-023-11743-0.
37. Sun, Y., Wu, J., Shang, X., Xue, L., Ji, G., Chang, S., Niu, J., Emaneghemi, B., Screening of Siderophore-Producing Bacteria and their Effects on Promoting the Growth of Plants. *Curr. Microbiol.*, *79*, 5, 150, 2022, https://doi.org/10.1007/s00284-022-02777-w.
38. Etesami, H. and Maheshwari, D.K., Use of plant growth promoting rhizobacteria (PGPRs) with multiple plant growth promoting traits in stress agriculture: Action mechanisms and future prospects. *Ecotoxicol. Environ. Saf.*, *156*, 225–246, 2018, https://doi.org/10.1016/j.ecoenv.2018.03.013.
39. Yu, X., Ai, C., Xin, L., Zhou, G., The siderophore-producing bacterium, *Bacillus subtilis* CAS15, has a biocontrol effect on Fusarium wilt and promotes the growth of pepper. *Eur. J. Soil Biol.*, *47*, 2, 138–145, 2011, https://doi.org/10.1016/j.ejsobi.2010.11.001.

40. Dimkpa, C.O., Merten, D., Svatoš, A., Büchel, G., Kothe, E., Metal-induced oxidative stress impacting plant growth in contaminated soil is alleviated by microbial siderophores. *Soil Biol. Biochem.*, *41*, 1, 154–162, 2009, https://doi.org/10.1016/j.soilbio.2008.10.010.
41. Karuppiah, V., Natarajan, S., Gangatharan, M., Aldayel, M.F., Alsowayeh, N., Thangavel, K., Development of siderophore-based rhizobacterial consortium for the mitigation of biotic and abiotic environmental stresses in tomatoes: An in vitro and in planta approach. *J. Appl. Microbiol.*, *133*, 6, 3276–3287, 2022, https://doi.org/10.1111/jam.15625.
42. Solanki, M.K., Singh, R.K., Srivastava, S., Kumar, S., Kashyap, P.L., Srivastava, A.K., Arora, D.K., Isolation and characterization of siderophore producing antagonistic rhizobacteria against *Rhizoctonia solani*. *J. Basic Microbiol.*, *54*, 6, 585–597, 2014, https://doi.org/10.1002/jobm.201200564.
43. Rani, N., Kaur, G., Kaur, S., Mutreja, V., Upadhyay, S.K., Tripathi, M., Comparison of diversity and zinc solubilizing efficiency of rhizobacteria obtained from solanaceous crops under polyhouse and open field conditions. *Biotechnol. Genet. Eng. Rev.*, *39*, 2, 527–548, 2023, https://doi.org/10.1080/02648725.2022.2157949.
44. Ta, L., Mai, N.T., To, H.T., Mai, C.D., Chu, H.H., Le, B.T., Nguyen, P.T., In vitro screening of siderophore-producers from upland rice root endophytic bacteria for antagonistic activities against *fusarium oxysporum*, and *xanthomonas oryzea* PV. ORYZAE. *Int. J. Adv. Res.*, 2020.
45. Singh, J.S., Kumar, A., Rai, A.N., Singh, D.P., Cyanobacteria: A Precious Bioresource in Agriculture, Ecosystem, and Environmental Sustainability. *Front. Microbiol.*, *7*, 2016, https://doi.org/10.3389/fmicb.2016.00529.
46. Sultana, S., Alam, S., Karim, M.M., Screening of siderophore-producing salt-tolerant rhizobacteria suitable for supporting plant growth in saline soils with iron limitation. *J. Agric. Food Res.*, *4*, 100150, 2021, https://doi.org/10.1016/j.jafr.2021.100150 z.
47. Eigharlou, M., Javidpoor, S., Mohammadi, A., Khelghatibana, F., Nami, Y., Ma'mani, L., Sadeghi, A., Evaluation of herbicidal potential of Siderophores produced by *Amycolatopsis lurida* strain 407. *Sci. Rep.*, *14*, 1, 26212, 2024, https://doi.org/10.1038/s41598-024-77843-3.

14

Siderophores in Various Agroecosystems: From Theory to Practice

Rahul Gogoi[1], Abhisek Rath[2], Ratrismita Chetia[3], Jitul Bora[1], Madhurjya Ranjan Sharma[1] and Sudipta Sankar Bora[4*]

[1]Department of Agricultural Biotechnology, Assam Agricultural University, Jorhat, Assam, India
[2]Department of Plant Pathology, Assam Agricultural University, Jorhat, Assam, India
[3]Dr. D.Y. Patil Biotechnology and Bionformatics Institute, Pune, Maharashtra, India
[4]Multi-Disciplinary Research Unit, Jorhat Medical College and Hospital, Jorhat, Assam, India

Abstract

Siderophores, biologically active iron-chelating molecules, play a pivotal role in microbial interactions and nutrient cycling within agroecosystems. By modulating iron availability, siderophores not only mitigate iron deficiency in plants but also suppress phytopathogens, enhancing overall plant resilience. Additionally, their role in fostering beneficial plant–microbe interactions highlights their utility as biofertilizers and biocontrol agents. This chapter aims to advance the understanding and deployment of siderophores in promoting environmentally sustainable agricultural practices. It discusses the diverse functions of siderophores in various agricultural contexts, emphasizing their potential in improving soil fertility, plant health, and crop productivity. The chapter also synthesizes current theoretical frameworks and practical applications of siderophores, offering insights into their mechanisms of action, ecological implications, and contributions to sustainable agriculture. Future perspectives are outlined, advocating for integrative approaches that leverage siderophore-producing microbes to address the challenges of modern agroecosystems.

***Keywords*:** Biofertilizers, microbial interactions, plant resilience, nutrient cycling, siderophores

**Corresponding author*: sudip.asm@gmail.com

Debasis Mitra, Ayush Madan and Anju Rani (eds.) Microbial Siderophores for Agricultural Productivity: Perspectives of Use in Agriculture, (343–366) © 2026 Scrivener Publishing LLC

14.1 Introduction

Iron (Fe), the most common element on Earth and the fourth most common in the crust, may be found in a range of oxidation states, from -2 to +6. Being a catalyst in enzymatic reactions, oxygen metabolism, electron transport, and DNA and RNA synthesis, it is a necessary component for the development and operation of almost all living microbes [1]. Iron (Fe) typically exists in two oxidation states: Fe(III) (ferric) and Fe(II) (ferrous). Its capacity to cycle between both states allows it to play a vital part in oxidation–reduction reactions. Ferric iron is abundant, generally non-reactive, and insoluble, while ferrous iron is less prevalent, highly reactive, and soluble. Iron is required for various metabolic activities, including the tricarboxylic acid (TCA) cycle, electron transport chain, oxidative phosphorylation, and photosynthesis. Additionally, it is vital in controlling the formation of porphyrins, vitamins, toxins, cytochromes, antibiotics, siderophores, pigments, aromatic compounds, and nucleic acids [2]. According to recent research, iron controls the surface motility of bacteria, which is crucial for the production of microbial biofilms. According to a number of investigations, the dissolved ferrous iron concentrations under these circumstances are between 10^{-10} and 10^{-9} M, which is significantly less than the minimum ferrous iron levels needed for living things, which are thought to be between 10^{-7} and 10^{-5} M [3]. Microbes build specialized low-molecular-weight organic compounds called siderophores to survive in iron-depleted environments. Specialized receptors on the cell membrane allow these siderophores to firmly attach to ferric iron (Fe^{3+}) and create a siderophore–iron complex, which is then carried into the cell. Siderophore-binding proteins, permeases, and ATPases are involved in this process in Gram-positive bacteria. They cooperate to help move the siderophore–iron (Fe^{3+}) combination across the cell membrane [4]. Gram-negative bacteria have a very different membrane network than gram-positive bacteria. The siderophore iron (Fe^{+3}) complex is transported by the ATP-binding cassette transporter (ABC-transporter), which includes an outer membrane receptor, a periplasmic binding proteins, and a cytoplasmic membrane protein in gram-negative bacteria. Hydroxamates, catecholates, and carboxylates are the three forms of siderophores that are distinguished by the oxygen ligands that are utilized for Fe (III) coordination [5]. Despite being a micronutrient, iron is necessary for redox reactions, the manufacture of chlorophyll, and a number of other critical physiological processes in plants. Iron deficiency severely impacts crop output and quality, resulting to a reduction in agricultural productivity.

This reduction in agricultural output disturbs the ecosystem's natural food web, hurting both plant and animal populations that depend upon these resources for sustenance. Plants require around 10^{-17} mol/L of iron at neutral pH, while microorganisms require 10^{-6} mol/L under the same conditions [6]. Plant roots produce siderophores to regulate iron levels required for metabolic and physiological activities in iron-stressed, degraded soils, albeit these levels may still be inadequate. Siderophore-producing microorganisms (SPM) help to alleviate plant stress in such soils by creating a variety of iron-chelating chemicals. These bacteria serve as biofertilizers and biocontrol agents, encouraging sustainable agriculture and environmentally friendly crop production on degraded land. Furthermore, SPM promote plant growth by displaying plant growth-promoting (PGP) characteristics such as plant hormone synthesis, phosphate solubilization, and secondary metabolites. They produce favorable circumstances in strained soils, promoting plant development. Notably, two rhizobacterial species that produce siderophores, *Pseudomonas putida* and *Bacillus amyloliquefaciens*, are resistant to drought stress. This is due to their ability to unleash PGP features like as siderophore generation, hormone synthesis, mineral solubilization, biofilm development, and ACC deaminase activity, making them useful for stimulating crop growth under challenging situations [7]. Siderophores bind to harmful heavy metals, ensuring that they do not impair plant cell function. The toxic heavy metal detoxification and binding characteristics of siderophores play a critical role in sustaining plant growth on heavy-metal-polluted soils. By binding to hazardous metals, siderophores diminish their bioavailability and lessen their damaging effects on plants. In addition to microbial siderophores, plants create phyto-siderophores, which have the potential to directly chelate iron, which assists in maintaining critical iron levels and boosting plant health in tough situations. With the quick consumption of phyto-siderophore, the symptom of iron deficiency in some plants disappeared entirely. Thus, siderophores produced by microbes or plants are recognized as critical suppliers of iron, facilitating their survival and growth, particularly in iron-deficient or stressed conditions.

14.2 Siderophore Production and Regulation

Siderophores are low-molecular-weight chelating compounds produced by microorganisms to address the limited bioavailability of iron (Fe) in many environments. Iron, despite being a critical nutrient for most living organisms, is often inaccessible due to its tendency to form insoluble hydroxides under aerobic conditions. Siderophores are specialized secondary metabolites with high affinity for iron ($Kf > 10^{30}$) and molecular weight that ranged from 200 to 2,000 Da. They have developed to successfully overcome iron restriction [8]. To effectively bind iron, siderophores depend on particular organic functional groups, such phenolic hydroxyls, carboxyl groups, hydroxamates, and sulfhydryl groups. Because they form thermodynamically resistant hexacoordinate complexes with Fe(III), which is chosen over Fe(II) because of its higher stability constants, they are effective at binding iron. This makes it possible for siderophores to efficiently sequester iron in settings where it is either limited or bonded to other substances [9]. The stability and efficiency of these complexes make siderophores vital tools for microbial survival and adaptation in nutrient-poor environments as well as significant players in nutrient cycling and plant–microbe interactions within agroecosystems. The acid–base properties of siderophores further contribute to their chelation capabilities. Their acidic functional groups, such as hydroxyl (-OH) and carboxyl (-COOH), dissociate in aqueous solutions, releasing H_3O^+ ions and reducing pH. These dissociation properties, characterized by low pKa values, determine the conditions under which siderophores act most effectively to chelate iron [10].

14.2.1 Types of Siderophores

The functional groups of siderophores that are engaged in Fe(III) coordination are used to classify them into different types, namely:

a. Carboxylates:
Carboxylate siderophores contain α-hydroxycarboxylic acid groups, enabling the chelation of Fe(III) even at low pH levels. For example, citric acid serves as a model carboxylate, with pKa values ranging between 3.13 and 6.40, making it effective in acidic microbial environments. Although carboxylate siderophores generally form less stable complexes ($\log\beta \approx$ 20–30) than other types, their ecological success is often supported by high concentrations in the rhizosphere [9].

b. Catecholates (phenolates):
Catecholate siderophores, derived from catechols or 1,2-dihydroxybenzenes, exhibit high specificity for Fe(III) due to their acidic phenolic groups. These siderophores, such as enterobactin produced by *Escherichia coli*, form exceptionally stable Fe(III) complexes. Catecholates are notable for their dual antioxidant and pro-oxidant properties, contributing to

microbial competition and even antimicrobial activity under specific conditions [11].

c. Hydroxamates:

Hydroxamate siderophores are among the most efficient in iron chelation. Their hydroxamic acid groups exhibit significant acidity, forming stable complexes with Fe(III). An example is ferrichrome, produced by fungi and certain bacteria, which demonstrates high stability constants ($\log\beta >$ 28) [12].

d. Mixed-type siderophores:

Many siderophores possess a combination of functional groups, enhancing their versatility. For instance, pyoverdines produced by *Pseudomonas* spp. feature carboxylate and hydroxamate groups, enabling their activity across a broad pH range and diverse environmental conditions. These structural variations provide adaptability and robustness in natural settings [5].

14.3 Microorganisms Involved in Siderophore Production

Siderophore production is a fundamental trait among microorganisms, aiding their survival under iron-limiting conditions and contributing to diverse ecological roles. These microorganisms, which include numerous bacterial, fungal, and actinomycete taxa, use different processes to synthesize and utilize siderophores. Bacterial genera that produce siderophores include *Bacillus, Pseudomonas, Enterobacter, Klebsiella, Kosakonia, Methylobacterium, Nocardia, Pantoea, Paenibacillus, Rhodococcus, Serratia*, and others. Siderophores are produced by at least 20 different bacterial taxa, according to research. For example, *Azotobacter* species have been intensively researched for their plant-growth-promoting qualities, specifically their ability to create siderophores that promote iron intake and plant health [13]. *Bacillus* spp., notable for their robust environmental adaptability, produce siderophores with plant-growth-promoting effects [14]. Siderophore production in *Enterobacter* spp. has been linked to their ability to promote plant growth under iron deficiency. *Klebsiella* spp. similarly exhibit siderophore synthesis that contributes to their adaptability in plant and human environments [15]. Other notable genera, such as *Pantoea, Paenibacillus, Pseudomonas, Rhodococcus*, and *Serratia*, showcase diverse siderophore production pathways with implications for both agriculture and biocontrol. The genus *Pantoea* exemplifies the multi-functionality of siderophore producers. Species like *Pseudomonas ananatis*, *Pseudomonas agglomerans*, and *Pseudomonas diversa* not only synthesize siderophores but can also solubilize critical elements like phosphate and zinc. Furthermore, many species produce phytohormones including indole-3-acetic acid (IAA), which encourages plant growth. These bacteria's combined properties make them great plant growth promoters, increasing nutrient availability and promoting plant health in a variety of situations [16].

Fungal siderophore producers are equally diverse, with filamentous fungi, for example, *Aspergillus* and *Penicillium*, being prominent contributors. These fungi primarily synthesize hydroxamate-type siderophores like ferrichromes, aiding iron acquisition in nutrient-deficient soils and fostering plant symbiosis. *Trichoderma* spp., widely known for their role in biocontrol, produce siderophores such as coprogen and fusigen, which exhibit both plant-growth-promoting and antifungal properties [17]. Fungi like *Mucorales* produce carboxylate-type siderophores such as rhizoferrin, while *Penicillium bilaii* synthesizes the catecholate siderophore pistillarin. The actinomycetes, particularly the genus *Streptomyces*, produce siderophores with both ecological and pharmacological significance. For example, *Streptomyces pilosus* synthesizes desferrioxamine B, widely used as a chelation therapy drug for iron overload conditions [18].

14.4 Regulation of Siderophore Production

Siderophore production is intricately regulated by a network of genetic and environmental factors, with intracellular iron concentration playing a central role. In most bacteria, siderophore synthesis is significantly upregulated under iron-deficient conditions and suppressed when iron is abundant. This dynamic regulation ensures efficient resource utilization and adapts to the fluctuating iron availability in microbial environments [19]. At the genetic level, the regulation of siderophore biosynthesis and transport is mediated by transcriptional regulators such as the Ferric Uptake Regulator (Fur). Fur functions as a molecular "switch", repressing siderophore gene expression under iron-rich conditions by binding to promoter regions of target genes. When intracellular iron levels are low, Fur releases the DNA, allowing the transcription of siderophore biosynthesis genes [20]. In *Pseudomonas aeruginosa*, the Fur (ferric uptake regulator) protein regulates iron homeostasis *via* modulating the expression of the sigma factor PvdS. PvdS, in turn, controls the formation of pyoverdine (PVD), a siderophore that the bacteria secretes to absorb iron from its surroundings. This system enables *P. aeruginosa* to adapt to iron-limited circumstances while retaining enough iron for growth and metabolic functions. Under iron-deficient conditions, PvdS activation leads to robust PVD production, essential for iron acquisition. In addition to Fur, other regulatory mechanisms, such as quorum sensing, play a role in siderophore production. Quorum sensing allows bacteria to coordinate siderophore secretion within a population, ensuring synchronized responses to iron scarcity. This has been particularly noted in *Pseudomonas aeruginosa*, where quorum sensing tightly integrates siderophore production with other metabolic pathways [21]. Environmental factors, including nutrient availability, pH, and oxygen levels, further influence siderophore biosynthesis. For instance, sufficient glucose or ammonium may suppress siderophore production in some species, as these nutrients impact cellular metabolism and iron requirements. Competing microorganisms also modulate siderophore production, as siderophores often mediate microbial competition by sequestering iron from shared environments [22]. Advances in molecular biology have enabled the engineering of regulatory pathways to optimize siderophore production. For example, researchers have modified *Pseudomonas aeruginosa* by replacing the Fur-regulated promoter of the *pvdS* gene with an arabinose-inducible promoter. This modification allows for pyoverdine production independent of iron availability, enhancing bioweathering processes like asbestos alteration. Similarly, in *Pseudomonas putida*, enhanced siderophore production was accomplished by introducing a plasmid that carried an extra copy of the *pvdS* gene under a constitutive promoter. This genetic alteration led to higher expression of the sigma factor PvdS, leading in elevated synthesis of pyoverdine (PVD), the siderophore important for iron acquisition in iron-limited conditions. This technique demonstrates the potential of genetic tinkering to raise siderophore synthesis and improve the bacterium's ability to assist plant development or bioremediation efforts [23]. Another strategy involves the deletion of Fur in certain species. For instance, *Bacillus licheniformis* mutants lacking Fur displayed an increased production of schizokinen and bacillibactin, demonstrating the potential to boost siderophore yields through genetic modification [24]. Understanding these regulatory networks is pivotal for leveraging siderophores in applications such as agriculture and bioremediation. For example, modulating siderophore biosynthesis can improve plant growth under nutrient-poor conditions or enhance the degradation of environmental contaminants. Exploring the interplay between iron-sensing, quorum sensing, and environmental factors will further enable the targeted manipulation of siderophore production for sustainable practices.

14.5 Siderophores' Function in Plant–Microbe Communication

In soils, siderophore-producing microbes can function as facilitators or suppressors, depending upon the availability of compatible receptors in other organisms for siderophore uptake. Strain-specific receptors that identify and promote the absorption of the siderophore–iron complexes regulate iron acquisition through iron-loaded siderophores. Microbial siderophores supply iron nutrition to crop plants and facilitate their growth in conditions of iron bioavailability deficiency during salinity stress. Two possible processes for plants to absorb iron from microbial siderophores were put forth: (a) to provide Fe(II) to the plant's transport system, microbial siderophores with a high redox potential can be reduced. According to this method, microbial Fe(III)–siderophores are transported to the plant root's apoplast, where siderophore reduction may occur. Consequently, the apoplast sequesters Fe(II), which raises the concentration of Fe in the root; (b) After chelating iron from soils, microbial siderophores can exchange ligands with phytosiderophores [25]. This mechanism is influenced by a number of parameters, particularly the stability constants and concentrations of microbial and phytosiderophores, in addition to the pH and redox conditions of the roots. For example, bacterial siderophores, such as catechol and hydroxamate, obtained from *Arabidopsis thaliana*, *Festuca rubra*, and *Agrostis capillaris*, grown in heavy metal-contaminated regions, markedly increased the growth rate of *Festuca rubra* and *Brassica napus* [26].

Similarly, these microbial siderophores act as bioprotectants against specific phytopathogens by chelating ferric ions, thereby making them inaccessible to the pathogens and indirectly influencing their coexistence in the same niche, potentially suppressing the growth of these resident pathogens. By upsetting organelle iron homeostasis, triggering transcriptional pathways, and changing cellular behavior, these microbial siderophores can significantly affect infections. Furthermore, iron chelation may induce mitophagy, the regulated autophagy of mitochondria [27]. For instance, pyoverdine, the siderophore produced by *Pseudomonas aeruginosa*, is detrimental to mitochondria in various pathogens, causing mitochondrial fragmentation and a decline in mitochondrial membrane potential, ultimately resulting in cellular and, if sufficiently severe, organismal mortality. Siderophores have been shown to be efficient biostimulants and biocontrol agents in a number of studies. One study found that applying *Alcaligenes faecalis*, which produces siderophores, improved seed germination and chlorophyll content in groundnut crops by 8.75% and 8.0%, respectively, and increased root and shoot growth by 9.35% and 16%, respectively [28]. *Pseudomonas aureofaciens* has shown its efficacy for disease control in rice (*Oryza sativa*) against fungal pathogens including *Alternaria* sp., *Fusarium oxysporum*, and *Pyricularia oryzae* by efficiently inhibiting their mycelial growth [29]. *Alcaligenes* sp., *Pseudomonas fluorescens*, and *P. aeruginosa* all successfully inhibited the growth of *Rhizoctonia solani*, with mycelial growth inhibition rates ranging from 61.7%, 67.8%, and 71%, respectively [30]. An essential biocontrol agent in agricultural settings, *Lysobacter enzymogenes* is recognized for generating the spermidine siderophore N1, N8-bis(2,3-dihydroxybenzoyl). Furthermore, *Macrophomina phaseolina* (with growth inhibition rates of 37.3% to 57.5%) and *Sclerotium rolfsii* (with growth inhibition rates of 48.3% to 52.6%) were effectively inhibited by four siderophore-producing plant-growth-promoting rhizobacteria (PGPR): *Bacillus licheniformis, Bacillus subtilis*, and *Ochrobactrum rignonense* [31]. In a similar way *Brevibacillus brevis*, a siderophore-producing bacteria, had antifungal efficacy against *Candida albicans*, indicating its potential as a biocontrol agent. *Acinetobacter calcoaceticus*, which produces the catecholate siderophore (similar to Acinetobactin), has shown antifungal action against *Fusarium oxysporum*. Interestingly, when the spermidine biosynthetic gene, which encodes arginine decarboxylase or S-adenosylmethionine (SAM) decarboxylase, was deleted, the production of lysochelin, antifungals, and their analogues—all of which are essential for disease control—was reduced. This suggests that this gene is necessary for the biocontrol process. As a result, utilizing siderophore-producing microorganisms could be a useful technique for reducing reliance on synthetic drugs for phytopathogen management. However, phytopathogen suppression is not primarily dependent on siderophore formation; other parameters, such as the ability to create secondary metabolites including acetyl phloroglucinols, phenazines, and pyrroles, also play an important role in phytopathogen inhibition. These findings highlight the importance of understanding ecological interactions in order to effectively control soil-borne diseases within microbial populations.

Induced systemic resistance (ISR) in plants can also be triggered by siderophore synthesis. For instance, a study discovered that the siderophore pseudobactin caused *P. fluorescens* WCS374r to produce ISR in rice (*Oryza sativa*) against the leaf blast pathogen *Magnaporthe oryzae* [32]. The signal transduction route that triggered the defense mechanism was controlled by ethylene and jasmonic acid rather than salicylic acid. Additionally, siderophores play a key role in helping bacteria overcome membrane-associated drug resistance by promoting iron transport, which can be used to deliver medications into cells using siderophore-antimicrobial conjugates. Furthermore, plants may develop induced systemic resistance (ISR) as a result of siderophore production. For example, using the siderophore pseudobactin, *Pseudomonas fluorescens* WCS374r promoted ISR in rice (*Oryza sativa*) against the leaves blast pathogen *Magnaporthe oryzae* [32]. The defense mechanism was activated *via* a signal transduction route that was controlled by ethylene and jasmonic acid rather than salicylic acid. By promoting iron transport, siderophores help bacteria overcome membrane-associated drug resistance. This transport can then be used to deliver medications into cells using siderophore–antimicrobial conjugates. Furthermore, plants may develop systemic resistance (ISR) as a result of siderophore synthesis. Moreover, siderophores help many bacteria overcome membrane-associated drug resistance by taking use of their iron transport ability, which can subsequently be utilized to carry medications into cells *via* siderophore–antimicrobial conjugates.

Numerous parameters, such as the number of siderophores generated, the siderophore's chemical characteristics (such as its stability constant for iron), and its resistance to degradation, might affect the potential buildup of the siderophore and allow microorganisms to compete for iron. The unique characteristics of the transport systems used for iron uptake and the capacity to use siderophores made by other microbes are further considerations. The ability to obtain iron through siderophores may eventually determine which bacterium dominates the competition for different carbon sources made available by rhizodeposition or root exudates in highly competitive environments [33].

Siderophores significantly impact the rhizospheric microbial community by regulating iron availability, which can influence the composition and abundance of microbes, primarily favoring the growth of siderophore-producing bacteria while potentially limiting the growth of competing or cheating microorganisms that lack the ability to acquire iron effectively and cooperating with its similar kind, thereby shaping the overall microbial community dynamics within the rhizosphere; this can have positive effects on plant health and other beneficial microbes by promoting beneficial microbes and potentially suppressing pathogens [34].

A mutually beneficial form of cooperation that directly benefits both the producers and the recipients is the norm for cooperative behaviors, which entail an actor creating benefits for another individual. Ornibactin from *B. cenocepacia* 69, enterochelin from *Escherichia coli*, and pyoverdine and pyochelin from *Pseudomonas aeruginosa* and *Burkholderia cenocepacia* are a few examples of this. One organism produces these siderophores to sequester iron, which is advantageous to the producer as well as other microbes that can use them [35]. Cooperation may not be evolutionarily stable because each cooperative act can be taken advantage of by "cheaters," even though it can provide substantial fitness benefits to all parties involved. Increased spatial structure can help with this problem by limiting siderophore dispersion and cell dispersal, making it more probable that siderophores will be shared between clone mates than with cheaters. Nonetheless, bacteria frequently form varied consortia with interacting strains that generate distinct siderophores, each of which needs a particular cognate receptor to absorb iron. According to evolutionary theory, siderophores might compete with other species in these situations. Siderophores can sequester iron from rivals who lack the proper receptor, conserving it for the homologous producers. This can be especially useful in an iron-limited environment. According to evolutionary theory, social interactions mediated by siderophores have significant fitness consequences and will lead to reciprocal evolutionary adaptations between interacting strains or species. In other words, we would anticipate evolutionary shifts between situations, such as the formation of non-transitive community structures, co-evolutionary arms races, and increased ecological linkages. This ultimately depends on soil conditions, plant characteristics and siderophore types. The recent findings on the antipathogenic effects of siderophores are comprehensively summarized in Table 14.1.

14.6 Siderophores in Diverse Agroecosystems

Siderophores are indispensable components of agroecosystems, facilitating plant iron acquisition, supporting nutrient cycling, and promoting plant health through biocontrol mechanisms (Figure 14.1). Their ecological functions vary across agricultural soils, forest ecosystems, grasslands, and farming systems, reflecting their adaptability and significance in diverse environmental contexts. Harnessing siderophore-producing microorganisms represents a promising approach to improve agroecosystem sustainability, ensuring both productivity and environmental integrity.

14.6.1 Agricultural Soils

Microbial production of siderophores is essential to meet the high iron demands of crops in intensively cultivated soils. *Pseudomonas*, *Bacillus*, and fungi including *Aspergillus* are dominant rhizosphere microorganisms in these systems. Siderophores enhance plant iron uptake by solubilizing ferric iron, which is otherwise inaccessible due to its low solubility in neutral and alkaline soils. Beyond iron acquisition, siderophores play a biocontrol

Table 14.1 Role of siderophore as biocontrol and disease suppression.

Siderophore-producing organism	Target pathogen	Mechanism of action	References
Bacillus subtilis	*Alternaria alternata*	Siderophores reduce iron availability, indirectly inhibiting fungal development	Harish *et al.*, [36]
Trichoderma harzianum	Various fungal pathogens	Indirect competition *via* siderophores, along with mycoparasitism	Tyśkiewicz *et al.*, [37]
Pseudomonas putida	*Phytophthoracapsici*	Chelates iron, limiting pathogen growth and improving plant health	Tyśkiewicz *et al.*, [37]
Azotobacter chroococcum	Soil-borne pathogens	Increases plant growth and lowers disease severity through siderophore-mediated iron uptake	Pandit *et al.*, [38]
Aspergillus flavus	Soil-borne pathogens	Produces hydroxamate siderophores that promote plant tolerance to diseases	Wu *et al.*, [39]
Bradyrhizobium japonicum	Root rot pathogens in soybeans	Improves plant iron availability while battling with pathogens for iron	Huynh *et al.*, [40]

(*Continued*)

Table 14.1 Role of siderophore as biocontrol and disease suppression. (*Continued*)

Siderophore-producing organism	Target pathogen	Mechanism of action	References
Trichoderma virens	*Sclerotinia sclerotiorum*	Indirectly suppresses fungal infections by producing siderophores	Silva *et al.*, [41]
Myxococcus xanthus	Various soil pathogens	Siderophores battle with pathogens for iron, limiting their growth	Contreras-Moreno *et al.*, [42]
Azotobacter chroococcum	Soil-borne pathogens	Enhances plant growth and lowers disease severity *via* siderophore-mediated iron absorption	Li *et al.*, [43]
Bacillus subtilis	*Alternaria alternata*	Siderophores reduce iron availability, indirectly inhibiting fungal development	Ahmad *et al.*, [44]
Streptomyces griseus	General soil pathogens	Produces siderophores and antibacterial substances that effectively reduce pathogens	Khan *et al.*, [45]
Enterobacter cloacae	*Fusarium oxysporum*	Siderophore-mediated iron competition reduces pathogen growth in agricultural roots	Sallam *et al.*, [46]

(*Continued*)

Table 14.1 Role of siderophore as biocontrol and disease suppression. (*Continued*)

Siderophore-producing organism	Target pathogen	Mechanism of action	References
Pseudomonas protegens	*Ralstonia solanacearum*	It produces siderophores, which limit the availability of iron to bacterial pathogens	Clough *et al.*, [47]
Bacillus licheniformis	*Sclerotium rolfsii*	Produces bacillibactin, a siderophore that suppresses fungal growth and spore generation	Panichikkal *et al.*, [48]
Pseudomonas alcaligenes	*Alternaria solani*	Produces siderophores that hinder fungal hyphal extension and spore germination	Jia *et al.*, [49]

role by limiting the availability of iron to phytopathogens, thereby reducing disease incidence. The utilization of siderophore-producing microorganisms (SPMs) as bioinoculants has been proposed as a sustainable strategy to reduce dependency on chemical fertilizers and pesticides, supporting the development of environmentally friendly agricultural practices. In cereal crops, siderophores belonging to the family of mugineic acid (MA) mobilize and transport iron from rhizosphere to the plant. Since iron is poorly soluble in alkaline and calcareous soils, siderophores enhance its solubility by chelating ferric ions (Fe^{3+}) which are further taken up *via* specific transporters in plant roots, such as *Yellow Stripe1-like* (YSL) transporters. Problems associated with iron deficiency like poor plant growth, chlorosis and reduced photosynthetic activity can be solved by microbial siderophore production into the rhizosphere. Pyoverdine siderophores produced by *Pseudomonas* species have been demonstrated to promote plant growth; hence, these bacteria are classified as plant-growth-promoting bacteria (PGPB).

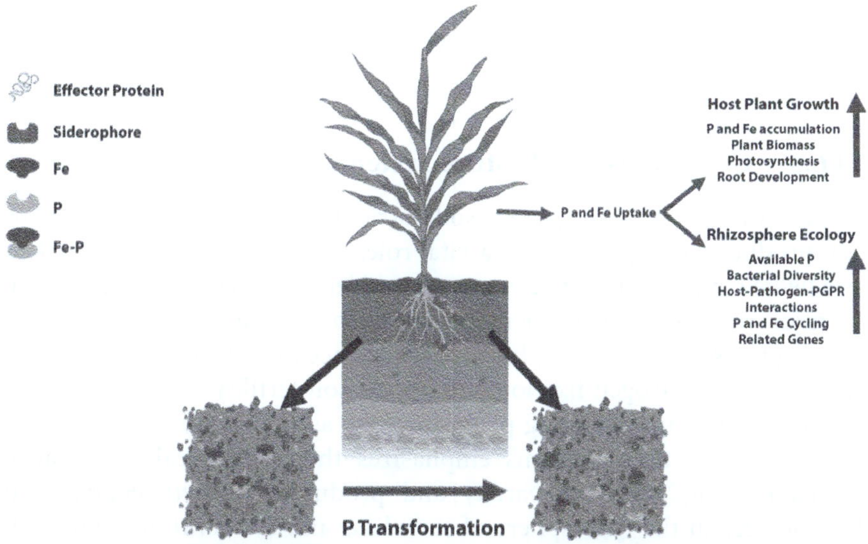

Figure 14.1 Role of siderophores in phosphorus and iron cycling for plant growth and rhizosphere ecology. The diagram highlights siderophores' role in soil P and Fe cycling, enhancing nutrient uptake, plant growth, and rhizosphere ecological balance.

Pyoverdine siderophores assist by chelating iron from the soil, thereby rendering it available to the plant, which then promotes numerous physiological processes required for growth.

14.6.2 Forest Ecosystems

In forest ecosystems, the typically acidic soil with limited iron availability necessitates the production of siderophores by ectomycorrhizal (ECM) and arbuscular mycorrhizal (AM) fungi. These fungi establish symbiotic relationships with tree species like pines (*Pinus* spp.) and oaks (*Quercus* spp.), generating a wide array of siderophores that enable the uptake of iron and other trace elements. The acquisition of iron through siderophores plays a vital role in sustaining forest productivity, particularly during periods of nutrient deficiency. Additionally, siderophores aid in the dissolution of other crucial minerals, including phosphate and zinc, thus supporting the ongoing nutrient cycles within these ecosystems. *Pseudomonas* strains (positive for siderophore production) are promising candidates as biofertilizers to improve shisham plantation in iron-deficient Tarai region of western Himalayas. Furthermore, siderophores play a crucial role in maintaining ecological equilibrium by reducing the toxicity of heavy metals and boosting the diversity of soil microbes. This microbial activity is essential for sustaining forest productivity, particularly in environments under stress.

14.6.3 Grasslands and Pasture Systems

In grasslands and pastures, where soil nutrients are scarce, microorganisms that produce siderophores play a vital role. These ecosystems are marked by a fierce competition between grasses and legumes, with iron often being a scarce resource. Rhizosphere microbes, particularly those associated with legumes, secrete siderophores that enhance iron availability. This process facilitates nitrogen fixation and boosts soil fertility. The involvement of siderophores in mediating plant–microbe as well as microbe–microbe competition in these systems emphasizes their ecological significance in maintaining grassland variety and productivity. *Pseudomonas* and *Enterobacter* in the rhizosphere of *S. indicus* and *P. coloratum* improve P acquisition through PGP traits like siderophore and IAA production in the alkaline-sodic lowlands of flooding pampa [50]. Research has shown that introducing plant-growth-promoting microorganisms (PGPMs), including those that produce siderophores, into pastures can improve the quality of forage and increase biomass yield. These applications are especially beneficial for rehabilitating deteriorated grasslands and fostering sustainable animal husbandry.

14.6.4 Organic and Integrated Farming Systems

In organic farming systems, where synthetic fertilizers and pesticides are minimized, siderophore-producing microorganisms thrive due to favorable soil microbiomes enriched by organic matter inputs. Siderophores improve iron uptake in crops and aid in disease prevention by reducing the growth of iron-dependent diseases [51]. In integrated farming systems, siderophores act as a critical component of nutrient cycling, reducing dependency on external chemical inputs. The integration of compost and biofertilizers containing SPMs, such as *Azotobacter* and *Pseudomonas*, has shown to enhance crop yields while improving soil health and microbial diversity. Various *Trichoderma* species and other endophytic organisms have been observed to boost disease resistance and tolerance to abiotic stress factors, in part through their ability to produce siderophores [52]. Maize harboring siderophore-producing strains of *Pseudomonas* and *Bacillus* enhances the suppression of *Fusarium oxysporum* and improves resistance in faba bean when grown in an intercropping system [53].

Siderophore production by *Pseudomonas* spp. consortia protects tomato plants against *Ralstonia solanacearum* by competing with the pathogen for iron utilization [54]. This underscores the potential of siderophore-producing microbial consortia as an eco-friendly strategy for managing soil-borne bacterial diseases in crops.

14.7 Applications of Siderophores in Agriculture

Over the past few decades, agriculture has been found to be profoundly dependent on chemical fertilizers to supplement plant growth for an increase in crop production and yield in order to fulfil the growing food demands of the accelerating population. The dependence, however, conversely affects the environment. The continuous and incessant application of chemical fertilizers damages the soil fertility, disturbs the soil microbial population, and most significantly, causes environmental pollution. Over-consumption of these chemical fertilizers leads to the accumulation of heavy metals by plants that could pose a threat when entering the food chain. Thus, in the recent years, environmentalists and agricultural scientists are directed toward the use of bio-fertilizers and organics to enhance soil fertility and plant productivity. Microbial inoculants, such as plant-growth-promoting rhizobacteria (PGPR), are essential alternatives to chemical fertilizers since they boost soil fertility while also contributing to crop improvement. The application of *Pseudomonas corrugate* as a bio-fertilizer against phytopathogens *Alternaria alternata* and *Fusarium oxysporum* is a classic example [55].

Iron is an important micronutrient, and its limitation causes problems for several physiological activities, chlorophyll synthesis, and certain redox reactions in plants. The iron sequestering ability establishes siderophores as a friendly alternative to harmful pesticides. In nature, there are different classes of siderophores produced by different microorganisms. Pyroverdines generated by *Pseudomonas* species enhance plant growth and give protection against diseases, therefore classifying them as plant-growth-promoting bacteria [56]. In an iron-deficient environment, other bacteria like *Serratia*, *Bacillus*, *Azotobacter*, *Enterobacterium*, *Azospirillum*, and *Rhizobium* chelate iron and form complexes with other metals like cadmium, manganese, cobalt, and nickel to make it easily available to the microbes for use. Crop production quality and quantity are problematic under any nutrient-deficit state, let alone iron deficiency. Significant work to mitigate such problems has been contributed by several researchers through plant–microbial interaction, with siderophore–plant interactions being listed as one of them.

A combination of *Bradyrhizobium* and *Pseudomonas* sp., which produces hydroxamate siderophores, improved nitrogen fixation and root nodulation in mung beans [57]. Siderophores can inhibit the growth of pathogenic microbes by outcompeting them for available iron, which is critical for most pathogens. This biocontrol mechanism is important in preventing plant diseases, including those caused by soil-borne fungi and bacteria. Siderophore-producing bacteria scavenge iron, making it unavailable for pathogens to feed on. Siderophore-producing bacteria not only suppress pathogens but also trigger systemic resistance in plants. Research has indicated that siderophores induce the expression of defense-related genes, activating pathways associated with systemic acquired resistance (SAR). This dual function is particularly effective in crops like wheat and rice, where increased resistance against bacterial blight and fungal infections has been observed [58].

Heavy metal contamination from rapidly growing petrochemical industries, mine tilling, waste water irrigation, and nuclear power is a serious concern for the environment. Bioremediation of toxic metal by employing biological environmentally sustainable methods is the need of the hour. Siderophore use is renowned for being economical, highly effective, and environmentally friendly. Siderophore's affinity toward metals (like Cr^{3+}, Al^{3+}, Cu^{2+}, Eu^{3+}, and Pb^{2+}) other than Fe^{3+} has been reported by several studies. The amount of different siderophores like pyoverdines, ferrioxamines, azotochelin, and schizokinen produced depends on the concentration of other metals in the medium [59]. Desferrioxamine B is more effective at chelating Co^{3+} than Fe^{3+} in alkaline environments. *Agrobacterium radiobacter* removes approximately 54% pollutants from arsenic-contaminated soil, and pyoverdine siderophore produced mobilize U6+, Np5+, and other metals from uranium mine waste. The combination of phytoremediation with siderophore-producing bacteria will increase the bioavailability of a range of metals required for the healthy growth of any plant. *Bacillus subtilis* is known to collect cadmium (Cd), whereas mangrove fungus and *Fusarium solani* accumulate copper (Cu) and zinc (Zn) in their biomass. Additionally, *Pseudomonas aeruginosa* and *Pseudomonas fluorescens*, which produce siderophores, have been shown to increase the absorption of chromium (Cr) and lead (Pb) in maize. These microbes can help plants absorb metals by increasing their bioavailability in the soil. This is especially advantageous in areas where metal contamination is present [60]. In soil contaminated with heavy metals, PGPB *Streptomyces tenae* F4 acts as a Cd phytoremediator and promotes the uptake of other metals [59]. Given all of the evidence shown above, siderophore has the potential to be an effective metal bioremediation agent. The versatility of siderophores in enhancing nutrient availability, protecting against pathogens, and managing environmental challenges underscores their potential in sustainable and innovative solutions across disciplines.

14.8 Conclusion

Iron is an essential metal needed for many cellular functions in all living things. Microorganisms' growth is hampered when iron levels are low. Microorganisms thrive in such iron-limited environments by secreting siderophores. Siderophores' numerous uses demonstrate that they have the potential to be used as a potential agent in a variety of fields, including ecology, agriculture, bioremediation, and biosensor. Plants interact with both beneficial and pathogenic microbes capable of producing siderophores. In agriculture, siderophore-producing bacteria in the rhizosphere play an important role. Siderophores are low-molecular-weight phenolic compounds that can chelate Fe^{3+} ions, convert them to Fe^{2+}, and provide essential iron to crops. In addition to providing iron to plants, siderophores also restrict the growth of soil-borne phytopathogens, many of which are iron-dependent. Research on the mechanisms of siderophore synthesis and its effects on plant growth and development is essential for the development of new, more efficient farming practices. Therefore, it is intended to use bacteria that produce siderophores in agricultural fields to boost plant growth and productivity.

AI disclosure statement: The author acknowledges the use of AI-based tools for minor language editing or grammatical correction. However, all intellectual contributions, analyses, and conclusions presented in this chapter are entirely the author's own.

References

1. Pandey, S.S., The Role of Iron in Phytopathogenic Microbe–Plant Interactions: Insights into Virulence and Host Immune Response. *Plants*, 12, 17, 3173, Sep. 2023, doi: 10.3390/plants12173173.
2. Saha, M., Sarkar, S., Sarkar, B., Sharma, B.K., Bhattacharjee, S., Tribedi, P., Microbial siderophores and their potential applications: a review. *Environ. Sci. Pollut. Res.*, 23, 5, 3984–3999, Mar. 2016, doi: 10.1007/s11356-015-4294-0.

3. Mahender, A., Swamy, B.P.M., Anandan, A., Ali, J., Tolerance of Iron-Deficient and -Toxic Soil Conditions in Rice. *Plants*, 8, 2, 31, Jan. 2019, doi: 10.3390/plants8020031.
4. Page, M.G.P., The Role of Iron and Siderophores in Infection, and the Development of Siderophore Antibiotics. *Clin. Infect. Dis.*, 69, Supplement_7, S529–S537, Nov. 2019, doi: 10.1093/cid/ciz825.
5. Miethke, M. and Marahiel, M.A., Siderophore-Based Iron Acquisition and Pathogen Control. *Microbiol. Mol. Biol. Rev.*, 71, 3, 413–451, Sep. 2007, doi: 10.1128/MMBR.00012-07.
6. Cheng, Y., Narayanan, M., Shi, X., Chen, X., Li, Z., Ma, Y., Phosphate-solubilizing bacteria: Their agroecological function and optimistic application for enhancing agro-productivity. *Sci. Total Environ.*, 901, 166468, Nov. 2023, doi: 10.1016/j.scitotenv.2023.166468.
7. Kumar, M., *et al.*, Synergistic effect of Pseudomonas putida and Bacillus amyloliquefaciens ameliorates drought stress in chickpea (Cicer arietinum L.). *Plant Signal. Behav.*, 11, 1, e1071004, 2016, doi: 10.1080/15592324.2015.1071004.
8. Khan, A., Singh, P., Srivastava, A., Synthesis, nature and utility of universal iron chelator – Siderophore: A review. *Microbiol. Res.*, 212–213, 103–111, Jul. 2018, doi: 10.1016/j.micres.2017.10.012.
9. Kraemer, S.M., Iron oxide dissolution and solubility in the presence of siderophores. *Aquat. Sci. - Res. Across Boundaries*, 66, 1, 3–18, Mar. 2004, doi: 10.1007/s00027-003-0690-5.
10. Reichard, P.U., Kretzschmar, R., Kraemer, S.M., Dissolution mechanisms of goethite in the presence of siderophores and organic acids. *Geochim. Cosmochim. Acta*, 71, 23, 5635–5650, Dec. 2007, doi: 10.1016/j.gca.2006.12.022.
11. Romero, R., Salgado, P.R., Soto, C., Contreras, D., Melin, V., An Experimental Validated Computational Method for pKa Determination of Substituted 1,2-Dihydroxybenzenes. *Front. Chem.*, 6, Jul. 2018, doi: 10.3389/fchem.2018.00208.
12. Boukhalfa, H. and Crumbliss, A.L., Chemical aspects of siderophore mediated iron transport. *Biometals*, 15, 4, 325–339, 2002, doi: 10.1023/A:1020218608266.
13. Baars, O., Zhang, X., Morel, F.M.M., Seyedsayamdost, M.R., The Siderophore Metabolome of Azotobacter vinelandii. *Appl. Environ. Microbiol.*, 82, 1, 27–39, Jan. 2016, doi: 10.1128/AEM.03160-15.
14. Kesaulya, H., Hasinu, J.V., Tuhumury, G.N., Potential of Bacillus spp produces siderophores insuppressing thewilt disease of banana plants. *IOP Conf. Ser. Earth Environ. Sci.*, 102, 012016, Jan. 2018, doi: 10.1088/1755-1315/102/1/012016.
15. Zhang, W., *et al.*, Siderophores in clinical isolates of Klebsiella pneumoniae promote ciprofloxacin resistance by inhibiting the oxidative stress. *Biochem. Biophys. Res. Commun.*, 491, 3, 855–861, Sep. 2017, doi: 10.1016/j.bbrc.2017.04.108.

16. Singh, P., *et al.*, Diazotrophic Bacteria Pantoea dispersa and Enterobacter asburiae Promote Sugarcane Growth by Inducing Nitrogen Uptake and Defense-Related Gene Expression. *Front. Microbiol.*, 11, Jan. 2021, doi: 10.3389/fmicb.2020.600417.
17. Nakkeeran, S., Rajamanickam, S., Karthikeyan, M., Mahendra, K., Renukadevi, P., Johnson, I., Antimicrobial secondary metabolites from Trichoderma spp. as next generation fungicides, in: *Biocontrol Agents and Secondary Metabolites*, pp. 257–282, Elsevier, Amsterdam, Netherlands, 2021, doi: 10.1016/B978-0-12-822919-4.00011-9.
18. Nakouti, I., Sihanonth, P., Hobbs, G., A new approach to isolating siderophore-producing actinobacteria. *Lett. Appl. Microbiol.*, 55, 1, 68–72, Jul. 2012, doi: 10.1111/j.1472-765X.2012.03259.x.
19. Puja, H., Mislin, G.L.A., Rigouin, C., Engineering Siderophore Biosynthesis and Regulation Pathways to Increase Diversity and Availability. *Biomolecules*, 13, 6, 959, Jun. 2023, doi: 10.3390/biom13060959.
20. Xie, P., Xu, Y., Tang, J., Wu, S., Gao, H., Multifaceted regulation of siderophore synthesis by multiple regulatory systems in Shewanella oneidensis. *Commun. Biol.*, 7, 1, 498, Apr. 2024, doi: 10.1038/s42003-024-06193-7.
21. McRose, D.L., Baars, O., Seyedsayamdost, M.R., Morel, F.M.M., Quorum sensing and iron regulate a two-for-one siderophore gene cluster in Vibrio harveyi. *Proc. Natl. Acad. Sci.*, 115, 29, 7581–7586, Jul. 2018, doi: 10.1073/pnas.1805791115.
22. David, S.R., Ihiawakrim, D., Regis, R., Geoffroy, V.A., Iron removal from raw asbestos by siderophores-producing Pseudomonas. *J. Hazard. Mater.*, 385, 121563, Mar. 2020, doi: 10.1016/j.jhazmat.2019.121563.
23. Becker, F., Wienand, K., Lechner, M., Frey, E., Jung, H., Interactions mediated by a public good transiently increase cooperativity in growing Pseudomonas putida metapopulations. *Sci. Rep.*, 8, 1, 4093, Mar. 2018, doi: 10.1038/s41598-018-22306-9.
24. Akhtar, S.S., *et al.*, Bacillus licheniformis FMCH001 Increases Water Use Efficiency *via* Growth Stimulation in Both Normal and Drought Conditions. *Front. Plant Sci.*, 11, Apr. 2020, doi: 10.3389/fpls.2020.00297.
25. Ahmed, E. and Holmström, S.J.M., Siderophores in environmental research: roles and applications. *Microb. Biotechnol.*, 7, 3, 196–208, May 2014, doi: 10.1111/1751-7915.12117.
26. Grobelak, A. and Hiller, J., Bacterial siderophores promote plant growth: Screening of catechol and hydroxamate siderophores. *Int. J. Phytorem.*, 19, 9, 825–833, Sep. 2017, doi: 10.1080/15226514.2017.1290581.
27. Wilson, B.R., Bogdan, A.R., Miyazawa, M., Hashimoto, K., Tsuji, Y., Siderophores in Iron Metabolism: From Mechanism to Therapy Potential. *Trends Mol. Med.*, 22, 12, 1077–1090, Dec. 2016, doi: 10.1016/j.molmed.2016.10.005.
28. Sayyed, R.Z., Gangurde, N.S., Patel, P.R., Joshi, S., Chincholkar, S.B., Siderophore production by Alcaligenes faecalis and its application for growth

promotion in Arachis hypogaea. *Indian J. Biotechnol.*, 9, 302–307, 2010, [Online]. Available: https://api.semanticscholar.org/CorpusID:34473728.
29. Chaiharn, M., Chunhaleuchanon, S., Lumyong, S., Screening siderophore producing bacteria as potential biological control agent for fungal rice pathogens in Thailand. *World J. Microbiol. Biotechnol.*, 25, 11, 1919–1928, Nov. 2009, doi: 10.1007/s11274-009-0090-7.
30. Solanki, M.K., *et al.*, Isolation and characterization of siderophore producing antagonistic rhizobacteria against Rhizoctonia solani. *J. Basic Microbiol.*, 54, 6, 585–597, Jun. 2014, doi: 10.1002/jobm.201200564.
31. Kalyan, V.S.R.K., Meena, S., Karthikeyan, S., Jawahar, D., Isolation, screening, characterization, and optimization of bacteria isolated from calcareous soils for siderophore production. *Arch. Microbiol.*, 204, 12, 721, Dec. 2022, doi: 10.1007/s00203-022-03322-w.
32. De Vleesschauwer, D., Djavaheri, M., Bakker, P.A.H.M., Höfte, M., Pseudomonas fluorescens WCS374r-Induced Systemic Resistance in Rice against Magnaporthe oryzae Is Based on Pseudobactin-Mediated Priming for a Salicylic Acid-Repressible Multifaceted Defense Response. *Plant Physiol.*, 148, 4, 1996–2012, Dec. 2008, doi: 10.1104/pp.108.127878.
33. Schalk, I.J., Bacterial siderophores: diversity, uptake pathways and applications. *Nat. Rev. Microbiol.*, 23, 1, 24–40, Jan. 2025, doi: 10.1038/s41579-024-01090-6.
34. Solomon, W., Janda, T., Molnár, Z., Unveiling the significance of rhizosphere: Implications for plant growth, stress response, and sustainable agriculture. *Plant Physiol. Biochem.*, 206, 108290, Jan. 2024, doi: 10.1016/j.plaphy.2023.108290.
35. Kramer, J., Özkaya, O., Kümmerli, R., Bacterial siderophores in community and host interactions. *Nat. Rev. Microbiol.*, 18, 3, 152–163, Mar. 2020, doi: 10.1038/s41579-019-0284-4.
36. Harish, B.N., *et al.*, Molecular characterization and antifungal activity of lipopeptides produced from Bacillus subtilis against plant fungal pathogen Alternaria alternata. *BMC Microbiol.*, 23, 1, 179, Jul. 2023, doi: 10.1186/s12866-023-02922-w.
37. Tyśkiewicz, R., Nowak, A., Ozimek, E., Jaroszuk-Ściseł, J., Trichoderma: The Current Status of Its Application in Agriculture for the Biocontrol of Fungal Phytopathogens and Stimulation of Plant Growth. *Int. J. Mol. Sci.*, 23, 4, 2329, Feb. 2022, doi: 10.3390/ijms23042329.
38. Pandit, M.A., *et al.*, Major Biological Control Strategies for Plant Pathogens. *Pathogens*, 11, 2, 273, Feb. 2022, doi: 10.3390/pathogens11020273.
39. Wu, Q., *et al.*, Differential Expression of Genes Related to Growth and Aflatoxin Synthesis in Aspergillus flavus When Inhibited by Bacillus velezensis Strain B2. *Foods*, 11, 22, 3620, Nov. 2022, doi: 10.3390/foods11223620.
40. Huynh, T., Navi, S.S., Yang, X.B., Efficacy of Seed Treatments with Bradyrhizobium japonicum to Reduce Occurrence of Soybean Sudden Death

Syndrome in Early-planted Soybeans, pp. 415–437, 2022, doi: 10.1007/978-981-16-8877-5_20.

41. Silva, L.G., Camargo, R.C., Mascarin, G.M., de O. Nunes, P.S., Dunlap, C., Bettiol, W., Dual functionality of Trichoderma: Biocontrol of Sclerotinia sclerotiorum and biostimulant of cotton plants. *Front. Plant Sci.*, 13, Oct. 2022, doi: 10.3389/fpls.2022.983127.
42. Contreras-Moreno, F.J., Pérez, J., Muñoz-Dorado, J., Moraleda-Muñoz, A., Marcos-Torres, F.J., Myxococcus xanthus predation: an updated overview. *Front. Microbiol.*, 15, Jan. 2024, doi: 10.3389/fmicb.2024.1339696.
43. Li, S., *et al.*, Vermicompost and Azotobacter chroococcum increase nitrogen retention in saline-alkali soil and nitrogen utilization of maize. *Appl. Soil Ecol.*, 201, 105512, Sep. 2024, doi: 10.1016/j.apsoil.2024.105512.
44. Ahmad, T., *et al.*, Biocontrol potential of lipopeptides produced by the novel Bacillus subtilis strain Y17B against postharvest Alternaria fruit rot of cherry. *Front. Microbiol.*, 14, Mar. 2023, doi: 10.3389/fmicb.2023.1150217.
45. Khan, S., Srivastava, S., Karnwal, A., Malik, T., Streptomyces as a promising biological control agents for plant pathogens. *Front. Microbiol.*, 14, Nov. 2023, doi: 10.3389/fmicb.2023.1285543.
46. Sallam, A.A., Haroun, S.A., Aboulnaga, E.A., Mowafy, A.M., Enterobacter cloacae Induces SA-Dependent Systemic Acquired Resistance of Zea mays Against Fusarium oxysporum. *J. Plant Growth Regul.*, 43, 8, 2536–2554, Aug. 2024, doi: 10.1007/s00344-024-11280-4.
47. Clough, S.E., Jousset, A., Elphinstone, J.G., Friman, V., Combining *in vitro* and *in vivo* screening to identify efficient Pseudomonas biocontrol strains against the phytopathogenic bacterium Ralstonia solanacearum. *Microbiologyopen*, 11, 2, Apr. 2022, doi: 10.1002/mbo3.1283.
48. Panichikkal, J., Manu, S., Krishnankutty, R.E., Bacillus licheniformis and Bacillus velezensis from Rhizosphere of Clerodendrum infortunatum L. Promote Plant Growth and Resistance to Sclerotium rolfsii in Vigna unguiculata (L.) Walp. *Curr. Microbiol.*, 81, 8, 238, Aug. 2024, doi: 10.1007/s00284-024-03749-y.
49. Jia, Y., *et al.*, Synergistic antagonism mechanism of Bacillus-Pseudomonas consortium against Alternaria solani. *Eur. J. Plant Pathol.*, 167, 4, 715–726, Dec. 2023, doi: 10.1007/s10658-023-02747-3.
50. Dip, D.P., Sannazzaro, A., II, Otondo, J., Pistorio, M., Estrella, M.J., Exploring Phosphate Solubilizing Bacterial Communities in Rhizospheres of Native and Exotic Forage Grasses in Alkaline-Sodic Soils of the Flooding Pampa. *Curr. Microbiol.*, 81, 7, 189, Jul. 2024, doi: 10.1007/s00284-024-03704-x.
51. Timofeeva, A.M., Galyamova, M.R., Sedykh, S.E., Bacterial Siderophores: Classification, Biosynthesis, Perspectives of Use in Agriculture. *Plants*, 11, 22, 3065, Nov. 2022, doi: 10.3390/plants11223065.
52. Harman, G.E., Integrated Benefits to Agriculture with Trichoderma and Other Endophytic or Root-Associated Microbes. *Microorganisms*, 12, 7, 1409, Jul. 2024, doi: 10.3390/microorganisms12071409.

53. Sun, X., *et al.*, High bacterial diversity and siderophore-producing bacteria collectively suppress Fusarium oxysporum in maize/faba bean intercropping. *Front. Microbiol.*, 13, Aug. 2022, doi: 10.3389/fmicb.2022.972587.
54. Shao, Z., *et al.*, Siderophore interactions drive the ability of Pseudomonas spp. consortia to protect tomato against Ralstonia solanacearum. *Hortic. Res.*, 11, 9, Sep. 2024, doi: 10.1093/hr/uhae186.
55. Trivedi, P., Pandey, A., Palni, L.M.S., *In vitro* evaluation of antagonistic properties of Pseudomonas corrugata. *Microbiol. Res.*, 163, 3, 329–336, May 2008, doi: 10.1016/j.micres.2006.06.007.
56. Gamalero, E. and Glick, B.R., Mechanisms Used by Plant Growth-Promoting Bacteria, in: *Bacteria in Agrobiology: Plant Nutrient Management*, pp. 17–46, Springer Berlin Heidelberg, Berlin, Heidelberg, 2011, doi: 10.1007/978-3-642-21061-7_2.
57. Garg, G., Kumar, S., Bhati, S., Siderophore in Plant Nutritional Management: Role of Endophytic Bacteria, pp. 315–329, 2021, doi: 10.1007/978-3-030-65447-4_14.
58. Backer, R., *et al.*, Plant Growth-Promoting Rhizobacteria: Context, Mechanisms of Action, and Roadmap to Commercialization of Biostimulants for Sustainable Agriculture. *Front. Plant Sci.*, 9, Oct. 2018, doi: 10.3389/fpls.2018.01473.
59. Dimkpa, C.O., Svatoš, A., Dabrowska, P., Schmidt, A., Boland, W., Kothe, E., Involvement of siderophores in the reduction of metal-induced inhibition of auxin synthesis in Streptomyces spp. *Chemosphere*, 74, 1, 19–25, Dec. 2008, doi: 10.1016/j.chemosphere.2008.09.079.
60. Ullah, A., Heng, S., Munis, M.F.H., Fahad, S., Yang, X., Phytoremediation of heavy metals assisted by plant growth promoting (PGP) bacteria: A review. *Environ. Exp. Bot.*, 117, 28–40, Sep. 2015, doi: 10.1016/j.envexpbot.2015.05.001.

15

Emerging Trends and Future Directions in Siderophore Research

Devi Chhatrodiya[1,2], Bhagyashri Lalwani[1,2], Mital D. Jasani[1] and Jignesh H. Kamdar[1*]

[1]*Department of Microbiology, School of Science, RK University, Rajkot, Gujarat, India*
[2]*JM Science College & PG Center, Veraval, Gir Somnath, Gujarat, India*

Abstract
The increasing awareness of siderophores' potential uses in biotechnology, agriculture, medicine, and environmental sustainability has led to notable advances in siderophore research in recent years. Emerging trends and future goals in siderophore research are highlighted in this chapter. These include the identification of new siderophores, the creation of treatments inspired by siderophores, and the investigation of siderophore-mediated interactions between microbes and plants. We also go over how siderophore research and discovery can be accelerated by integrating state-of-the-art technologies like machine learning, genomics, and synthetic biology. We further stress that in order to fully utilize siderophores, interdisciplinary cooperation and creative methods are required. With an emphasis on the fascinating potential and difficulties that lie ahead, this chapter offers a thorough summary of the state of siderophore research as it is today and its prospects for the future.

***Keywords*:** Siderophores, future directions, siderophore-inspired therapeutics, plant–microbe interactions, synthetic biology, medicine, environmental sustainability

**Corresponding author*: jignesh.kamdar@rku.ac.in; ORCID: 0000-0003-4047-5687

Debasis Mitra, Ayush Madan and Anju Rani (eds.) Microbial Siderophores for Agricultural Productivity: Perspectives of Use in Agriculture, (367–380) © 2026 Scrivener Publishing LLC

15.1 Introduction

15.1.1 Significance and Importance of Siderophore Research

Siderophores are secondary metabolites and low molecular weight molecules that are secreted by bacteria to scavenge iron from the surrounding environment and transport it to the bacterial cell wall through certain receptors. Siderophores are essential to microbial ecology because they affect plant–microbe interactions, microbial competition, and nutrient cycling [2]. Additionally, siderophores have attracted a lot of interest due to their prospective uses in biotechnology, medicine, and agriculture, such as antibacterial agents, iron supplements, and instruments for environmental remediation [1]. In recent years, siderophores have attracted a lot of attention because of their potential uses in a variety of sectors. Siderophores are being investigated in biotechnology as new antibacterial agents and biocatalysts [3]. In the medical field, siderophores have demonstrated potential in the treatment of infectious diseases and iron-related conditions [4]. In agriculture, siderophores are being used to develop sustainable fertilizers and to enhance crop yields [5]. Furthermore, the potential of siderophores in environmental remediation, including the removal of heavy metals and the biodegradation of contaminants, is being investigated [1]. This chapter aims to give a summary of the new developments and potential paths in siderophore research. Because of their potential uses in biotechnology, agriculture, medicine, and environmental sustainability, siderophores as iron-chelating compounds made by microorganisms have attracted a lot of attention lately [48]. The present status of siderophore research, including new developments in siderophore synthesis, purification, and characterization, as well as their applications across a range of industries, is intended to be highlighted in this chapter.

The objective of this chapter is to give a thorough overview of the new developments and potential paths in siderophore research. We will talk about the most recent advancements in siderophore research, such as the identification of new siderophores, the creation of medicines inspired by siderophores, and the investigation of siderophore-mediated interactions between microbes and plants [17]. The potential uses of siderophores in a variety of industries, such as agriculture, medicine, and environmental sustainability, will also be highlighted. This chapter seeks to drive innovation in this fascinating topic and inspire new research approaches by offering a thorough overview of the current trends and future directions in siderophore research.

15.2 Current State of Siderophore Research

15.2.1 Overview of Current Understanding of Siderophore Biology and Chemistry

Microorganisms use siderophores, which are complex, low-molecular-weight compounds, to chelate iron from their surroundings. The production, regulation, and transport of siderophores have been clarified by recent research, which has shown a high level of complexity and diversity [6]. The presence of functional groups like hydroxamates, catecholates, and carboxylates in siderophores' chemistry allows them to bind iron with a high affinity [7]. The identification and characterization of novel siderophores has been made easier by developments in analytical techniques like nuclear magnetic resonance spectroscopy and mass spectrometry, which have increased our knowledge of siderophore biology and chemistry [8]. The chemical structures of various important siderophores are shown in Figure 15.1.

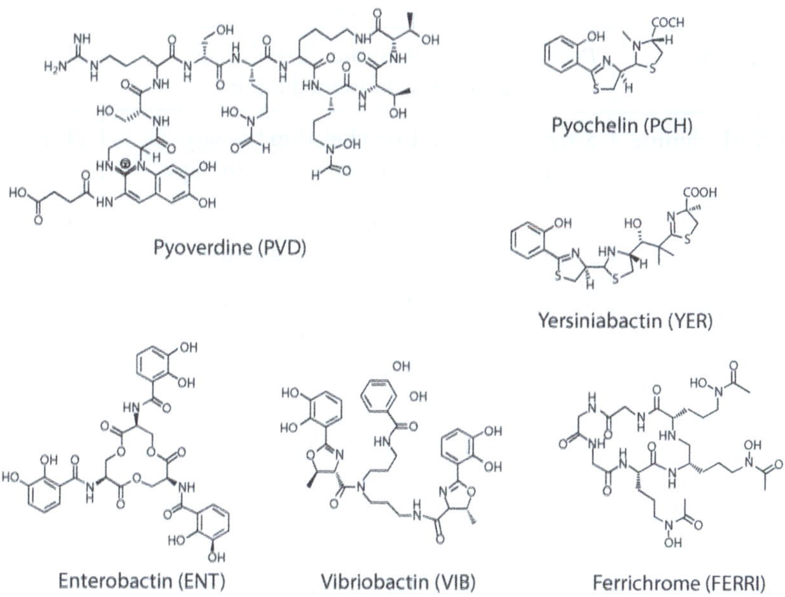

Figure 15.1 Chemical structures of siderophores [47].

15.2.2 Siderophores: Production, Characterization, and Applications

Our knowledge of siderophores has greatly increased as a result of recent developments in their synthesis, purification, and characterization. Innovative fermentation techniques have improved siderophore yields and efficiency, including solid-state fermentation and bioreactor-based production [9]. The purification and separation of siderophores has been made easier by developments in chromatographic techniques, such as high-performance liquid chromatography (HPLC) and ultra-performance liquid chromatography (UPLC) [10]. Additionally, the use of spectroscopic methods like mass spectrometry (MS) and nuclear magnetic resonance (NMR) spectroscopy has made it possible to thoroughly characterize siderophore structures and their iron-chelating characteristics [10, 11]. Many biotechnological applications have found siderophores to be useful tools (Table 15.1). Siderophores' capacity to chelate heavy metals and make it easier

Table 15.1 Example of siderophore with potential applications.

Siderophore	Producing organism	Application	Reference
Enterobactin	*E. coli*	Antimicrobial therapy, iron supplementation	[42]
Pyoverdine	*Pseudomonas aeruginosa*	Antimicrobial therapy, bioremediation	[43]
Desferrioxamine	*Streptomyces pilosus*	Iron chelation therapy, antimicrobial therapy	[44]
Albomycin	*Streptomyces griseus*	Antimicrobial therapy, cancer treatment	[45]
Yersiniabactin	*Yersinia pestis*	Antimicrobial therapy, bioterrorism agent detection	[46]
Azotobactin	*Azotobacter vinelandii*	Plant growth promotion, biofertilizer	[43]
Pseudobactin	*Pseudomonas fluorescens*	Plant growth promotion, biofertilizer	[44]
Ferrioxamine	*Streptomyces* spp.	Iron supplementation, antimicrobial therapy	[42]

to remove them from contaminated environments has made them useful in bioremediation [11]. Furthermore, siderophores have demonstrated potential as antibacterial agents by securing vital iron and preventing the growth of infections. The potential of siderophores in agriculture as biofertilizers and plant growth boosters has also been studied [1]. Additionally, siderophores have been investigated for their potential as a treatment for iron-related conditions such as iron insufficiency and overload [12, 13].

15.3 Emerging Trends in Siderophore Research

15.3.1 Siderophore Research: Genomics, Bioinformatics, and Synthetic Biology Approaches

The study of siderophores has advanced recently as new research directions have been made available. Investigating siderophore-inspired treatments for iron-related illnesses, like cancer and neurological diseases, is one new development [14]. Growing interest is also being shown in the creation of siderophore-based biosensors for the detection and measurement of heavy metals in environmental samples [15]. Additionally, the use of computational tools and machine learning to forecast siderophore shapes and functions is becoming more popular, making it possible to quickly identify new siderophores with possible uses [16]. These new developments highlight the enormous potential of siderophore research to tackle urgent issues in biotechnology, environmental sustainability, and human health. The identification of new siderophores has been transformed by the development of bioinformatics tools and high-throughput sequencing technology. Siderophore biosynthesis gene clusters have been identified and new siderophore structures have been predicted thanks to genomic and bioinformatics techniques including genome mining and comparative genomics [17]. Additionally, the quick identification and characterization of bacteria that produce siderophores has been made possible by the application of machine learning techniques and bioinformatics tools like antiSMASH and PRISM [17, 18]. These methods have significantly sped up the identification of new siderophores, increasing our knowledge of their variety, role, and possible uses.

15.3.2 Siderophore-Inspired Materials and Microbe-Host Interactions

Genetic engineering and synthetic biology have become effective methods for increasing and diversifying the production of siderophores. Recent developments in gene cluster assembly and CRISPR-Cas9 genome editing have made it possible to create new siderophore biosynthesis pathways and enhance preexisting ones [19]. Furthermore, siderophore production levels have been fine-tuned and new siderophore-based biotechnological applications have been developed thanks to the use of synthetic biology techniques including promoter engineering and gene regulation [20]. These developments have greatly increased siderophores' potential in industries like environmental remediation, agriculture, and medicine. Siderophore-inspired materials and technologies, which take advantage of the special qualities of siderophores to provide novel solutions, have become a promising field of study. Nanomaterials based on siderophores have been the subject of recent research because they have demonstrated promise for use in targeted medication delivery, imaging, and cancer treatment [21]. Siderophore-inspired materials have been developed using biomimetic techniques for environmental remediation, including the removal of contaminants and heavy metals from water [22]. Additionally, the development of innovative antimicrobial coatings and wound healing applications has been investigated by the inclusion of siderophores into biomaterials [23].

Interactions mediated by siderophores are essential in determining how human–microbe and plant–microbe systems behave. Recent research has demonstrated the significance of siderophores in enhancing plant immunity and growth through rhizosphere microbiome modulation [24, 25]. Similarly, the composition and function of the gut microbiota have been demonstrated to be influenced by siderophores in human–microbe systems, which has implications for human health and disease [26]. Studies have also looked into siderophores' potential as innovative therapeutic agents for treating infectious disorders and modifying the microbiome [25].

15.3.3 Siderophore-Based Therapeutics and Diagnostics

The potential of siderophore-based therapies and diagnostics to treat a wide range of illnesses and ailments has drawn a lot of attention in recent years. Studies have shown that siderophores are effective against bacteria that are resistant to antibiotics, leading to their exploration as potential antimicrobial agents [27]. As possible diagnostic instruments for identifying and tracking infectious diseases like tuberculosis, siderophores have been studied [28]. Additionally, scientists have looked into using siderophores as medicinal agents to address conditions connected to iron, like iron overload and deficiency [13].

15.4 Future Directions in Siderophore Research

15.4.1 Siderophore Regulation, Signaling, and Environmental Sustainability

Research on siderophores is expected to continue growing quickly in the future due to developments in synthetic biology, genomics, and analytical methods. New research topics include how siderophores influence the human microbiota [26] and the development of siderophore-based therapeutics for infectious diseases [13], hoping to provide fresh perspectives on the biology and uses of these fascinating molecules. Additionally, it is anticipated that the combination of machine learning and computational modeling techniques would speed up the identification of new siderophores and their optimization for a range of uses [16]. Determining the functions of siderophores in microbial physiology and pathogenicity requires an understanding of their regulation and signaling mechanisms. Recent research has clarified the complex regulatory networks that control the synthesis of siderophores, including the role of transcriptional regulators like Fur and Irr [29]. Studies have also looked at the signaling pathways that siderophores mediate, including how they function in biofilm development and quorum sensing [30]. Additionally, the discovery of new regulatory elements and signaling pathways involved in siderophore regulation has been made possible by the application of biological techniques like transcriptomics and proteomics [20].

15.4.2 Siderophore-Based Innovations for Bioremediation and Antibiotic Resistance

The biogeochemical cycling of iron and other vital nutrients is influenced by siderophores, which are important for environmental sustainability. The role of siderophores in enhancing soil fertility and plant growth, especially in iron-limited conditions, has been emphasized by recent studies [31]. It has been demonstrated that siderophores accelerate the breakdown of heavy metals and organic contaminants in contaminated soils and water [11, 22]. The potential of siderophores as biofertilizers and bioremediation agents has also been investigated, providing a sustainable substitute for conventional chemical-based methods [1, 11]. Researchers looked into the potential of microorganisms that produce siderophores for the biodegradation of organic contaminants, including pesticides and polycyclic aromatic hydrocarbons (PAHs) [32]. Additionally, the development of biosensors and bioassays based on siderophores has made it possible to recognize and monitor environmental contaminants, offering a useful instrument for pollution prevention and bioremediation projects [15].

The investigation of new antimicrobial tactics has become necessary due to the increase in antibiotic resistance, and siderophore-inspired remedies present an attractive strategy. Recent research has shown that siderophore-based antimicrobials have the ability to target and stop the growth of bacteria that are resistant to antibiotics, such as carbapenem-resistant Enterobacteriaceae (CRE) and methicillin-resistant *Staphylococcus aureus* (MRSA) [33]. Moreover, scientists have looked into the use of siderophore-conjugated antibiotics, which have demonstrated lower toxicity and increased antibacterial activity in comparison to conventional antibiotics [34]. These results demonstrate how siderophore-inspired approaches may be used to fight increasing risks of antibiotic resistance.

15.4.3 Integrating Siderophore Research with Other Disciplines

New opportunities for innovation and discovery have been made possible by the merging of siderophore research with other fields—for instance, the development of new biological systems for the synthesis of siderophores and siderophore-inspired molecules has been made possible by the convergence of siderophore research with synthetic biology [22]. The development of innovative biomaterials and nanomaterials with potential uses in biomedicine, environmental remediation, and energy storage has also resulted from the integration of siderophore research and materials science [23, 35]. Moreover, siderophore research has been integrated with computational modeling and machine learning to make it easier to predict siderophore shapes and functions. This has sped up the process of finding new siderophores with possible uses [16].

15.5 Challenges and Opportunities

15.5.1 Scaling Siderophore Production: Overcoming Challenges and Regulatory Concerns

If we are going to properly utilize the advantages of siderophores, we must overcome a number of obstacles that arise as we continue to investigate their enormous potential. The development of economical and scalable processes for siderophore manufacturing and purification is one of the main obstacles [36]. Furthermore, the complex nature of siderophore activities and structures is a major obstacle to their application and characterization [37]. However, these difficulties also offer chances for creativity and learning, such as the creation of innovative synthetic biology techniques and the investigation of fresh uses in industries like environmental remediation, agriculture, and medicine [1, 38]. Siderophores are becoming more and more popular, yet there are still significant challenges to their scalability and manufacture. To increase siderophore yields, recent research has emphasized the significance of regulating fermentation conditions, including temperature, pH, and nutrient availability [9, 36]. Furthermore, the creation of new expression systems, such *Pichia pastoris* and recombinant *Escherichia coli*, has demonstrated potential for increasing siderophore production [38]. The production and scaling issues around siderophores may be resolved by these developments, opening the door for their broad use across a range of sectors.

Addressing safety and regulatory issues becomes more crucial as siderophores move from lab research to industrial and medical uses. Comprehensive risk assessments and safety evaluations of siderophores are necessary, according to recent studies, especially in view of their possible effects on the environment and human health [11]. To guarantee adherence to current rules and regulations, regulatory frameworks, such as those set up by the European Medicines Agency (EMA) and the US Environmental Protection Agency (EPA), must also be negotiated. In order to resolve these issues and eventually enable the responsible development and implementation of siderophore-based solutions, researchers and industry stakeholders must collaborate to create standardized procedures for siderophore safety testing and regulatory compliance.

15.5.2 Translating Siderophore Research: Collaborative Approaches and Practical Applications

There are many chances for collaborative research and interdisciplinary methods in the quickly developing field of siderophore research. In order to create innovative siderophore-based solutions, recent research has shown the benefits of combining knowledge from microbiology, biochemistry, and materials science [39]. Additionally, partnerships between government, business, and academia have made it easier to translate fundamental research findings into useful applications, including the creation of bioremediation agents and antimicrobials inspired by siderophores [13]. It is crucial to encourage ongoing interaction and knowledge exchange between researchers from many fields in order to fully realize the potential of siderophores. Additionally, creative collaborations and financing programs that support interdisciplinary research and development must be established. A rapidly growing area of study, siderophore translation into practical applications, has the potential to affect a number of sectors, including environmental cleanup, agriculture, and medicine. According to recent research, siderophores can be used as biofertilizers to improve plant nutrition and crop output [40]. Furthermore, siderophore-inspired antimicrobials have demonstrated promise in the treatment of illnesses resistant to antibiotics, underscoring siderophores' potential to address urgent global health issues [14]. Additionally, there has been a lot of interest lately in the development of siderophore-based technologies for environmental remediation, including wastewater treatment and heavy metal removal [41].

15.6 Conclusion

Research on siderophores is developing quickly, and new trends and possibilities for future study hold enormous potential for creative uses. Finding new siderophores and optimizing them for a range of applications could be sped up by combining siderophore research with state-of-the-art technologies including synthetic biology, genomics, and machine learning. Furthermore, novel approaches to urgent global issues are anticipated as a result of the investigation of siderophores in a variety of domains, such as biotechnology, medicine, agriculture, and environmental sustainability. Fostering cooperation, promoting creativity, and tackling the difficulties of increasing siderophore production while guaranteeing their safe and efficient use are all crucial as this field of study develops. In the end, siderophore research has enormous potential to change a variety of industries and enhance human health, environmental sustainability, and agricultural productivity. Research on siderophores has a wide-ranging and complex potential impact on numerous sectors and fields. Siderophores have the potential to transform industries like biotechnology, agriculture, medicine, and environmental sustainability as long as research in this area continues to progress. Addressing urgent global issues like antibiotic resistance, food security, and environmental pollution may be possible with the development of siderophore-based antimicrobials, biofertilizers, and bioremediation agents. Additionally, it is anticipated that siderophore research will advance innovation and discovery in this area by integrating with cutting-edge technologies, including synthetic biology, genomics, and machine learning. Ultimately, siderophore research has the potential to have a huge impact on a variety of sectors and domains, and sustained investment in this subject is probably going to produce game-changing answers for a healthier and more sustainable future. Siderophores are an intriguing class of chemicals that have enormous potential in a number of industries, such as biotechnology, medicine, agriculture, and environmental sustainability. As our knowledge of siderophore biology, chemistry, and control has grown recently, new siderophore-based solutions have been made possible. However, in order to fully realize the promise of siderophores, issues related to production, scalability, and regulatory frameworks must be resolved. It is crucial to promote interdisciplinary cooperation, stimulate creativity, and set up standardized protocols for siderophore safety testing and regulatory compliance as this field of study develops. Finally, siderophore research and uses could transform a number of industries and enhance agricultural output, environmental sustainability, and human health.

AI disclosure statement: The author confirms that no artificial intelligence (AI)-generated content or AI-assisted writing tools were used in the preparation of this chapter (including summaries and the abstract). All text, analysis, and conclusions are the result of the author's original work.

References

1. Singh, P., Chauhan, P.K., Upadhyay, S.K., Singh, R.K., Dwivedi, P., Wang, J., Jiang, M., Mechanistic insights and potential use of siderophores producing microbes in rhizosphere for mitigation of stress in plants grown in degraded land. *Front. Microbiol.*, *13*, 898979, 2022.
2. Ahmed, E. and Holmström, S.J., Siderophores in environmental research: roles and applications. *Microb. Biotechnol.*, *7*, 3, 196–208, 2014.
3. Terra, L., Ratcliffe, N., Castro, H.C., Vicente, A.C., Dyson, P., Biotechnological potential of Streptomyces siderophores as new antibiotics. *Curr. Med. Chem.*, *28*, 7, 1407–1421, 2021.
4. Khasheii, B., Mahmoodi, P., Mohammadzadeh, A., Siderophores: Importance in bacterial pathogenesis and applications in medicine and industry. *Microbiol. Res.*, *250*, 126790, 2021.
5. Timofeeva, A.M., Galyamova, M.R., Sedykh, S.E., Plant growth-promoting soil bacteria: nitrogen fixation, phosphate solubilization, siderophore production, and other biological activities. *Plants*, *12*, 24, 4074, 2023.
6. Schalk, I.J., Bacterial siderophores: diversity, uptake pathways and applications. *Nat. Rev. Microbiol.*, 1–17, 2024.

7. Kumar, A., Chakravorty, S., Yang, T., Russo, T.A., Newton, S.M., Klebba, P.E., Siderophore-mediated iron acquisition by Klebsiella pneumoniae. *J. Bacteriol.*, *206*, 5, e00024–24, 2024.
8. Grosse, C., Brandt, N., Van Antwerpen, P., Wintjens, R., Matthijs, S., Two new siderophores produced by Pseudomonas sp. NCIMB 10586: The anti-oomycete non-ribosomal peptide synthetase-dependent mupirochelin and the NRPS-independent triabactin. *Front. Microbiol.*, *14*, 1143861, 2023.
9. Soares, E.V., Perspective on the biotechnological production of bacterial siderophores and their use. *Appl. Microbiol. Biotechnol.*, *106*, 11, 3985–4004, 2022.
10. Rivaro, P., Vivado, D., Ianni, C., Salis, A., Parodi, A., Millo, E., A New Approach to Characterize Siderophore-Type Ligands in Seawater by Solid Phase Synthesis and SPE-HPLC-ESI-MS/MS Analysis. *J. Mar. Sci. Eng.*, *12*, 1, 110, 2024.
11. Gomes, A.F., Sousa, E., Resende, D., II, A Practical Toolkit for the Detection, Isolation, Quantification, and Characterization of Siderophores and Metallophores in Microorganisms. *ACS Omega*, *9*, 25, 26863–26877, 2024.
12. Sachla, A.J., Soni, V., Piñeros, M., Luo, Y., Im, J.J., Rhee, K.Y., Helmann, J.D., The Bacillus subtilis yqgC-sodA operon protects magnesium-dependent enzymes by supporting manganese efflux. *J. Bacteriol.*, e00052–24, 2024.
13. Ribeiro, M., Sousa, C.A., Simões, M., Harnessing microbial iron chelators to develop innovative therapeutic agents. *J. Adv. Res.*, *39*, 89–101, 2022.
14. Fan, D. and Fang, Q., Siderophores for medical applications: imaging, sensors, and therapeutics. *Int. J. Pharm.*, *597*, 120306, 2021.
15. Prabhakar, P.K., Bacterial siderophores and their potential applications: a review. *Curr. Mol. Pharmacol.*, *13*, 4, 295–305, 2020.
16. Shyam, M., Thakur, A., Velez, C., Daniel, C., Acevedo, O., Bhakta, S., Jayaprakash, V., Mycobactin analogue interacting with siderophore efflux-pump protein: insights from molecular dynamics simulations and whole-cell assays. *Front. Antibiot.*, *3*, 1362516, 2024.
17. Reitz, Z.L. and Medema, M.H., Genome mining strategies for metallophore discovery. *Curr. Opin. Biotechnol.*, *77*, 102757, 2022.
18. Cavas, L. and Kirkiz, I., Characterization of siderophores from Escherichia coli strains through genome mining tools: an antiSMASH study. *AMB Express*, *12*, 1, 74, 2022.
19. Sharma, A., Gupta, A.K., Devi, B., Current trends in management of bacterial pathogens infecting plants. *Antonie van Leeuwenhoek*, *116*, 4, 303–326, 2023.
20. Pecoraro, L., Wang, X., Shah, D., Song, X., Kumar, V., Shakoor, A., Rani, R., Biosynthesis pathways, transport mechanisms and biotechnological applications of fungal siderophores. *J. Fungi*, *8*, 1, 21, 2021.
21. Dell'Anno, F., Vitale, G.A., Buonocore, C., Vitale, L., Palma Esposito, F., Coppola, D., de Pascale, D., Novel insights on pyoverdine: from biosynthesis to biotechnological application. *Int. J. Mol. Sci.*, *23*, 19, 11507, 2022.

22. Nosrati, R., Dehghani, S., Karimi, B., Yousefi, M., Taghdisi, S.M., Abnous, K., Ramezani, M., Siderophore-based biosensors and nanosensors; new approach on the development of diagnostic systems. *Biosens. Bioelectron.*, *117*, 1–14, 2018.
23. McHendrie, R., Nguyen, N.H., Nguyen, M.T., Fallahnezhad, K., Vasilev, K., Truong, V.K., Hashemi, R., Development of Novel Antibacterial Ti-Nb-Ga Alloys with Low Stiffness for Medical Implant Applications. *J. Funct. Biomater.*, *15*, 6, 167, 2024.
24. Kamath, A., Patel, D., Shukla, A., Traversing through the intricate interplay of plant microbiome front runners. *Eur. J. Plant Pathol.*, *165*, 3, 407–419, 2023.
25. Ellermann, M. and Arthur, J.C., Siderophore-mediated iron acquisition and modulation of host-bacterial interactions. *Free Radical Biol. Med.*, *105*, 68–78, 2017.
26. Jaswal, K., Todd, O.A., Behnsen, J., Neglected gut microbiome: interactions of the non-bacterial gut microbiota with enteric pathogens. *Gut Microbes*, *15*, 1, 2226916, 2023.
27. Rayner, B., Verderosa, A.D., Ferro, V., Blaskovich, M.A., Siderophore conjugates to combat antibiotic-resistant bacteria. *RSC Med. Chem.*, *14*, 5, 800–822, 2023.
28. Gokarn, K. and Pal, R.B., Preliminary evaluation of anti-tuberculosis potential of siderophores against drug-resistant Mycobacterium tuberculosis by mycobacteria growth indicator tube-drug sensitivity test. *BMC Complement. Altern. Med.*, *17*, 1–8, 2017.
29. Xie, P., Xu, Y., Tang, J., Wu, S., Gao, H., Multifaceted regulation of siderophore synthesis by multiple regulatory systems in Shewanella oneidensis. *Commun. Biol.*, *7*, 1, 498, 2024.
30. Law, S.K.K. and Tan, H.S., The role of quorum sensing, biofilm formation, and iron acquisition as key virulence mechanisms in Acinetobacter baumannii and the corresponding anti-virulence strategies. *Microbiol. Res.*, *260*, 127032, 2022.
31. Pandey, S., Investigating the impact of siderophore producing bacteria on iron bioavailability in tomato (Solanum lycopersicum L.), Doctoral dissertation, GB Pant University of Agriculture & Technology, Pantnagar-263145, 2023.
32. Behera, S. and Das, S., Potential and prospects of Actinobacteria in the bioremediation of environmental pollutants: Cellular mechanisms and genetic regulations. *Microbiol. Res.*, *273*, 127399, 2023.
33. Endicott, N.P., Siderophore-Dependent Transport Paradigms for Iron Across the Bacterial Cell Envelope in the Human Pathogen Staphylococcus aureus, Washington University in St. Louis, 2019.
34. Negash, K.H., Norris, J.K., Hodgkinson, J.T., Siderophore–antibiotic conjugate design: New drugs for bad bugs? *Molecules*, *24*, 18, 3314, 2019.

35. Rahim, M.A., Kristufek, S.L., Pan, S., Richardson, J.J., Caruso, F., Phenolic building blocks for the assembly of functional materials. *Angew. Chem. Int. Ed.*, *58*, 7, 1904–1927, 2019.
36. Kumar, S., Kumar, N., Islam, S.U. (Eds.), *Role of Microbes in Industrial Products and Processes*, John Wiley & Sons, Hoboken, NJ, 2022.
37. Passari, A.K., Ruiz-Villafán, B., Cruz-Bautista, R., Díaz-Domínguez, V., Rodríguez-Sanoja, R., Sanchez, S., Opportunities and challenges of microbial siderophores in the medical field. *Appl. Microbiol. Biotechnol.*, *107*, 22, 6751–6759, 2023.
38. Cui, F., Fan, R., Wang, D., Li, J., Li, T., Research progress on iron uptake pathways and mechanisms of foodborne microorganisms and their application in the food sector. *Crit. Rev. Food Sci. Nutr.*, *64*, 24, 8892–8910, 2024.
39. Maier, A., Mügge, C., Tischler, D., Biotechnological Aspects of Siderophore Biosynthesis by Actinobacteria, in: *Natural Products from Actinomycetes: Diversity, Ecology and Drug Discovery*, pp. 445–474, Springer Singapore, Singapore, 2022.
40. Aloo, B.N., Tripathi, V., Makumba, B.A., Mbega, E.R., Plant growth-promoting rhizobacterial biofertilizers for crop production: The past, present, and future. *Front. Plant Sci.*, *13*, 1002448, 2022.
41. Sarvepalli, M., Velidandi, A., Ramachandravarapu, A.K., Korrapati, N., Marine Actinomycetes Siderophores: Types, High Throughput Characterization Techniques, Applications, and Their Association with Nanotechnology: A Comprehensive Review. *Nanoworld J.*, *10*, 1, 1–21, 2024.
42. Krewulak, K.D. and Vogel, H.J., Structural biology of bacterial iron uptake. *Biochim. Biophys. Acta (BBA)-Biomembr.*, *1778*, 9, 1781–1804, 2008.
43. Harrisson, O., Developing and Assessing Efficient Hydrocarbon Bioremediation Strategies for Cold-Temperature Soils and Sediments: Two Case Studies In Northern Canada, McGill University Canada, 2023.
44. Sandy, M. and Butler, A., Microbial iron acquisition: marine and terrestrial siderophores. *Chem. Rev.*, *109*, 10, 4580–4595, 2009.
45. Braun, V. and Hantke, K., Recent insights into iron import by bacteria. *Curr. Opin. Chem. Biol.*, *15*, 2, 328–334, 2011.
46. Heffernan, J.R., Wildenthal, J.A., Tran, H., Katumba, G.L., McCoy, W.H., Henderson, J.P., Yersiniabactin is a quorum-sensing autoinducer and siderophore in uropathogenic Escherichia coli. *Mbio*, *15*, 2, e00277–23, 2024.
47. Perraud, Q., Cantero, P., Roche, B., Gasser, V., Normant, V.P., Kuhn, L., Schalk, I.J., Phenotypic adaption of Pseudomonas aeruginosa by hacking siderophores produced by other microorganisms. *Mol. Cell. Proteomics*, *19*, 4, 589–607, 2020.
48. Luo, V.C. and Peczuh, M.W., Location, Location, Location: Establishing Design Principles for New Antibacterials from Ferric Siderophore Transport Systems. *Molecules*, *29*, 16, 3889, 2024.

16

Siderophores as Key Players in Sustainable Agriculture

Ibrahim Isse Ali[1]* and Ali Hussein Ahmed[2]

[1]*Department of Plant Protection, Faculty of Agriculture, Selçuk University, Konya, Türkiye*
[2]*Department of Agribusiness Management, Faculty of Agriculture, Zamzam University, Mogadishu, Somali*

Abstract
Siderophores, which are microbial metabolites with a high affinity for iron, are essential tools in sustainable agriculture that enhance plant growth, disease resistance, and nutrient uptake. Catecholates, hydroxamates, and carboxylate siderophores are three primary classes of siderophores identified based on their functional groups responsible for iron chelation. In this chapter, we highlight the significance of siderophores in soil health and fertility, focusing on their capacity to reduce reliance on pesticides and chemical fertilizers. Additionally, we discuss their prospective application as biological control agents to combat phytopathogens, including viruses, bacteria, and fungi, which are destructive phytopathogens for agricultural production. Beyond biological roles, we also demonstrate that siderophores can be widely used to mitigate climate change challenges, which are major challenges confronting agriculture worldwide through the climate-smart agricultural system. Finally, this chapter underscores the relevance of integrating siderophore-based strategies into sustainable farming practices to achieve food security globally by highlighting recent advances in siderophore research and innovation requirements.

***Keywords*:** Microbial siderophores, sustainable farming practices, environmental sustainability, climate-smart agriculture, biological control

**Corresponding author*: ibraisse252@gmail.com

Debasis Mitra, Ayush Madan and Anju Rani (eds.) Microbial Siderophores for Agricultural Productivity: Perspectives of Use in Agriculture, (381–418) © 2026 Scrivener Publishing LLC

16.1 Introduction

Sustainable agriculture refers to methods that fulfill the community's current fiber and food needs without sacrificing the future generations' capacity to satisfy their requirements. Despite numerous attempts to define sustainable agriculture, its contested nature and the complexity of sustainable agriculture in the nation continue to make it seem ambiguous and vague. Since 1987, when the Brundtland Report was released, the concept of sustainable agriculture has become increasingly significant [1]. One of its common definitions, as described by the FAO, is "the management and conservation of the natural resource base and the orientation of technological and institutional change in such a manner as to ensure the attainment and continued satisfaction of human needs for present and future generations. Such development ... conserves land, water, plant, and animal genetic resources; is environmentally non-degrading, technically appropriate, economically viable, and socially acceptable" [2]. Conventional farming, which relies heavily on high-input, synthetic chemical pesticides, mechanization, extensive land use, and monoculture farming and often prioritizes quantity over quality, is the vice versa of sustainable agriculture [3–6].

Environmental, social, and economic factors, as shown in Figure 16.1, are three main key dimensions to ensuring long-term agricultural viability, reducing reliance on pesticides and synthetic fertilizers and protecting natural resources [7–9]. Sustainable agriculture not only enhances production

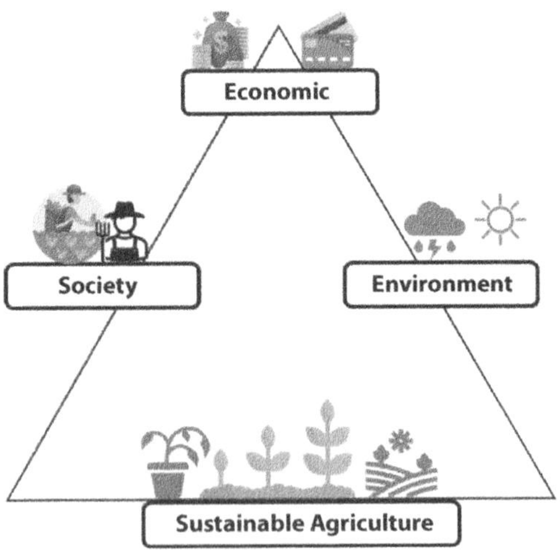

Figure 16.1 Dimensions of sustainable agriculture (created by the authors).

and promotes biodiversity while preserving soil health but also reduces reliance on petroleum-based inputs and non-renewable sources, mitigating their environmental impacts and high costs by substituting them with renewable ones [10, 11].

In terrestrial ecosystems, the interactions between microbes and plants profoundly influence biogeochemical cycling, community formation, and individual growth. Furthermore, plant–microbe interactions have been proven crucial in the advancement of biofertilizers, biological control agents, plant growth promoters, soil conditioners, and bioremediation for sustainable agriculture by tolerating abiotic and biotic stresses [12–14]. Employing microbe-derived biological control agents, including fungi and bacteria, as well as natural substances, predators, and parasitoids against phytopathogens is a well-known principle of agricultural sustainability by avoiding pesticides and the use of chemicals [15–17]. These microbes can produce siderophores, one of the key mechanisms through which beneficial microbes compete against plant pathogens.

This chapter aims to discuss the significance of siderophores in enhancing plant growth, nutrient uptake, soil health, and disease suppression, which are essential components of sustainable agriculture practices. Additionally, it describes the primary types of siderophores and their mechanisms of action in soil and plant health, focusing on reducing reliance on chemical fertilizers. The chapter also investigates how siderophores help mitigate climate change challenges through climate-smart agriculture. Finally, it highlights recent advances in siderophore research and their applications in sustainable agriculture, emphasizing the need for innovation in this field to support global food security.

16.2 Siderophore Production: A Key Factor in Plant-Pathogen Growth and Virulence

Siderophores are compound Greek terms, sidero, meaning "iron," and phore, meaning "carriers," therefore signifying "iron carriers" [18, 19]. Microbes produce these low-molecular-weight secondary metabolites in response to iron deprivation, facilitating the organism's iron acquisition. They are the most potent Fe^{3+} binding agents known, even though a growing variety of siderophore activities are increasingly being recognized [20]. As shown in Table 16.1, siderophores are generated by several microbes, including *Trichoderma*, *Bacillus*, *Rhizobium*, some *Pseudomonas* species, and others [21–23]. Previous studies also declared that some plant species,

Table 16.1 Common bioagents used in plant disease control.

Bioagent	Diseases/target pathogens	Crop	Effectiveness	References
Bacterial bioagents				
Bacillus subtilis	Root rot, common bacterial blight, fire blight, leaf blight, gray mold, bacterial leaf spot, powdery mildew, rice blast, black Sigatoka, walnut blight, and white mold	Various crops	Reduced disease incidence and severity, promoted plant development and seed germination	[119–124]
B. amyloliquefaciens	Anthracnose rot, bacterial wilt, blackleg, soft rot, *Botrytis cinerea*, *Penicillium expansum*, *Rhizopus stolonifer* *Moniliophthora racemosa*, *Aspergillus niger*, *Fusarium oxysporum*, *Botryosphaeria dothidea*, and *Candida albicans*	Fruit, cereal, legumes, and other crops	Hindered mycelial and conidia development, suppresses disease; also induces plant resistance genes	[88, 125, 126]
Bacillus thuringiensis (*Bt*)	Fall armyworm, European corn borer, caterpillars, pea aphids, butterflies, beetles, and various Lepidoptera insect larvae	Maize, soybean, potato, cotton, rice, eggplant, and tomato	Demonstrated growth inhibition and high mortality rates in multiple insect larval hosts	[127–130]

(*Continued*)

Table 16.1 Common bioagents used in plant disease control. (*Continued*)

Bioagent	Diseases/target pathogens	Crop	Effectiveness	References
B. velezensis	*Botrytis cinerea, Ralstonia solanacearum, Fusarium oxysporum, Trichoderma aggressivum f. europaeum, Bipolaris sorokiniana, X. oryzae* pv. *oryzicola*, and *X. oryzae* pv. *oryzae*	Potato, eggplant, banana, and tomato	Suppressed the disease incidence and severity of these pathogens with a control efficiency	[131–134]
B. pumilus	*Pseudomonas syringae* pv. *lachrymans, Xanthomonas axonopodis, Arthrobotrys conoides, Rhizoctonia solani, Fusarium solani, Fagopyrum esculentum, Erwinia tracheiphila, Sclerotinia sclerotiorum, Cronartium quercuum,* and *Fusarium oxysporum*	Several crops	Promoted plant growth and triggered plant systemic resistance	[91, 135–137]
B. licheniformis	Anthracnose, vascular wilt, early blight, damping off, and dollar spot	Chilli, lawns, sports turf, golf courses, and arboretum	Inhibited disease by 81.18% compared to the control group; decreased disease severity and incidence; also induced plant resistance genes	[89, 138–140]

(*Continued*)

Table 16.1 Common bioagents used in plant disease control. (*Continued*)

Bioagent	Diseases/target pathogens	Crop	Effectiveness	References
Pseudomonas fluorescens	*Erwinia amylovora, Plasmodiophora brassicae, X. oryzae* pv. *oryzae, G. graminis* var. *tritici, Thielaviopsis basicola, Macrophomira phaseolina, Xanthomonas axonopodis* pv. *phaceoli,*	Rice, pear, wheat, beans, black gram, and Chinese cabbages	Mycelial growth and spore germination were reduced, as well as increased plant resistance	[92, 141–143]
P. aeruginosa	*Pythium splendens, Pythium myriotylum,* and *Fusarium oxysporum*	Chickpeas, pigeonpeas, and banana	Reduced disease incidence and severity; also inhibited the growth of mycelial and spore germination	[93, 144]
Rhizobium spp.	*Cylindrocladium parasiticum, Macrophomina phaseolina, Fusarium solani* f. sp. *phaseoli, Xanthomonas phaseoli* pv. *phaseoli, Xanthomonas campestris* pv. *glycine,* and *Pseudomonas syringae* pv. *phaseolicola*	Several crops	Plant growth has been increased significantly by suppressing disease severity	[145–147]
Certain *R. radiobacter* strains	*Rhizobium rhizogenes*	Apple	Completely inhibited hairy root disease	[96]

(*Continued*)

Table 16.1 Common bioagents used in plant disease control. (*Continued*)

Bioagent	Diseases/target pathogens	Crop	Effectiveness	References
Paenibacillus polymyxa	*Phytophthora capsici* and *Fusarium oxysporum* f. sp. *cucumerinum*	Several crops	Decreased disease index and symptoms, as well as promoted plant growth	[97]
Fungal bioagents				
Trichoderma harzianum	*Botrytis, Rhizoctonia, Fusarium, Pythium* spp., or other foliar pathogenic, soil fungi, and bacterial diseases	Various crops	Protected plants against diseases and stimulates the growth of roots and root hairs	[148]
Trichoderma viride	Potato virus Y and root rot diseases	Potato, strawberry, and soybean	Increased defense-related genes and the productivity of the crop	[149–151]
Trichoderma/ Gliocladium virens	Shallot wilt and vascular wilt diseases	Shallots and eggplants	Promoted seedling growth, reduced disease severity, and increased stem and root length, showcasing how well plant development may be encouraged	[152, 153]
Beauveria bassiana	Powdery mildew, aphids, whiteflies, thrips, brown marmorated stink bugs, yellow mealworm beetles, and tea pests	Tomato, cucumber, and strawberry crops	Induced a hypersensitive response-like cell death against disease and reduced the insect population	[154–156]

particularly the grass family, such as wheat, barley, maize, and rice, can generate siderophores, which are called phytosiderophores or plant siderophores, from roots to chelate and dissolve Fe^{3+} in the soil [24–26]. When iron availability is limited, especially in calcareous (high pH) soils, these plants can convert iron into insoluble forms that plant pathogens cannot take. Deoxymugineic acid (DMA), mugineic acid (MA), and epihydroxy mugineic acid (epiHMA) are well-known examples of plant siderophores [26–28]. Plant siderophores solve iron deficiency and support nutrient uptake, stress tolerance, and sustainable agricultural practices. Non-grass crops also utilize strategy I, which involves acidifying, reducing, and transporting the rhizosphere to mobilize and acquire iron [29].

Iron is one of the micronutrients that are physiologically necessary for plant development and growth, especially in metabolic processes like respiration, cell proliferation, chlorophyll synthesis, electron transfer, photosynthesis, and DNA synthesis [30]. Iron, the lithosphere's fourth most abundant metal, is typically found in large amounts in soils. Nevertheless, its bioavailability is restricted in surroundings with neutral pH and aerobic conditions [31, 32]. It is also important for numerous functions related to proteins, especially enzymes engaged in redox processes, which necessitate iron for their functions. This element is not only crucial for plants but also has key importance for plant pathogens' growth. Sufficient iron acquisition is linked to the virulence of many plant pathogens by supporting the growth and reproduction and effectively colonizing their host of plant pathogens. Additionally, as iron is an indispensable component of enzymes like catalase and peroxidase, it protects pathogens from oxidative stress during host invasion.

Phytopathogenic fungi and bacteria acquire iron through two distinct pathways: low-affinity and high-affinity uptake systems [33, 34]. As is common, plant photosynthesis and chlorophyll production are reduced by iron deficiency, which causes chlorosis and serious growth problems [35, 36] as well as affects the proliferation and virulence of diseases. Thus, there is competition among plant pathogens and plants with help from beneficial microbes for iron elements.

16.3 Role of Siderophores in Soil Health and Fertility

Soil plays a major role in maintaining the equilibrium of ecosystems on Earth, with fertility and productivity of the soil crucial for agricultural sustainability [37]. Like humans and animals, plants need food and nutrients to grow properly. Internal signals that rely on the proper movement of mineral-rich nutrients from the soil to the roots and from the roots to the shoots control the developmental processes of plants [38, 39]. A total of 18 critical elements are among the several chemical components, referred to as plant nutrients, divided into macronutrients and micronutrients. The whole life cycle of a plant, including germination, productivity, metabolism, and defense, depends on these nutrients. Potassium (K), phosphorus (P), and nitrogen (N) are the primary macronutrients necessary for plant development and growth, while the pace of nutrient requirements varies. Compared to K, P, and N, the secondary macronutrients sulfur (S), magnesium (Mg), and calcium (Ca) are required in smaller quantities [40]. For crop growth and maximum production, micronutrients such as nickel (Ni), cobalt (Co), chlorine (Cl), copper (Cu), boron (B), molybdenum (Mo), manganese (Mn), zinc (Zn), and iron (Fe) are needed in far lesser amounts than the macronutrients [41]. Therefore, if plants are provided enough of these nutrients, more yields can be harvested, and the income of farmers can be increased. Unfortunately, soil faces several challenges, including soil erosion [42], salinization [43], acidification [44], compaction [45], nutrient imbalance [46, 47], loss of biodiversity [48, 49], and contamination [50, 51] from pesticide residue, industrial pollutants, and heavy metals, posing serious obstacles to sustainable agriculture. Hence, achieving food security and sustainable agriculture is a daydream until the abovementioned issues are solved. Using siderophore-producing microbes and beneficial communities can be a solution to many of the problems mentioned above.

Soil microbial communities are important for controlling soil-related activities including organic matter decomposition and nutrient cycling; hence, it is essential to comprehend the variables that influence their size, activity, and structure [52]. They exhibit significant spatial organization, influenced partially by the soil's structure. Comprehending the spatial organization of microorganisms is essential for grasping soil microbial ecology and its influence on ecosystem services and biogeochemical processes [53]. The soil microbial community's composition may be significantly modified by its interaction with certain plant species, and these alterations can have crucial repercussions for plant ecology. Negative plant–soil response is crucial for sustaining plant community variety, whereas positive plant–soil feedback may lead to community transformation [54].

Siderophore-producing microbial communities not only enhance soil fertility, structure, nutrient uptake, water retention, and carbon sequestration but also exhibit tolerance to plant salinity, thereby enhancing crop productivity in degraded soils [55–58]. Additionally, they stimulate root elongation and branching, thereby increasing root systems and reducing soil compaction. Furthermore, they naturally supply iron and other nutrients, and some are resistant to heavy metals, even chelating heavy metals such as cadmium, arsenic, and lead, which reduces the need for synthetic fertilizers and minimizes the entry of toxins into food chains [59–61]. Therefore, utilizing siderophore-producing microbes as biofertilizers and biological control agents shapes a cornerstone of sustainable agricultural practices by helping to restore degraded soil, improve crop productivity, and promote long-term sustainability.

16.4 Biological Control of Plant Diseases Through Siderophores

Plant diseases are triggered by phytopathogens such as viruses, fungi, and bacteria, which disrupt plants' normal physiological processes. These diseases are characterized by slowed plant growth and improvement, reduced quality and production, and, in extreme cases, plant mortality [62, 63]. Plant diseases are a major factor preventing the attainment of food security in developed and developing countries [62, 64, 65], as they not only cause lower yields but also decrease the value, quantity, and quality of agricultural commodities, raise the cost of inputs and labor, and have detrimental effects on farmers' incomes and livelihoods [66]. Plant diseases are believed to account for approximately 30% of global crop yield losses yearly, which amounts to losses of hundreds of billions of dollars [67, 68]. Several phytopathogens, such as *Claviceps* spp., *Alternaria* spp., *Penicillium* spp., *Aspergillus* spp., *Fusarium* spp., and others, produce mycotoxins that are dangerous to both animals and human health [69–72].

A common method for controlling diseases in plants involves the application of chemical and pesticide treatments, but unfortunately, they have many negative impacts on human health, biodiversity, and the environment, as shown in Figure 16.2. The utilization of pesticides against plant diseases and pests not only pollutes soil, air, and water but also harms non-target species, including bees, birds, fishes, and natural pest predators, which results in an increase in pest populations and a local ecosystem's imbalance [73–79]. In recent years, biological control and biopesticides, which are eco-friendly alternative methods against plant pests and diseases, have been increasingly employed due to the growing knowledge of the risks posed by pesticides [80–84]. These methods lower the severity and incidence of diseases and help preserve natural enemies and biodiversity.

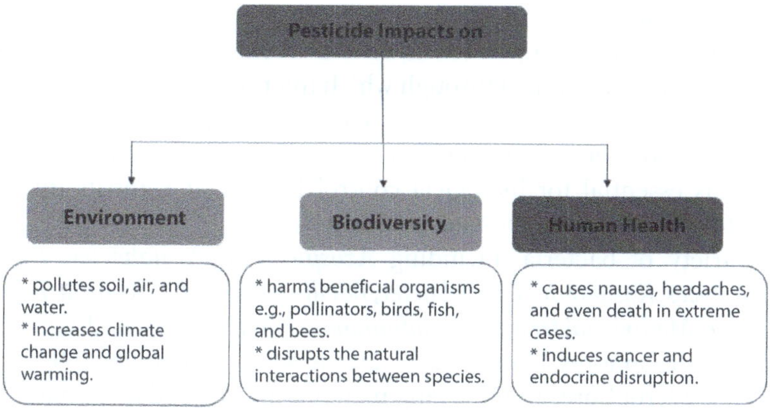

Figure 16.2 Pesticide usage impacts the environment, biodiversity, and human health (created by the authors).

Many species of bacteria, including *Bacillus subtilis* [85, 86], *B. thuringiensis* [85, 87], *B. amyloliquefaciens* [88], *B. licheniformis* [89], *B. velezensis* [90], *B. pumilus* [91], *Pseudomonas fluorescens* [92], *P. aureofaciens* [93], *Rhizobium* spp. [94, 95], *R. radiobacter* [96], and *Paenibacillus polymyxa* [97], are known as plant-growth-promoting bacteria and are effective against plant pathogens. *Trichoderma harzianum*, *Trichoderma viride*, *Gliocladium virens*, and *Beauveria bassiana* are beneficial fungi used as bioagents against plant diseases [98–100]. Table 16.1 shows how these bacteria and fungi be against common pests and phytopathogens such as *Aspergillus niger*, *Penicillium expansum*, *Botrytis cinerea*, *Rhizopus stolonifera*, *Botryosphaeria dothidea*, *Candida albicans*, *Ralstonia solanacearum*, *Fusarium oxysporum*, *Trichoderma aggressivum f. europaeum*, *Bipolaris sorokiniana*, *X. oryzae* pv. *oryzae*, and *X. oryzae* pv. *oryzicola* [101–108]. There are two mechanisms: direct, which includes nutrient fixation, solubilization, phytohormones, siderophores, hydrolytic and lytic enzyme production; and indirect, which provides for the release of volatile organic compounds (VOCs), the activation of systemic resistance, biofilm formation, and detoxification of pathogen toxins [55, 109–117].

Competing for essential resources such as oxygen, nutrients, and space is common among soil microbes. The competition between pathogenic and non-pathogenic microbes for nutrients and space surrounding host plants may lead to biological control [118]. The most common nutrient element over which beneficial microbes, including *Trichoderma* spp., *Pseudomonas* spp., *Rhizobium* spp., *Bacillus* spp., *Streptomyces* spp., and others, and pathogenic microbes compete is iron (Fe). Because of insufficient bio-available iron in the soil environment, there is intense competition. The process responsible for this competition is known as siderophores. Siderophores represent a key mechanism through which microbes compete against plant pathogens. Their primary function is to chelate iron, making it soluble and extractable from organic materials and minerals [157]. Siderophore production is essential for biological control because it combats infectious agents for iron, limiting their accessibility [158].

A variety of bacteria, including *Azospirillum*, *Azotobacter*, *Dickeya*, *Streptomyces*, *Kosakonia*, *Enterobacter*, *Bacillus*, *Klebsiella*, *Serratia*, *Pantoea*, *Nocardia*, *Methylobacterium*, *Pseudomonas*, *Paenibacillus*, and others, can produce siderophores. Numerous researchers have documented the significance of siderophores in inhibiting plant pathogens and lowering disease severity and incidence. Kloepper *et al.* [159] investigated the importance of siderophores in suppressing pathogens like *Pectobacterium carotovorum*, showcasing the capability of PGPR to biologically control plant diseases.

The effectiveness of *Bacillus amyloliquefaciens* strain S1 against *C. michiganensis* spp. *Michiganensis*, which causes canker disease in tomatoes, was reported. The effectiveness of strain S1 was linked to the synthesis of siderophores, antibacterial metabolites, and lytic enzymes in *in vitro* conditions [160]. *Fusarium oxysporum*, *Botrytis cinerea*, *Nigrospora oryzae*, *Rhizoctonia solani* AG1-IA, and *Cochliobolus heterostrophus* are common fungi that induce diseases in several crops. A *Bacillus subtilis* strain 330-2, obtained from rapeseed, has been shown to generate siderophores, lytic enzymes, and indole-3-acetic acid, inhibiting these fungal pathogens and promoting maize and rice seedling growth by 14% to 37% in an *in vitro* study [161]. The three main kinds of siderophores, catecholate [162, 163], hydroxamates [164, 165], and carboxylate siderophores [166, 167], are distinguished by the functional groups that enable iron chelation. As a fourth siderophore type, some researchers include mixed-type siderophores [168, 169]. All of these siderophore types have features that can protect plants from phytopathogens, diseases, and pests.

As shown in Table 16.2, multiple microorganisms produce siderophores to safeguard plants against pests, pathogens, and diseases. *Bacillus subtilis*, one of the most frequently used bacteria as a bioagent, can produce a siderophore called bacillibactin, which belongs to the catecholate group [170, 171]. This beneficial bacterium, according to previous studies, has become successful in controlling *Cephalosporium maydis*, *Erwinia amylovora*, *Fusarium oxysporum* Schl. f. sp. *capsica*, and *Xanthomonas axonopodis* pv. *phaseoli*, which harm various crops, including maize, pepper, apple, beans, sesame, and others [120, 121, 170, 172]. Shizokinen and bacillibactin siderophores are named in the hydroxamate and catecholate groups,

Table 16.2 Siderophore-producing microbes against plant diseases.

Microbe	Pathogen/disease	Siderophore type (name)	Crop	Reference
Bacillus subtilis	*Cephalosporium maydis* and *Fusarium oxysporum* Schl. f.sp. *capsici*	Catecholate (bacillibactin)	Maize, pepper, and sesame	[170–173]
Bacillus megaterium	*Meloidogyne graminicola* and *Mycosphaerella graminicola*	Hydroxamate (shizokinen and N-deoxyshizokinen)	Rice and wheat	[174, 175]
Bacillus amyloliquefaciens	*Pseudomonas syringae* pv. *tomato* and Fusarium wilt	Catecholate (bacillibactin)	Tomato and brinjal plants	[176, 177]
Azotobacter brasilense	*Colletotrichum acutatum*	Catecholate (spirilobactin)	Strawberry and sugarcane crop	[178, 179]
Pseudomonas fluorescens	*Pythium ultimum*	Mixed (ferribactin)	Cotton	[180, 181]
Pseudomonas putida	*Fusarium oxysporum* f. sp. *melonis* and *Glomerella tucumensis*	Mixed (pyoverdine)	Sugarcane and muskmelon	[182–184]
Pseudomonas aeruginosa	*Sclerotinia sclerotiorum*, *Colletotrichum gloeosporioides*, and *Rhizoctonia solani*	Catecholate (pyoverdin and pyochelin)	Tomato and chili	[185–187]
Trichoderma harzianum	*Pythium* spp., *Sclerotinia sclerotiorum*, *Pseuperonospora cubensis*, *Sphaerotheca fusca* (*syn. S. fuliginea*), and *Botrytis cinerea*	Hydroxamate (harzianic acid)	Cucumber, radish, and tomato	[188–191]
Enterobacter cloacae	*Pythium ultimum*, *F. solani var coeruleum*, and/or *F. sambucinum*	Catecholate and hydroxamate (enterobactin and aerobactin)	Cucumber and potato	[192– 194]

(*Continued*)

Table 16.2 Siderophore-producing microbes against plant diseases. (*Continued*)

Microbe	**Pathogen/disease**	**Siderophore type (name)**	**Crop**	**Reference**
Rhizobium meloti	*Fusarium* spp. and *Pythium* spp.	Catecholate (rhizobactin)	Alfalfa	[195, 196]
Rhizobium radiobacter K84	*Rhizobium radiobacter* (formerly known as *Agrobacterium tumefaciens*)	Hydroxamate (agrobactin)	Fruit crops such as stone, pome, cane fruit, nuts, and blueberries	[197, 198]
Burkholderia cepacia	*Pythium aphanidermatum* and *Aphanomyces euteiches*	Hydroxamate (ornibactin)	Peas	[199, 200]
Rhizobium leguminosarum	*Fusarium solani* and *Sclerotium rolfsii*	Hydroxamate (schizokinen)	Common bean	[201–203]
Pseudomonas syringae	*Penicillium digitatum, Botrytis cinerea, Rhizopus stolonifera*, and *Rhizoctonia solani*	Mixed (pyoverdine)	Citrus, stone fruits, and potatoes.	[204, 205]
Streptomyces spp.	*Fusarium oxysporum, Agrobacterium tumefaciens, Xanthomonas campestris* pv. *vesicatoria, Erwinia amylovora, Magnaporthe oryzae, Rhizoctonia solani*, and *Pyrrhoderma noxium*	Hydroxymate (desferrioxamine B)	Sunflower, potato, and several crops	[206–213]

produced by *Bacillus megaterium* and *Bacillus amyloliquefaciens*, respectively, against *Meloidogyne graminicola*, *Mycosphaerella graminicola*, and *Pseudomonas syringae* pv. *tomato* [174, 175].

Pseudomonas aeruginosa, *Pseudomonas fluorescens*, and *Pseudomonas putida* are beneficial *Pseudomonas* strains that can also generate siderophores named ferribactin, pyoverdine, and pyochelin from the mixed and catecholate groups, respectively. These siderophores play a crucial role in managing *Rhizoctonia solani*, *Pythium ultimum*, *Fusarium oxysporum* f. sp. *melonis*, *Sclerotinia sclerotiorum*, *Glomerella tucumensis*, and *Colletotrichum gloeosporioides* [181, 183, 186]. Additionally, these strains support the soil microbial community, improving nutrient availability and enhancing stress resistance.

Jia *et al.* [214] evaluated the potential of *Pseudomonas fluorescens* J8 and *Bacillus subtilis* J3 as components of a synthetic microbial community (BCA) against *A. solani*, which induces early blight in tomatoes. They found that BCA increased the biomass growth by 66.91% to 489.21%, decreased the disease incidence by 38.26% to 42.87%, and increased the tomato enzyme activities (peroxidase, catalase, and superoxide dismutase) by 73.11% to 90.22%. Furthermore, BCA increased beneficial phyllosphere and rhizosphere microbial communities by 6.58%–192.77%, contributing to disease control.

Trichoderma species are also one of the most common advantageous microbes utilized as biocontrol for seed and soil-borne phytopathogens [215, 216]. One of the *Trichoderma* species, *Trichoderma harzianum*, generates a siderophore called harzianic acid, which belongs to the hydroxamate group in controlling *Colletotrichum gloeosporioides*, *Ralstonia solanacearum*, *Rhizoctonia solani*, *Fusarium oxysporum* f. sp. *lycopersicum*, and *Pythium* spp., deteriorating chili, tomato, pea, and brinjal production [189, 190, 217]. A study conducted by Freitas *et al.* [218] demonstrated a notable suppressive effect against *Fusarium oxysporum* on tomatoes when cocoa pod husk extract and harzianic acid were incorporated into hydrocolloid-based coating formulations. Based on their findings, the coated tomatoes reduced degradation (33.33%) and weight loss (23.26%) while maintaining key quality attributes such as β-carotene, lycopene, antioxidant capacity, and acidity. Thus, the utilization of beneficially siderophore-producing microbes is essential for organic agricultural sustainability and achieving food security globally.

16.5 The Importance of Siderophores in Advancing Climate-Smart Agriculture

Climate change is a critical issue confronting agriculture worldwide. Alterations in rainfall patterns, rising temperatures, and increased carbon dioxide levels, which negatively impact staple cereal and legume crops, have been observed in both developing and developed nations, contributing to the spread of diseases and pests [219–227]. To mitigate these issues and achieve food security, scientists and environmentalists have suggested adopting a climate-smart agriculture (CSA) strategy.

Climate-smart agriculture is a system designed to reduce greenhouse gas emissions, lessen the consequences of climate change, and increase agricultural productivity [228–231]. Drought-resistant crop varieties, integrated pest management (IPM), efficient water management, agroforestry, conservation tillage, crop rotation, and diversification are the most common applications in climate-smart agriculture, as shown in Figure 16.3, contributing to the achievement of several SDGs [232–239].

Droughts and water shortages caused by climate change are two of the most prestigious challenges in our world. Therefore, beneficial microbes, including the *Bacillus* and *Trichoderma* species, not only manage diseases but also can boost the soil's capacity to retain moisture, enhancing the bacteria community in the rhizosphere, fixing nitrogen and resisting stresses. Bi *et al.* [240] tested *Bacillus subtilis*' influences on the growth and physiology of cotton under salt and water stress. Based on their result, the yield was increased by 3.32%–54.67%, and water use efficiency improved

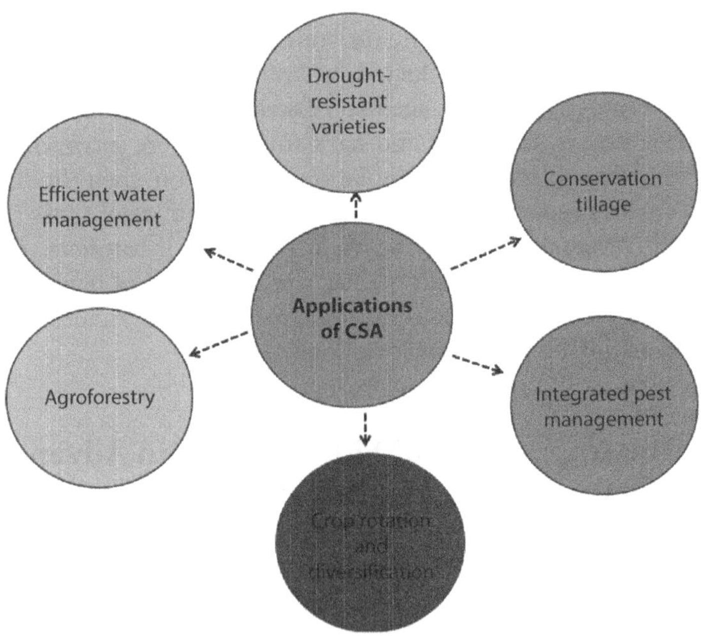

Figure 16.3 Applications of climate-smart agriculture strategy (created by the authors).

by 1.68%–41.07% compared to the control group. They also highlighted that this bacterium significantly promoted soil desalination, enhanced soil water holding capacity, and improved cotton growth indices such as stem diameter, dry matter accumulation, leaf area index, and plant height under water and salt stress. Another study done under drought-stressed conditions evaluated the effectiveness of some *Bacillus* spp., including Bacillus *megaterium* and *Bacillus thuringiensis*, in increasing potassium content, reducing stomatal conductance, and controlling proline accumulation in *Salvia officinalis* and *Lavandula dentata* [241].

Rice is cultivated as a staple food across various continents; unfortunately, drought affects nearly 50% of global rice production due to the crop's sensitivity [242–245]. Many researchers have emphasized that using microbes can increase production and reduce the impact of drought on rice. As mentioned earlier, *Trichoderma harzianum*, a siderophore-producing fungus, is identified as drought-tolerant because of its capacity to effectively colonize neem de-oiled cake at low moisture levels, improving rice production at the tillering stage, reducing leaf rolling, minimizing proline content, and enhancing membrane stability index, leaf chlorophyll content, relative water content, and leaf area index under water shortage conditions [246]. Another work done by Quazi *et al.* [247] in Bangladesh investigated the effects of *Trichoderma brevicompactum* and *Trichoderma asperelloides* on rice crops subjected to drought stress. They reported that these species enhanced drought tolerance in drought-susceptible (*IR64*) and drought-tolerant (*BRRI dhan56*) rice varieties by improving root length and altering biochemical markers, including total phenolic content (TFC), malondialdehyde (MDA), proline, and H_2O_2 levels. Hence, the integration of siderophores into agricultural practices not only controls phytopathogens but also provides an innovative solution to the issues posed by climate change, improving nutrient availability, reducing chemical input dependency, and promoting sustainable soil health suitable perfectly with the goals of climate-smart agriculture.

16.6 Conclusion and Future Directions

In sustainable agriculture, siderophores are a potential but little-studied technique. Beneficial microorganisms produce these natural compounds by playing an essential role in encouraging plant development, improving the accessibility of nutrients, and controlling pathogens while minimizing the need for pesticides and synthetic fertilizers which harm biodiversity and the environment. Additionally, utilizing siderophore-produced microbes not only protects against plant diseases but also addresses global food security challenges by mitigating the consequences of environmental stress like soil salinity, drought, and nutrient deficiencies as well as building a more sustainable future for agriculture. However, further investigation, collaboration, and innovation are needed to maximize the effectiveness of siderophores.

The investigation of new and diverse sources of siderophore-producing microbes that can adapt to challenging conditions is essential. Furthermore, studies focused on understanding the genetic and metabolic pathways of these microbes in different environments are required to achieve agricultural sustainability. While modern research is conducted in laboratories, evaluating the scalability and practical applications of siderophore-based technologies in real-world farming systems is also crucial. Finally, with continued research and collaboration, siderophores can revolutionize sustainable agriculture and support global food security in an increasingly changing world.

AI disclosure statement: The author acknowledges the use of AI-based tools for minor language editing or grammatical correction. However, all intellectual contributions, analyses, and conclusions presented in this chapter are entirely the author's own.

References

1. Velten, S., Leventon, J., Jager, N., Newig, J., What Is Sustainable Agriculture? A Systematic Review. *Sustainability*, 7, 6, 7833–7865, 2015, https://doi.org/10.3390/su7067833.
2. FAO, Sustainable development and natural resources management. *Twenty-fifth Conference, paper C 89/2, Supp. 2*, Rome, Food and Agriculture Organization, 1989, https://openknowledge.fao.org/items/3d4031a5-393d-49a2-a644-8889afda1d2b.
3. Aslam, W., Shahzad Noor, R., Ullah, S., Chen, H., Comparative economic analysis of crop yield under organic and conventional farming systems in Punjab, Pakistan. *Asian J. Agric. Biol.*, 8, 2, 2020.
4. Dantsis, T., Loumou, A., Giourga, C., Organic Agriculture's Approach towards Sustainability; Its Relationship with the Agro-Industrial Complex, A Case Study in Central Macedonia, Greece. *J. Agric. Environ. Ethics*, 22, 197–216, 2009, https://doi.org/10.1007/s10806-008-9139-0.
5. Mäder, P., Edenhofer, S., Boller, T., Wiemken, A., Niggli, U., Arbuscular mycorrhizae in a long-term field trial comparing low-input (organic,

biological) and high-input (conventional) farming systems in a crop rotation. *Biol. Fertil. Soils*, 31, 150–156, 2000, https://doi.org/10.1007/s003740050638.

6. Średnicka-Tober, D., Obiedzińska, A., Kazimierczak, R., Rembiałkowska, E., Environmental impact of organic vs. conventional agriculture-a review. *J. Res. Appl. Agric. Eng.*, 61, 4, 204–211, 2016.
7. Dönmez, D., Isak, M.A., İzgü, T., Şimşek., O., Green Horizons: Navigating the Future of Agriculture through Sustainable Practices. *Sustainability*, 16, 8, 3505–3505, 2024, https://doi.org/10.3390/su16083505.
8. Bathaei, A. and Štreimikienė, D., A Systematic Review of Agricultural Sustainability Indicators. *Agriculture*, 13, 2, 241, 2023, https://doi.org/10.3390/agriculture13020241.
9. Sahraei, S., Pakravan-Charvadeh, M.R., Gholamrezai, S., Rahimian, M., Assessing the association of sustainable agriculture with rural household food security (considering ecological, economic, and social aspects). *Front. Nutr.*, 9, 2022, https://doi.org/10.3389/fnut.2022.899427.
10. Bathaei, A. and Štreimikienė, D., Renewable Energy and Sustainable Agriculture: Review of Indicators. *Sustainability*, 15, 19, 14307, 2023, https://doi.org/10.3390/su151914307.
11. Majeed, Y., Khan, M.U., Waseem, M., Zahid, U., Mahmood, F., Majeed, F., Sultan, M., Raza, A., Renewable energy as an alternative source for energy management in agriculture. *Energy Rep.*, 10, 344–359, 2023, https://doi.org/10.1016/j.egyr.2023.06.032.
12. Singh, P.P., Kujur, A., Yadav, A., Kumar, A., Singh, S.K., Prakash, B., Mechanisms of Plant-Microbe Interactions and its Significance for Sustainable Agriculture, in: *PGPR Amelioration in Sustainable Agriculture*, A.K. Singh, A. Kumar, P.K. Singh (Eds.), pp. 17–39, Woodhead Publishing, 2019, https://doi.org/10.1016/B978-0-12-815879-1.00002-1.
13. Fadiji, A.E., Yadav, A.N., Santoyo, G., Babalola, O.O., Understanding the plant-microbe interactions in environments exposed to abiotic stresses: An overview. *Microbiol. Res.*, 271, 127368, 2023, https://doi.org/10.1016/j.micres.2023.127368.
14. Cao, Y., Shen, Z., Zhang, N., Deng, X., Thomashow, L.S., Lidbury, I., Liu, H., Li, R., Shen, Q., Kowalchuk, G.A., Phosphorus availability influences disease-suppressive soil microbiome through plant-microbe interactions. *Microbiome*, 12, 1, 185, 2024, https://doi.org/10.1186/s40168-024-01906-w.
15. Lee, J., Kim, S., Jung, H., Koo, B., Han, J.A., Lee, H.S., Exploiting Bacterial Genera as Biocontrol Agents: Mechanisms, Interactions and Applications in Sustainable Agriculture. *J. Plant Biol.*, 66, 6, 485–498, 2023, https://doi.org/10.1007/s12374-023-09404-6.
16. Woo, S.L., Hermosa, R., Lorito, M., Monte, E., *Trichoderma*: a multipurpose, plant-beneficial microorganism for eco-sustainable agriculture. *Nat. Rev. Microbiol.*, 21, 5, 312–326, 2023, https://doi.org/10.1038/s41579-022-00819-5.

17. Ayaz, M., Li, C.-H., Ali, Q., Zhao, W., Chi, Y.-K., Shafiq, M., Ali, F., Yu, X.-Y., Yu, Q., Zhao, J.-T., Yu, J.-W., Qi, R.-D., Huang, W.-K., Bacterial and Fungal Biocontrol Agents for Plant Disease Protection: Journey from Lab to Field, Current Status, Challenges, and Global Perspectives. *Molecules*, 28, 18, 6735, 2023, https://doi.org/10.3390/molecules28186735.
18. Neilands, J.B., Siderophores: structure and function of microbial iron transport compounds. *J. Biol. Chem.*, 270, 45, 26723–26726, 1995, https://doi.org/10.1074/jbc.270.45.26723.
19. Saha, R., Saha, N., Donofrio, R.S., Bestervelt, L.L., Microbial siderophores: a mini review. *J. Basic Microbiol.*, 53, 4, 303–317, 2013, https://doi.org/10.1002/jobm.201100552.
20. Johnstone, T.C. and Nolan, E.M., Beyond iron: Non-classical biological functions of bacterial siderophores. *Dalton Trans.*, 44, 14, 6320–6339, 2015, https://doi.org/10.1039/C4DT03559C.
21. Srivastava, P., Sahgal, M., Sharma, K., Enshasy, H.A.E., Gafur, A., Alfarraj, S., Ansari, M.J., Sayyed, R.Z., Optimization and identification of siderophores produced by *Pseudomonas monteilii* strain MN759447 and its antagonism toward fungi associated with mortality in *Dalbergia sissoo* plantation forests. *Front. Plant Sci.*, 13, 2022, https://doi.org/10.3389/fpls.2022.984522.
22. Schalk, I.J., Bacterial siderophores: diversity, uptake pathways and applications. *Nat. Rev. Microbiol.*, 23, 1, 24–40, 2025, https://doi.org/10.1038/s41579-024-01090-6.
23. Barman, M., Shah, M.H., Samanta, S., Dutta, S., Panda, M., Thakur, H., Pramanik, K., Islam, S., Bacillus secondary metabolites and their applications in agriculture, in: *Bacterial Secondary Metabolites*, K.A. Abd-Elsalam and H. Mohamed II (Eds.), pp. 239–258, Elsevier, 2024, https://doi.org/10.1016/B978-0-323-95251-4.00017-X.
24. Nozoye, T., Nagasaka, S., Kobayashi, T., Takahashi, M., Sato, Y., Sato, Y., Uozumi, N., Nakanishi, H., Nishizawa, N.K., Phytosiderophore efflux transporters are crucial for iron acquisition in graminaceous plants. *J. Biol. Chem.*, 286, 7, 5446–5454, 2011, https://doi.org/10.1074/jbc.M110.180026.
25. Römheld, V. and Marschner, H., Evidence for a specific uptake system for iron phytosiderophores in roots of grasses. *Plant Physiol.*, 80, 1, 175–180, 1986.
26. Kobayashi, T., Nakanishi, H., Takahashi, M., Kawasaki, S., Nishizawa, N.K., Mori, S., *In vivo* evidence that Ids3 from *Hordeum vulgare* encodes a dioxygenase that converts 2′-deoxymugineic acid to mugineic acid in transgenic rice. *Planta*, 212, 5, 864–871, 2001, https://doi.org/10.1007/s004250000453.
27. Bashir, K., Inoue, H., Nagasaka, S., Takahashi, M., Nakanishi, H., Mori, S., Nishizawa, N.K., Cloning and Characterization of Deoxymugineic Acid Synthase Genes from Graminaceous Plants. *J. Biol. Chem.*, 281, 43, 32395–32402, 2006, https://doi.org/10.1074/jbc.M604133200.

28. Takagi, S., Nomoto, K., Takemoto, T., Physiological aspect of mugineic acid, a possible phytosiderophore of graminaceous plants. *J. Plant Nutr.*, 7, 1–5, 469–477, 1984, https://doi.org/10.1080/01904168409363213.
29. Santi, S. and Schmidt, W., Dissecting iron deficiency-induced proton extrusion in Arabidopsis roots. *New Phytol.*, 183, 4, 1072–1084, 2009, https://doi.org/10.1111/j.1469-8137.2009.02908.x.
30. Sun, J., Xiao, S., Xue, C., The tug-of-war on iron between plant and pathogen. *Phytopathol. Res.*, 5, 1, 61, 2023, https://doi.org/10.1186/s42483-023-00215-8.
31. Rout, G.R. and Sahoo, S., Role of iron in plant growth and metabolism. *Rev. Agric. Sci.*, 3, 1–24, 2015, https://doi.org/10.7831/ras.3.1.
32. Mori, S., Iron acquisition by plants. *Curr. Opin. Plant Biol.*, 2, 3, 250–253, 1999, https://doi.org/10.1016/s1369-5266(99)80043-0.
33. Chu, B.C., Garcia-Herrero, A., Johanson, T.H., Krewulak, K.D., Lau, C.K., Peacock, R.S., Slavinskaya, Z., Vogel, H.J., Siderophore uptake in bacteria and the battle for iron with the host; a bird's eye view. *Biometals*, 23, 4, 601–611, 2010, https://doi.org/10.1007/s10534-010-9361-x.
34. Haas, H., Fungal siderophore metabolism with a focus on *Aspergillus fumigatus. Nat. Prod. Rep.*, 31, 10, 1266–1276, 2014.
35. Mahawar, L., Ramasamy, K.P., Pandey, A., Prasad, S.M., Iron deficiency in plants: an update on homeostasis and its regulation by nitric oxide and phytohormones. *Plant Growth Regul.*, 100, 2, 283–299, 2023, https://doi.org/10.1007/s10725-022-00853-6.
36. Ahmadi, H., Motesharezadeh, B., Dadrasnia, A., Iron chlorosis in fruit stone trees with emphasis on chlorosis correction mechanisms in orchards: A review. *J. Plant Nutr.*, 46, 5, 782–800, 2023, https://doi.org/10.1080/01904167.2022.2087088.
37. Ali, I.I., Yusuf, G.B., Jama, A.M., Nutrient-Pathogen Interactions in Agriculture: Mechanisms, Relationships, and Sustainable Practices. *Int. J. Adv. Nat. Sci. Eng. Res.*, 8, 10, 153–162, 2024.
38. Maathuis, F.J.M. and Diatloff, E., Roles and Functions of Plant Mineral Nutrients, in: *Plant Mineral Nutrients: Methods and Protocols*, F.J.M. Maathuis (Ed.), pp. 1–21, Humana Press, Totowa, New Jersey, United States, 2023, https://doi.org/10.1007/978-1-62703-152-3_1.
39. Bollard, E.G. and Butler, G.W., Mineral Nutrition of Plants. *Annu. Rev. Plant Physiol.*, 17, 77–112, 1996, Medium: X, https://doi.org/10.1146/annurev.pp.17.060166.000453.
40. Monib, A.W., Alimyar, O., Mohammad, M.U., Akhundzada, M.S., Niazi, P., Macronutrients for Plants Growth and Humans Health. *J. Res. Appl. Sci. Biotechnol.*, 2, 2, 268–279, 2023, https://doi.org/10.55544/jrasb.2.2.38.
41. Ahmed, N., Zhang, B., Chachar, Z., Li, J., Xiao, G., Wang, Q., Hayat, F., Deng, L., Narejo, M., Bozdar, B., Tu, P., Micronutrients and their effects on Horticultural crop quality, productivity and sustainability. *Sci. Hortic.*, 323, 112512, 2024, https://doi.org/10.1016/j.scienta.2023.112512.

42. Boardman, J. and Vandaele, K., Soil erosion and runoff: The need to rethink mitigation strategies for sustainable agricultural landscapes in western Europe. *Soil Use Manage.*, 39, 2, 673–685, 2023, https://doi.org/10.1111/sum.12898.
43. Tarolli, P., Luo, J., Park, E., Barcaccia, G., Masin, R., Soil salinization in agriculture: Mitigation and adaptation strategies combining nature-based solutions and bioengineering. *iScience*, 27, 2, 2024, https://doi.org/10.1016/j.isci.2024.108830.
44. Du, L., Zhang, Z., Chen, Y., Wang, Y., Zhou, C., Yang, H., Zhang, W., Heterogeneous impact of soil acidification on crop yield reduction and its regulatory variables: A global meta-analysis. *Field Crops Res.*, 319, 109643, 2024, https://doi.org/10.1016/j.fcr.2024.109643.
45. Zhang, B., Jia, Y., Fan, H., Guo, C., Fu, J., Li, S., Li, M., Liu, B., Ma, R., Soil compaction due to agricultural machinery impact: A systematic review. *Land Degrad. Dev.*, 35, 10, 3256–3273, 2024, https://doi.org/10.1002/ldr.5144.
46. Sun, Z., Yang, R., Wang, J., Zhou, P., Gong, Y., Gao, F., Wang, C., Effects of Nutrient Deficiency on Crop Yield and Soil Nutrients Under Winter Wheat–Summer Maize Rotation System in the North China Plain. *Agronomy*, 14, 11, 2690, 2024, https://doi.org/10.3390/agronomy14112690.
47. Ghimirey, V., Chaurasia, J., Marahatta, S., Plant Nutrition Disorders: Insights from Clinic Analyses and Their Impact on Plant Health. *Agric. Ext. Dev. Countries*, 2, 1, 09–17, 2024.
48. Hald-Mortensen, C., The Main Drivers of Biodiversity Loss: A Brief Overview. *J. Ecol. Nat. Resour.*, 7, 3, 000346, 2023.
49. Tekwa, E., Gonzalez, A., Zurell, D., O'Connor, M., Detecting and attributing the causes of biodiversity change: needs, gaps and solutions. *Philos. Trans. R. Soc. B*, 378, 1881, 20220181, 2023, https://doi.org/doi:10.1098/rstb.2022.0181.
50. Sánchez-Castro, I., Molina, L., Prieto-Fernández, M.-A., Segura, A., Past, present and future trends in the remediation of heavy-metal contaminated soil - Remediation techniques applied in real soil-contamination events. *Heliyon*, 9, 6, 2023, https://doi.org/10.1016/j.heliyon.2023.e16692.
51. Poggere, G., Gasparin, A., Barbosa b, J.Z., Melo c, W.G., Corrêa d, R.S., Motta, V.A.C., Soil contamination by copper: Sources, ecological risks, and mitigation strategies in Brazil. *J. Trace Ele. Miner.*, 4, 100059, 2023, https://doi.org/10.1016/j.jtemin.2023.100059.
52. Watts, D.B., Torbert, H.A., Feng, Y., Prior, S.A., Soil Microbial Community Dynamics as Influenced by Composted Dairy Manure, Soil Properties, and Landscape Position. *Soil Sci.*, 175, 10, 474–486, 2010, https://doi.org/10.1097/SS.0b013e3181f7964f.
53. Upton, R.N., Bach, E.M., Hofmockel, K.S., Spatio-temporal microbial community dynamics within soil aggregates. *Soil Biol. Biochem.*, 132, 58–68, 2019, https://doi.org/10.1016/j.soilbio.2019.01.016.

54. Bever, J.D., Platt, T.G., Morton, E.R., Microbial population and community dynamics on plant roots and their feedbacks on plant communities. *Annu. Rev. Microbiol.*, 66, 265–283, 2012, https://doi.org/10.1146/annurev-micro-092611-150107.
55. Zhang, S., Deng, Z., Borham, A., Ma, Y., Wang, Y., Hu., J., Wang, J., Bohu, T., Significance of Soil Siderophore-Producing Bacteria in Evaluation and Elevation of Crop Yield. *Horticulturae*, 9, 3, 370, 2023, https://doi.org/10.3390/horticulturae9030370.
56. O'Brien, S., Culbert, C.T., Barraclough, T.G., Community composition drives siderophore dynamics in multispecies bacterial communities. *BMC Ecol. Evol.*, 23, 1, 45, 2023, https://doi.org/10.1186/s12862-023-02152-8.
57. Sultana, S., Alam, S., Karim, M.M., Screening of siderophore-producing salt-tolerant rhizobacteria suitable for supporting plant growth in saline soils with iron limitation. *J. Agric. Food Res.*, 4, 100150, 2021, https://doi.org/10.1016/j.jafr.2021.100150.
58. Sadeghi, A., Karimi, E., Dahaji, P.A., *et al.*, Plant growth promoting activity of an auxin and siderophore producing isolate of *Streptomyces* under saline soil conditions. *World J. Microbiol. Biotechnol.*, 28, 4, 1503–1509, 2012, https://doi.org/10.1007/s11274-011-0952-7.
59. Rajkumar, M., Ae, N., Prasad, M.N.V., Freitas, H., Potential of siderophore-producing bacteria for improving heavy metal phytoextraction. *Trends Biotechnol.*, 28, 3, 142–149, 2010, https://doi.org/10.1016/j.tibtech.2009.12.002.
60. Hesse, E., O'Brien, S., Tromas, N., Bayer, F., Luján, A.M., van Veen, E.M., Hodgson, D.J., Buckling, A., Ecological selection of siderophore-producing microbial taxa in response to heavy metal contamination. *Ecol. Lett.*, 21, 1, 117–127, 2018, https://doi.org/10.1111/ele.12878.
61. Wang, Y., Huang, W., Li, Y., Yu, F., Penttinen, P., Isolation, characterization, and evaluation of a high-siderophore-yielding bacterium from heavy metal–contaminated soil. *Environ. Sci. Pollut. Res.*, 29, 3, 3888–3899, 2022, https://doi.org/10.1007/s11356-021-15996-8.
62. Strange, R.N. and Scott, P.R., Plant disease: a threat to global food security. *Annu. Rev. Phytopathol.*, 43, 1, 83–116, 2005, https://doi.org/10.1146/annurev.phyto.43.113004.133839.
63. Fones, H.N., Bebber, D.P., Chaloner, T.M., Kay, W.T., Steinberg, G., Gurr, S.J., Threats to global food security from emerging fungal and oomycete crop pathogens. *Nat. Food*, 1, 6, 332–342, 2020, https://doi.org/10.1038/s43016-020-0075-0.
64. Mwangi, R.W., Mustafa, M., Charles, K., Wagara, I.W., Kappel, N., Selected emerging and reemerging plant pathogens affecting the food basket: A threat to food security. *J. Agric. Food Res.*, 14, 100827, 2023, https://doi.org/10.1016/j.jafr.2023.100827.

65. Tian, B., *et al.*, A cosmopolitan fungal pathogen of dicots adopts an endophytic lifestyle on cereal crops and protects them from major fungal diseases. *ISME J.*, 14, 12, 3120–3135, 2020.
66. Gai, Y. and Wang, H., Plant Disease: A Growing Threat to Global Food Security. *Agronomy*, 14, 8, 1615, 20242024, https://doi.org/10.3390/agronomy14081615.
67. Oerke, E.C., Crop losses to pests. *J. Agric. Sci.*, 144, 1, 31–43, 2006, https://doi.org/10.1017/S0021859605005708.
68. Savary, S., Willocquet, L., Pethybridge, S.J., *et al.*, The global burden of pathogens and pests on major food crops. *Nat. Ecol. Evol.*, 3, 3, 430–439, 2019, https://doi.org/10.1038/s41559-018-0793-y.
69. Ma, L.J., Geiser, D.M., Proctor, R.H., Rooney, A.P., O'Donnell, K., Trail, F., Gardiner, D.M., Manners, J.M., Kazan, K., *Fusarium* pathogenomics. *Annu. Rev. Microbiol.*, 67, 1, 399–416, 2013, https://doi.org/10.1146/annurev-micro-092412-155650.
70. Wu, F., Measuring the economic impacts of *Fusarium* toxins in animal feeds. *Anim. Feed Sci. Technol.*, 137, 3, 363–374, 2007, https://doi.org/10.1016/j.anifeedsci.2007.06.010.
71. Richard, J.L., Some major mycotoxins and their mycotoxicoses—An overview. *Int. J. Food Microbiol.*, 119, 1, 3–10, 2007, https://doi.org/10.1016/j.ijfoodmicro.2007.07.019.
72. Sweeney, M.J. and Dobson, A.D.W., Mycotoxin production by *Aspergillus*, *Fusarium* and *Penicillium* species. *Int. J. Food Microbiol.*, 43, 3, 141–158, 1998, https://doi.org/10.1016/S0168-1605(98)00112-3.
73. Tudi, M., Daniel Ruan, H., Wang, L., Lyu, J., Sadler, R., Connell, D., Chu, C., Phung, D.T., Agriculture Development, Pesticide Application and Its Impact on the Environment. *Int. J. Environ. Res. Public Health*, 18, 3, 1112, 2021, https://doi.org/10.3390/ijerph18031112.
74. Khan, B.A., *et al.*, Pesticides: Impacts on Agriculture Productivity, Environment, and Management Strategies, in: *Emerging Contaminants and Plants: Interactions, Adaptations and Remediation Technologies*, T. Aftab (Ed.), pp. 109–134, Springer International Publishing, Cham, Switzerland, 2023, https://doi.org/10.1007/978-3-031-22269-6_5.
75. Pawan, K., Kumar, R., Thakur, K., *et al.*, Impact of Pesticides Application on Aquatic Ecosystem and Biodiversity: A Review. *Biol. Bull.*, 50, 6, 1362–1375, 2023, https://doi.org/10.1134/S1062359023601386.
76. Kim, K.-H., Kabir, E., Jahan, S.A., Exposure to pesticides and the associated human health effects. *Sci. Total Environ.*, 575, 525–535, 2017, https://doi.org/10.1016/j.scitotenv.2016.09.009.
77. Fantke, P. and Jolliet, O., Life cycle human health impacts of 875 pesticides. *Int. J. Life Cycle Assess.*, 21, 5, 722–733, 2016, https://doi.org/10.1007/s11367-015-0910-y.

78. Quarcoo, F., Bonsi, C., Tackie, N., Pesticides, the environment, and human health. Agricultural and Biological Sciences, in: *Pesticides—Toxic Aspects*, pp. 81–103, InTech Europe, Rijeka, 2014.
79. Arias-Estévez, M., López-Periago, E., Martínez-Carballo, E., Simal-Gándara, J., Mejuto, J.C., García-Río, L., The mobility and degradation of pesticides in soils and the pollution of groundwater resources. *Agric. Ecosyst. Environ.*, 123, 4, 247–260, 2008, https://doi.org/10.1016/j.agee.2007.07.011.
80. Etesami, H., Jeong, B.R., Glick, B.R., Biocontrol of plant diseases by *Bacillus* spp. *Physiol. Mol. Plant Pathol.*, 126, 102048, 2023, https://doi.org/10.1016/j.pmpp.2023.102048.
81. Abdelaziz, A.M., Hashem, A.H., El-Sayyad, G.S., *et al.*, Biocontrol of soil borne diseases by plant growth promoting rhizobacteria. *Trop. Plant Pathol.*, 48, 2, 105–127, 2023, https://doi.org/10.1007/s40858-022-00544-7.
82. van Loon, L.C., Bakker, P.A., Pieterse, C.M., Systemic resistance induced by rhizosphere bacteria. *Annu. Rev. Phytopathol.*, 36, 453–483, 1998, https://doi.org/10.1146/annurev.phyto.36.1.453.
83. Umarusman, M.A. and Baştaş, K.K., Potential Biological Control Agents against Soft Rot Diseases Caused by *Pectobacteria on* Some Sugar Beet Cultivars. *Turk. J. Agric. Food Sci. Technol.*, 11, s1, 2601–2608, 2023, https://doi.org/10.24925/turjaf.v11is1.2601-2608.6535.
84. Butt, H. and Bastas, K.K., Environment-Friendly Management of Plant Diseases by *Bacillus* Through Molecular Pathways, in: *Microbial Biocontrol: Molecular Perspective in Plant Disease Management*, K.K. Bastas, A. Kumar, U. Sivakumar (Eds.), pp. 217–241, Springer Nature Singapore, 2023, https://doi.org/10.1007/978-981-99-3947-3_11.
85. Ünlü, E., Çalış, O., Say, A., Karim, A.A., Yetişir, H., Yılmaz, S., Investigation of the effects of *Bacillus subtilis* and *Bacillus thuringiensis* as Bio-agents against powdery mildew (*Podosphaera xanthii*) disease in zucchini (*Cucurbita pepo* L.). *Microb. Pathogen.*, 185, 106430, 2023, https://doi.org/10.1016/j.micpath.2023.106430.
86. Abo-Elyousr, K.A.M., Sallam, N.M.A., Mousa, M.A.A., Imran, M., Abdel-Rahim, I.R., Synergistic effect of *Bacillus subtilis* and benzothiadiazole (Bion®) on the suppression of *Fusarium oxysporum* and the enhancement of disease resistance in *Capsicum annuum*. *J. Plant Pathol.*, 106, 1, 127–138, 2024, https://doi.org/10.1007/s42161-023-01527-6.
87. Dolzhenko, V.I., Shapoval, O.V., Pleskachev, Y.N., Astarkhanova, T.S., Efficiency of *Bacillus thuringiensis* bioinsecticide in grapevine protection. *RUDN J. Agron. Anim. Ind.*, 18, 4, 493–500, 2023, https://doi.org/10.22363/2312-797x-2023-18-4-493-500.
88. Marković, S., *et al.*, Biological control of major pathogenic bacteria of potato by *Bacillus amyloliquefaciens* strains SS-12.6 and SS-38.4. *Biol. Control*, 182, 105238, 2023, https://doi.org/10.1016/j.biocontrol.2023.105238.
89. Ravindran, A.P., *et al.*, Efficacy of *Bacillus licheniformis*: A biocontrol agent against *Colletotrichum gloeosporioides* Penz. (Penz. & Sacc.) causing

anthracnose in greater yam (*Dioscorea alata* L.). *Egypt. J. Biol. Pest Control*, 33, 1, 112, 2023, https://doi.org/10.1186/s41938-023-00755-3.

90. Keshmirshekan, A., de Souza Mesquita, L.M., Ventura, S.P.M., Biocontrol manufacturing and agricultural applications of *Bacillus velezensis*. *Trends Biotechnol.*, 42, 8, 986–1001, 2024, https://doi.org/10.1016/j.tibtech.2024.02.003.
91. Dobrzyński, J., Jakubowska, Z., Kulkova, I., Kowalczyk, P., Kramkowski, K., Biocontrol of fungal phytopathogens by *Bacillus pumilus*. *Front. Microbiol.*, 14, 20232023, https://doi.org/10.3389/fmicb.2023.1194606.
92. Panpatte, D.G., Jhala, Y.K., Shelat, H.N., Vyas, R.V., *Pseudomonas fluorescens*: A Promising Biocontrol Agent and PGPR for Sustainable Agriculture, in: *Microbial Inoculants in Sustainable Agricultural Productivity*, vol. 1: Research Perspectives, D.P. Singh, H.B. Singh, R. Prabha (Eds.), pp. 257–270, Springer, New Delhi, India, 2016, https://doi.org/10.1007/978-81-322-2647-5_15.
93. Anjaiah, V., *et al.*, Involvement of phenazines and anthranilate in the antagonism of *Pseudomonas aeruginosa* PNA1 and Tn5 derivatives toward *Fusarium* spp. and *Pythium* spp. *Mol. Plant-Microbe Interact.*, 11, 9, 847–854, 1998.
94. Gao, P., Guo, Y., Li, Y., Duan, T., Effects of dual inoculation of AMF and *rhizobium* on alfalfa (*Medicago sativa*) root rot caused by *Microdochium tabacinum*. *Australas. Plant Pathol.*, 47, 2, 195–203, 2018, https://doi.org/10.1007/s13313-018-0543-2.
95. Zhang, X.-X., *et al.*, *Rhizobium oryzicola* sp. nov., potential plant-growth-promoting endophytic bacteria isolated from rice roots. *Int. J. Syst. Evol. Microbiol.*, 65, Pt_9, 2931–2936, 2015.
96. Kumar Junta, M., Gupta, A.K., Mahajan, R., Biological control of hairy root (*Rhizobium rhizogenes*) in apple nurseries through *Rhizobium radiobacter* antagonists (strain K-84 and native strain UHFBA-218). *Biol. Control*, 164, 104762, 2021, https://doi.org/10.1016/j.biocontrol.2021.104762.
97. Xu, S. and Kim, B.-S., Evaluation of *Paenibacillus polymyxa* strain SC09-21 for biocontrol of Phytophthora blight and growth stimulation in pepper plants. *Trop. Plant Pathol.*, 41, 3, 162–168, 2016, https://doi.org/10.1007/s40858-016-0077-5.
98. Mukhopadhyay, R. and Kumar, D., *Trichoderma*: A beneficial antifungal agent and insights into its mechanism of biocontrol potential. *Egypt. J. Biol. Pest Control*, 30, 1, 133, 2020, https://doi.org/10.1186/s41938-020-00333-x.
99. Singh, V., Srivastava, S., Lal, R.J., Awasthi, S., Joshi, B., Biological control of red rot disease of sugarcane through *Trichoderma harzianum* and *Trichoderma viride*. *Indian Phytopathol.*, 61, 4, 486–491, 2008.
100. Sreenivasaprasad, S. and Sreenivasaprasad, M., Efficacy of *Gliocladium virens* and *Trichoderma longibrachiatum* as biological control agents of groundnut root and stem rot diseases. *Int. J. Pest Manage.*, 39, 2, 167–171, 1993, https://doi.org/10.1080/09670879309371784.
101. Williamson, B., Tudzynski, B., Tudzynski, P., Van Kan, J.A., *Botrytis cinerea*: the cause of grey mould disease. *Mol. Plant Pathol*, 8, 5, 561–580, 2007.

102. Yu, L., *et al.*, Postharvest control of *Penicillium expansum* in fruits: A review. *Food Biosci.*, 36, 100633, 2020, https://doi.org/10.1016/j.fbio.2020.100633.
103. Bautista-Baños, S., Bosquez-Molina, E., Barrera-Necha, L.L., Chapter 1 - *Rhizopus stolonifer* (Soft Rot), in: *Postharvest Decay*, S. Bautista-Baños (Ed.), pp. 1–44, Academic Press, Cambridge, Massachusetts, United States, 2014, https://doi.org/10.1016/B978-0-12-411552-1.00001-6.
104. Schuster, E., Dunn-Coleman, N., Frisvad, J., van Dijck, P., On the safety of *Aspergillus niger* – a review. *Appl. Microbiol. Biotechnol.*, 59, 4, 426–435, 2002, https://doi.org/10.1007/s00253-002-1032-6.
105. Ma, Z., Morgan, D.P., Michailides, T.J., Effects of water stress on Botryosphaeria blight of pistachio caused by *Botryosphaeria dothidea. Plant Dis.*, 85, 7, 745–749, 2001.
106. Al-Sadi, A.M., Bipolaris sorokiniana-Induced Black Point, Common Root Rot, and Spot Blotch Diseases of Wheat: A Review. *Front. Cell. Infect. Microbiol.*, 11, 2021, https://doi.org/10.3389/fcimb.2021.584899.
107. Shen, Y. and Ronald, P., Molecular determinants of disease and resistance in interactions of *Xanthomonas oryzae* pv. oryzae and rice. *Microbes Infect.*, 4, 13, 1361–1367, 2002, https://doi.org/10.1016/S1286-4579(02)00004-7.
108. Xu, Y., *et al.*, Status of *streptomycin* resistance development in *Xanthomonas oryzae* pv. oryzae and *Xanthomonas oryzae* pv. oryzicola in China and their resistance characters. *J. Phytopathol.*, 158, 9, 601–608, 2010.
109. Shahwar, D., *et al.*, Role of microbial inoculants as bio fertilizers for improving crop productivity: A review. *Heliyon*, 9, 6, 2023.
110. Dhiman, K., Sharma, D., Kumari, R., Tomar, P., Biofortification of crops using microbes–a promising sustainable agriculture strategy. *J. Plant Nutr.*, 46, 12, 2912–2935, 2023.
111. Tiwari, S., Sharma, B., Bisht, N., Tewari, L., Role of beneficial microbial gene pool in mitigating salt/nutrient stress of plants in saline soils through underground phytostimulating signalling molecules. *Pedosphere*, 33, 1, 153–171, 2023, https://doi.org/10.1016/j.pedsph.2022.06.029.
112. Chepsergon, J. and Moleleki, L.N., Rhizosphere bacterial interactions and impact on plant health. *Curr. Opin. Microbiol.*, 73, 102297, 2023.
113. Sarraf, M., *et al.*, Understanding the role of beneficial elements in developing plant stress resilience: Signalling and crosstalk with phytohormones and microbes. *Plant Stress*, 10, 100224, 2023, https://doi.org/10.1016/j.stress.2023.100224.
114. Deb, C.R. and Tatung, M., Siderophore producing bacteria as biocontrol agent against phytopathogens for a better environment: A review. *S. Afr. J. Bot.*, 165, 153–162, 2024, https://doi.org/10.1016/j.sajb.2023.12.031.
115. Ingole, A., Ismail, S., Dhamak, A., Evaluation of Siderophore Production by Different Promising Microbial Isolates. *Int. J. Plant Soil Sci.*, 35, 19, 1091–1096, 2023, https://doi.org/10.9734/ijpss/2023/v35i193647.

116. Srikamwang, C., *et al.*, Role of microbial volatile organic compounds in promoting plant growth and disease resistance in horticultural production. *Plant Signal. Behav.*, 18, 1, 2227440, 2023.
117. Ajijah, N., Fiodor, A., Pandey, A.K., Rana, A., Pranaw, K., Plant Growth-Promoting Bacteria (PGPB) with Biofilm-Forming Ability: A Multifaceted Agent for Sustainable Agriculture. *Diversity*, 15, 1, 112, 2023, https://doi.org/10.3390/d15010112.
118. Pal, K.K. and McSpadden Gardener, B., Biological Control of Plant Pathogens. *Plant Health Instr.*, 2, 1117–1142, 2006.
119. Qiao, J., Zhang, R., Liu, Y., Liu, Y., Evaluation of the Biocontrol Efficiency of *Bacillus subtilis* Wettable Powder on Pepper Root Rot Caused by *Fusarium solani*. *Pathogens*, 12, 2, 225, 2023, https://doi.org/10.3390/pathogens12020225.
120. Belete, T., Bastas, K.K., Francesconi, S., Balestra, G.M., Biological effectiveness of *Bacillus subtilis* on common bean bacterial blight. *J. Plant Pathol.*, 103, 1, 249–258, 2021, https://doi.org/10.1007/s42161-020-00727-8.
121. Gür, A. and Baştaş, K.K., Effectiveness of Phosphorous acid, *Bacillus subtilis* and Copper Compounds on Apple cv. Gala with M9 Rootstock in the Control of Fire Blight. *Turk. J. Agric. Food Sci. Technol.*, 11, s1, 2595–2600, 2023, https://doi.org/10.24925/turjaf.v11is1.2595-2600.6533.
122. Wang, J., *et al.*, Plant Growth Promotion and Biocontrol of Leaf Blight Caused by *Nigrospora sphaerica* on Passion Fruit by Endophytic *Bacillus subtilis* Strain GUCC4. *J. Fungi*, 9, 2, 132, 2023, https://doi.org/10.3390/jof9020132.
123. Harish, B.N., *et al.*, Molecular characterization and antifungal activity of lipopeptides produced from *Bacillus subtilis* against plant fungal pathogen *Alternaria alternata*. *BMC Microbiol.*, 23, 1, 179, 2023, https://doi.org/10.1186/s12866-023-02922-w.
124. Al-Mutar, D.M.K., *et al.*, The Extracellular Lipopeptides and Volatile Organic Compounds of *Bacillus subtilis* DHA41 Display Broad-Spectrum Antifungal Activity against Soil-Borne Phytopathogenic Fungi. *J. Fungi*, 9, 8, 797, 2023, https://doi.org/10.3390/jof9080797.
125. Arrebola, E., Sivakumar, D., Bacigalupo, R., Korsten, L., Combined application of antagonist *Bacillus amyloliquefaciens* and essential oils for the control of peach postharvest diseases. *Crop Prot.*, 29, 4, 369–377, 2010, https://doi.org/10.1016/j.cropro.2009.08.001.
126. Li, X., *et al.*, Antifungal Activity of Isolated *Bacillus amyloliquefaciens* SYBC H47 for the Biocontrol of Peach Gummosis. *PLoS One*, 11, 9, e0162125, 2016, https://doi.org/10.1371/journal.pone.0162125.
127. Li, H., Chougule, N.P., Bonning, B.C., Interaction of the *Bacillus thuringiensis* delta endotoxins Cry1Ac and Cry3Aa with the gut of the pea aphid, *Acyrthosiphon pisum* (Harris). *J. Invertebr. Pathol.*, 107, 1, 69–78, 2011, https://doi.org/10.1016/j.jip.2011.02.001.
128. Zhang, J., *et al.*, Transcriptome profiling analysis of the intoxication response in midgut tissue of Agrotis ipsilon larvae to *Bacillus thuringiensis*

Vip3Aa protoxin. *Pestic. Biochem. Physiol.*, 160, 20–29, 2019, https://doi.org/10.1016/j.pestbp.2019.06.001.

129. Karshanal, J. and Kalia, V.K., Efficacy of native *Bacillus* isolates against different larval instars of fall armyworm, *Spodoptera frugiperda* alone and in combination. *Egypt. J. Biol. Pest Control*, 33, 1, 102, 2023, https://doi.org/10.1186/s41938-023-00743-7.
130. Bravo, A., Likitvivatanavong, S., Gill, S.S., Soberón, M., *Bacillus thuringiensis*: A story of a successful bioinsecticide. *Insect Biochem. Mol. Biol.*, 41, 7, 423–431, 2011, https://doi.org/10.1016/j.ibmb.2011.02.006.
131. Cao, Y., *et al.*, Antagonism of Two Plant-Growth Promoting *Bacillus velezensis* Isolates Against *Ralstonia solanacearum* and *Fusarium oxysporum*. *Sci. Rep.*, 8, 1, 4360, 2018, https://doi.org/10.1038/s41598-018-22782-z.
132. Kang, X., *et al.*, *Bacillus velezensis* CC09: a potential 'vaccine'for controlling wheat diseases. *Mol. Plant-Microbe Interact.*, 31, 6, 623–632, 2018.
133. Ge, B., *et al.*, *Bacillus methylotrophicus* Strain NKG-1, Isolated from Changbai Mountain, China, Has Potential Applications as a Biofertilizer or Biocontrol Agent. *PLoS One*, 11, 11, e0166079, 2016, https://doi.org/10.1371/journal.pone.0166079.
134. Kosanovic, D., Dyas, M., Grogan, H., Kavanagh, K., Differential proteomic response of *Agaricus bisporus* and *Trichoderma aggressivum* f. europaeum to *Bacillus velezensis* supernatant. *Eur. J. Plant Pathol.*, 160, 2, 397–409, 2021, https://doi.org/10.1007/s10658-021-02252-5.
135. Enebak, S. and Carey, W., Evidence for induced systemic protection to fusiform rust in loblolly pine by plant growth-promoting rhizobacteria. *Plant Dis.*, 84, 3, 306–308, 2000.
136. Jeong, H., Choi, S.-K., Kloepper, J.W., Ryu, C.-M., Genome sequence of the plant endophyte *Bacillus pumilus* INR7, triggering induced systemic resistance in field crops. *Genome Announc.*, 2, 5, 2014, 10.1128/genomea.01093-01014.
137. Li, W., Lee, S.-Y., Cho, Y.-J., Ghim, S.-Y., Jung, H.-Y., Mediation of induced systemic resistance by the plant growth-promoting rhizobacteria *Bacillus pumilus* S2-3-2., 2020.
138. Mehetre, S.T. and Kale, S.P., Comparative efficacy of thermophilic bacterium, *Bacillus licheniformis* (NR1005) and antagonistic fungi, *Trichoderma harzianum* to control *Pythium aphanidermatum*-induced damping off in chilli (*Capsicum annuum* L.). *Arch. Phytopathol. Plant Prot.*, 44, 11, 1068–1074, 2011, https://doi.org/10.1080/03235401003755262.
139. Nawaz, H.H., *et al.*, Evaluation of antifungal metabolites activity from *bacillus licheniformis* OE-04 against *Colletotrichum gossypii*. *Pestic. Biochem. Physiol.*, 146, 33–42, 20182018, https://doi.org/10.1016/j.pestbp.2018.02.007.
140. Ramírez-Cariño, H.F., *et al.*, Biocontrol of *Alternaria alternata* and *Fusarium oxysporum* by *Trichoderma asperelloides* and *Bacillus paralicheniformis* in tomato plants. *Antonie van Leeuwenhoek*, 113, 9, 1247–1261, 2020, https://doi.org/10.1007/s10482-020-01433-2.

141. RB., M., Epiphytic bacteria from various bean genotypes and their potential for biocontrol of *Xanthomonas axonopodis* pv. Phaseoli. *Tanz J. Agric. Sci.*, 2, 19–26, 1999.
142. Elsherif, M. and F., G., Investigations on biological control of some plant pathogenic fungi by *Fluorescent Pseudomonads* using different method of application. *Z. Pflanzenkrankh. Pflanzenschutz*, 98, 236–249, 1991.
143. Jayashree, K., Shanmugam, V., Raguchander, T., Ramanathan, A., R., S., Evaluation of *Pseudomonas fluorescens* (pf1) against blackgram and sesame root-rot disease. *J. Biol. Control*, 14, 55–61, 2000.
144. Lu, J., *et al.*, Antimicrobial mechanisms and antifungal activity of compounds generated by banana rhizosphere *Pseudomonas aeruginosa* Gxun-2 against *fusarium oxysporum* f. sp. cubense. *Front. Microbiol.*, 15, 2024, https://doi.org/10.3389/fmicb.2024.1456847.
145. Gao, X., *et al.*, Co-inoculation with rhizobia and AMF inhibited soybean red crown rot: from field study to plant defense-related gene expression analysis. *PLoS One*, 7, 3, e33977, 2012.
146. Estevez de Jensen, C., Percich, J.A., Graham, P.H., Integrated management strategies of bean root rot with *Bacillus subtilis* and *Rhizobium* in Minnesota. *Field Crops Res.*, 74, 2, 107–115, 2002, https://doi.org/10.1016/S0378-4290(01)00200-3.
147. Al-Ani, R.A., Adhab, M.A., Mahdi, M.H., Abood, H.M., *Rhizobium japonicum* as a biocontrol agent of soybean root rot disease caused by *Fusarium solani* and *Macrophomina phaseolina*. *Plant Prot. Sci.*, 48, 4, 149–155, 2012, https://doi.org/10.17221/16/2012-PPS.
148. Xiao, Z., Zhao, Q., Li, W., Gao, L., Liu, G., Strain improvement of *Trichoderma harzianum* for enhanced biocontrol capacity: Strategies and prospects. *Front. Microbiol.*, 14, 2023, https://doi.org/10.3389/fmicb.2023.1146210.
149. Aseel, D.G., Soliman, S.A., Al-Askar, A.A., Elkelish, A., Elbeaino, T., Abdelkhalek, A., *Trichoderma viride* Isolate Tvd44 Enhances Potato Growth and Stimulates the Defense System against Potato Virus Y. *Horticulturae*, 9, 6, 716, 2023, https://doi.org/10.3390/horticulturae9060716.
150. Abied, M.A.E., Khalil, M.I., Awad, M.F., Youssef, S.A., Alleviation of black root rot symptoms and alteration of strawberry growth *via* modulating physiological and biochemical mechanisms using *Trichoderma viride* and *Bacillus subtilis*. *Eur. J. Plant Pathol.*, 167, 2, 235–250, 2023, https://doi.org/10.1007/s10658-023-02697-w.
151. Gao, P., *et al.*, Effect of *Trichoderma viride* on rhizosphere microbial communities and biocontrol of soybean root rot. *Front. Microbiol.*, 14, 2023, https://doi.org/10.3389/fmicb.2023.1204688.
152. Muhammad, F., Elkawakib, S., Katriani, M., Survival Rate, Disease Incidence, and Yield of Shallots by Seed Priming and Application of Tithonia Compost Enriched with *Gliocladium virens*. *Int. J. Life Sci. Agric. Res.*, 2, 05, 57–62, 2023, https://doi.org/10.55677/ijlsar/V02I05Y2023-03.

153. Tomah, A.A., Alamer, I.S.A., Khattak, A.A., Ahmed, T., Hatamleh, A.A., Al-Dosary, M.A., Ali, H.M., Wang, D., Zhang, J., Xu, L., Li, B., Potential of *Trichoderma virens* HZA14 in Controlling Verticillium Wilt Disease of Eggplant and Analysis of Its Genes Responsible for Microsclerotial Degradation. *Plants*, 12, 21, 3761, 2023, https://doi.org/10.3390/plants12213761.
154. Iida, Y., *et al.*, Entomopathogenic fungus *Beauveria bassiana*–based bioinsecticide suppresses severity of powdery mildews of vegetables by inducing the plant defense responses. *Front. Plant Sci.*, 14, 2023, https://doi.org/10.3389/fpls.2023.1211825.
155. Swathy, K., Parmar, M.K., Vivekanandhan, P., Biocontrol efficacy of entomopathogenic fungi *Beauveria bassiana* conidia against agricultural insect pests. *Environ. Qual. Manage.*, 34, 1, e22174, 2024.
156. Bhattacharyya, P.N., *et al.*, Perspectives of *Beauveria bassiana*, an entomopathogenic fungus for the control of insect-pests in tea [*Camellia sinensis* (L.) O. Kuntze]: opportunities and challenges. *Int. J. Trop. Insect Sci.*, 43, 1, 1–19, 2023, https://doi.org/10.1007/s42690-022-00932-1.
157. Miljaković, D., Marinković, J., Balešević-Tubić, S., The Significance of *Bacillus* spp. in Disease Suppression and Growth Promotion of Field and Vegetable Crops. *Microorganisms*, 8, 7, 2020, https://doi.org/10.3390/microorganisms8071037.
158. Beneduzi, A., Ambrosini, A., Passaglia, L.M.P., Plant growth-promoting rhizobacteria (PGPR): Their potential as antagonists and biocontrol agents. *Genet. Mol. Biol.*, 35, 4 suppl 1, 1044–1051, 2012, https://doi.org/10.1590/s1415-47572012000600020.
159. Kloepper, J.W., Ryu, C.-M., Zhang, S., Induced Systemic Resistance and Promotion of Plant Growth by *Bacillus* spp. *Phytopathology®*, 94, 11, 1259–1266, 2004, https://doi.org/10.1094/phyto.2004.94.11.1259.
160. Gautam, S., Chauhan, A., Sharma, R., Sehgal, R., Shirkot, C.K., Potential of *Bacillus amyloliquefaciens* for biocontrol of bacterial canker of tomato incited by *Clavibacter michiganensis* ssp. michiganensis. *Microb. Pathogen.*, 130, 196–203, 2019, https://doi.org/10.1016/j.micpath.2019.03.006.
161. Ahmad, Z., Wu, J., Chen, L., Dong, W., Isolated *Bacillus subtilis* strain 330-2 and its antagonistic genes identified by the removing PCR. *Sci. Rep.*, 7, 1, 1777, 2017, https://doi.org/10.1038/s41598-017-01940-9.
162. Moynié, L., *et al.*, Hijacking of the Enterobactin Pathway by a Synthetic Catechol Vector Designed for Oxazolidinone Antibiotic Delivery in *Pseudomonas aeruginosa*. *ACS Infect. Dis.*, 8, 9, 1894–1904, 2022, https://doi.org/10.1021/acsinfecdis.2c00202.
163. Liu, Z., Huang, T., Shi, Q., Deng, Z., Lin, S., Catechol siderophores framed on 2,3-dihydroxybenzoyl-L-serine from *Streptomyces varsoviensis*. *Front. Microbiol.*, 14, 2023, https://doi.org/10.3389/fmicb.2023.1182449.

164. Khilyas, I.V., *et al.*, Genomic insights and anti-phytopathogenic potential of siderophore metabolome of endolithic *Nocardia mangyaensis* NH1. *Sci. Rep.*, 14, 1, 5676, 2023, https://doi.org/10.1038/s41598-024-54095-9.
165. Abanoz-Seçgin, B., Otur, C., Okay, S., Kurt-Kızıldoğan, A., The regulatory role of Fur-encoding SCLAV_3199 in iron homeostasis in *Streptomyces clavuligerus*. *Gene*, 878, 147594, 2023, https://doi.org/10.1016/j.gene.2023.147594.
166. Will, V., *et al.*, Siderophore specificities of the *Pseudomonas aeruginosa* TonB-dependent transporters ChtA and ActA. *FEBS Lett.*, 597, 23, 2963–2974, 2023, https://doi.org/10.1002/1873-3468.14740.
167. Rehan, M., Barakat, H., Almami, I.S., Qureshi, K.A., Alsohim, A.S., Production and potential genetic pathways of three different siderophore types in *Streptomyces tricolor* strain HM10. *Fermentation*, 8, 8, 346, 2022.
168. Wilson, B.R., Bogdan, A.R., Miyazawa, M., Hashimoto, K., Tsuji, Y., Siderophores in Iron Metabolism: From Mechanism to Therapy Potential. *Trends Mol. Med.*, 22, 12, 1077–1090, 2016, https://doi.org/10.1016/j.molmed.2016.10.005.
169. Crichton, R.R., Chapter 4 - Biological Ligands for Metal Ions, in: *Biological Inorganic Chemistry (Second Edition)*, R.R. Crichton (Ed.), pp. 69–89, Elsevier, 2012, https://doi.org/10.1016/B978-0-444-53782-9.00004-8.
170. Yu, X., Ai, C., Xin, L., Zhou, G., The siderophore-producing bacterium, *Bacillus subtilis* CAS15, has a biocontrol effect on Fusarium wilt and promotes the growth of pepper. *Eur. J. Soil Biol.*, 47, 2, 138–145, 2011, https://doi.org/10.1016/j.ejsobi.2010.11.001.
171. Nithyapriya, S., Lalitha, S., Sayyed, R.Z., Reddy, M.S., Dailin, D.J., El Enshasy, H.A., Luh Suriani, N., Herlambang, S., Production, Purification, and Characterization of Bacillibactin Siderophore of *Bacillus subtilis* and Its Application for Improvement in Plant Growth and Oil Content in Sesame. *Sustainability*, 13, 10, 5394, 2021, https://doi.org/10.3390/su13105394.
172. Ghazy, N. and El-Nahrawy, S., Siderophore production by *Bacillus subtilis* MF497446 and *Pseudomonas koreensis* MG209738 and their efficacy in controlling *Cephalosporium maydis* in maize plant. *Arch. Microbiol.*, 203, 3, 1195–1209, 2021, https://doi.org/10.1007/s00203-020-02113-5.
173. May, J.J., Wendrich, T.M., Marahiel, M.A., The dhb operon of *Bacillus subtilis* encodes the biosynthetic template for the catecholic siderophore 2,3-dihydroxybenzoate-glycine-threonine trimeric ester bacillibactin. *J. Biol. Chem.*, 276, 10, 7209–7217, 2001, https://doi.org/10.1074/jbc.M009140200.
174. Hu, X. and Boyer, G.L., Isolation and characterization of the siderophore N-deoxyschizokinen from *Bacillus megaterium* ATCC 19213. *Biometals*, 8, 4, 357–364, 1995, https://doi.org/10.1007/BF00141610.
175. Kildea, S., *et al.*, *Bacillus megaterium* shows potential for the biocontrol of *septoria tritici* blotch of wheat. *Biol. Control*, 47, 1, 37–45, 2008, https://doi.org/10.1016/j.biocontrol.2008.07.001.

176. Dimopoulou, A., *et al.*, Direct Antibiotic Activity of Bacillibactin Broadens the Biocontrol Range of *Bacillus amyloliquefaciens* MBI600. *mSphere*, 6, 4, e0037621, 2021, https://doi.org/10.1128/mSphere.00376-21.
177. Pradhan, S., *et al.*, Siderophore-producing *Bacillus amyloliquefaciens* BM3 mitigate arsenic contamination and suppress Fusarium wilt in brinjal plants. *J. Appl. Microbiol.*, 134, 10, 2023, https://doi.org/10.1093/jambio/lxad217.
178. Bachhawat, A.K. and Ghosh, S., Iron Transport in *Azospirillum Brasilense*: Role of the Siderophore Spirilobactin. *Microbiology*, 133, 1759–1765, 1987, https://doi.org/10.1099/00221287-133-7-1759.
179. da Silva Viana, R., *et al.*, Morphological Changes in Sugarcane Crop Induced by the Plant Growth-Promoting Bacterium *Azospirillum brasilense. Sugar Tech*, 22, 2, 241–249, 2020, https://doi.org/10.1007/s12355-019-00773-2.
180. Philson, S.B. and Llinás, M., Siderochromes from *Pseudomonas Fluorescens.* I. Isolation and Characterization. *J. Biol. Chem.*, 257, 8081–8085, 1982.
181. Loper, J.E., Role of fluorescent siderophore production in biological control of *Pythium ultimum* by a *Pseudomonas fluorescens* strain. *Phytopathology*, 78, 2, 166–172, 1988.
182. Parker, D.L., *et al.*, Pyoverdine synthesis by the Mn(II)-oxidizing bacterium *Pseudomonas putida* GB-1. *Front. Microbiol.*, 5, 2014, https://doi.org/10.3389/fmicb.2014.00202.
183. Bora, T., Özaktan, H., Göre, E., Aslan, E., Biological control of *Fusarium oxysporum* f. sp. melonis by wettable powder formulations of the two strains of *Pseudomonas putida. J. Phytopathol.*, 152, 8–9, 471–475, 2004.
184. Hassan, M.N., Afghan, S., Hafeez, F.Y., Biological control of red rot in sugarcane by native pyoluteorin-producing *Pseudomonas putida* strain NH-50 under field conditions and its potential modes of action. *Pest. Manage. Sci.*, 67, 9, 1147–1154, 2011.
185. Briskot, G., Taraz, K., Budzikiewicz, H., Siderophore vom Pyoverdin-Typ aus *Pseudomonas aeruginosa* [1] / Pyoverdine Type Siderophores from *Pseudomonas aeruginosa. Z. für Naturforschung C*, 41, 5–6, 497–506, 1986, https://doi.org/doi:10.1515/znc-1986-5-601.
186. Deshwal, V.K., *Pseudomonas aeruginosa* as biological control agent against plant pathogenic fungus *Sclerotina sclerotiorum. Int. J. Plant Anim. Environ. Sci.*, 2, 1, 14–17, 2012.
187. Sasirekha, B. and Srividya, S., Siderophore production by *Pseudomonas aeruginosa* FP6, a biocontrol strain for *Rhizoctonia solani* and *Colletotrichum gloeosporioides* causing diseases in chilli. *Agric. Nat. Resour.*, 50, 4, 250–256, 2016, https://doi.org/10.1016/j.anres.2016.02.003.
188. Vinale, F., *et al.*, Harzianic acid: a novel siderophore from *Trichoderma harzianum. FEMS Microbiol. Lett.*, 347, 2, 123–129, 2013, https://doi.org/10.1111/1574-6968.12231.
189. Li, R.-X., *et al.*, Solubilisation of Phosphate and Micronutrients by *Trichoderma harzianum* and Its Relationship with the Promotion of Tomato

Plant Growth. *PLoS One*, 10, 6, e0130081, 2015, https://doi.org/10.1371/journal.pone.0130081.

190. Hadar, Y., Harman, G., Taylor, A., Evaluation of *Trichoderma koningii* and *T. harzianum* from New York soils for biological control of seed rot caused by *Pythium* spp. *Phytopathology*, 74, 1, 106–110, 1984.
191. Elad, Y., Biological control of foliar pathogens by means of *Trichoderma harzianum* and potential modes of action. *Crop Prot.*, 19, 8, 709–714, 2000, https://doi.org/10.1016/S0261-2194(00)00094-6.
192. Van Tiel-Menkveld, G.J., Mentjox-Vervuurt, J.M., Oudega, B., de Graaf, F.K., Siderophore production by *Enterobacter cloacae* and a common receptor protein for the uptake of aerobactin and cloacin DF13. *J. Bacteriol.*, 150, 2, 490–497, 1982, https://doi.org/10.1128/jb.150.2.490-497.1982.
193. Costa, J.M. and Loper, J.E., Characterization of siderophore production by the biological control agent *Enterobacter cloacae*. *MPMI-Molecular Plant Microbe Interact.*, 7, 4, 440–448, 1994.
194. Al-Mughrabi, K.I., Biological control of Fusarium dry rot and other potato tuber diseases using *Pseudomonas fluorescens* and *Enterobacter cloacae*. *Biol. Control*, 53, 3, 280–284, 2010, https://doi.org/10.1016/j.biocontrol.2010.01.010.
195. Persmark, M., *et al.*, Isolation and structure of rhizobactin 1021, a siderophore from the alfalfa symbiont *Rhizobium meliloti* 1021. *J. Am. Chem. Soc.*, 115, 10, 3950–3956, 1993, https://doi.org/10.1021/ja00063a014.
196. Smith, M.J. and Neilands, J.B., Rhizobactin, a siderophore from *Rhizobium meliloti*. *J. Plant Nutr.*, 7, 1–5, 449–458, 1984, https://doi.org/10.1080/01904168409363211.
197. Penyalver, R., Oger, P., López, M.M., Farrand, S.K., Iron-binding compounds from Agrobacterium spp.: Biological control strain Agrobacterium rhizogenes K84 produces a hydroxamate siderophore. *Appl. Environ. Microbiol.*, 67, 2, 654–664, 2001, https://doi.org/10.1128/aem.67.2.654-664.2001.
198. Kuzmanović, N., *et al.*, A Novel Group of *Rhizobium* tumorigenes-Like Agrobacteria Associated with Crown Gall Disease of Rhododendron and Blueberry. *Phytopathology*, 109, 11, 1840–1848, 2019, https://doi.org/10.1094/phyto-05-19-0167-r.
199. Sokol, P.A., Darling, P., Lewenza, S., Corbett, C.R., Kooi, C.D., Identification of a siderophore receptor required for ferric ornibactin uptake in *Burkholderia cepacia*. *Infect. Immun.*, 68, 12, 6554–6560, 2000, https://doi.org/10.1128/iai.68.12.6554-6560.2000.
200. Heungens, K. and Parke, J., Postinfection biological control of oomycete pathogens of pea by *Burkholderia cepacia* AMMDR1. *Phytopathology*, 91, 4, 383–391, 2001.
201. Storey, E.P., Boghozian, R., Little, J.L., Lowman, D.W., Chakraborty, R., Characterization of 'Schizokinen'; a dihydroxamate-type siderophore produced by *Rhizobium leguminosarum* IARI 917. *Biometals*, 19, 6, 637–649, 2006, https://doi.org/10.1007/s10534-006-9001-7.

202. Hassan Dar, G., Zargar, M.Y., Beigh, G.M., Biocontrol of Fusarium Root Rot in the Common Bean (*Phaseolus vulgaris* L.) by using Symbiotic *Glomus mosseae* and *Rhizobium leguminosarum*. *Microb. Ecol.*, 34, 1, 74–80, 1997, https://doi.org/10.1007/s002489900036.
203. Volpiano, C.G., *et al.*, *Rhizobium* strains in the biological control of the phytopathogenic fungi *Sclerotium* (Athelia) *rolfsii* on the common bean. *Plant Soil*, 432, 1, 229–243, 20182018, https://doi.org/10.1007/s11104-018-3799-y.
204. Bultreys, A., Gheysen, I., Schäfer, M., Budzikiewicz, H., Wathelet, B., The Pyoverdins of *Pseudomonas syringae* and *Pseudomonas cichorii*. *Z. für Naturforschung C*, 59, 9-10, 613–618, 2004, https://doi.org/doi:10.1515/znc-2004-9-1001.
205. Stockwell, V.O. and Stack, J.P., Using *Pseudomonas* spp. for integrated biological control. *Phytopathology*, 97, 2, 244–249, 2007.
206. Barona-Gómez, F., *et al.*, Multiple biosynthetic and uptake systems mediate siderophore-dependent iron acquisition in *Streptomyces coelicolor* A3(2) and *Streptomyces ambofaciens* ATCC 23877. *Microbiol. (Reading)*, 152, Pt 11, 3355–3366, 2006, https://doi.org/10.1099/mic.0.29161-0.
207. LeBlanc, N., Bacteria in the genus *Streptomyces* are effective biological control agents for management of fungal plant pathogens: A meta-analysis. *BioControl*, 67, 1, 111–121, 2022, https://doi.org/10.1007/s10526-021-10123-5.
208. Bibb, M., The regulation of antibiotic production in *Streptomyces coelicolor* A3 (2). *Microbiology*, 142, 6, 1335–1344, 1996.
209. Straight, P.D., Willey, J.M., Kolter, R., Interactions between *Streptomyces coelicolor* and *Bacillus subtilis*: Role of Surfactants in Raising Aerial Structures. *J. Bacteriol.*, 188, 13, 4918–4925, 2006, https://doi.org/doi:10.1128/jb.00162-06.
210. McManus, P.S., Stockwell, V.O., Sundin, G.W., Jones, A.L., Antibiotic use in plant agriculture. *Annu. Rev. Phytopathol.*, 40, 443–465, 2002, https://doi.org/10.1146/annurev.phyto.40.120301.093927.
211. Le, K.D., Yu, N.H., Park, A.R., Park, D.-J., Kim, C.-J., Kim, J.-C., *Streptomyces* sp. AN090126 as a Biocontrol Agent against Bacterial and Fungal Plant Diseases. *Microorganisms*, 10, 4, 2022, https://doi.org/10.3390/microorganisms10040791.
212. Panchalingam, H., *et al.*, Testing the Biocontrol Ability of a *Trichoderma-Streptomycetes Consortium* against *Pyrrhoderma noxium* (Corner) L.W. Zhou and Y.C. Dai in Soil. *J. Fungi (Basel)*, 9, 1, 2022, https://doi.org/10.3390/jof9010067.
213. Khan, S., Srivastava, S., Karnwal, A., Malik, T., *Streptomyces* as a promising biological control agents for plant pathogens. *Front. Microbiol.*, 14, 2023, https://doi.org/10.3389/fmicb.2023.1285543.
214. Jia, Y., *et al.*, Synergistic biocontrol of *Bacillus subtilis* and *Pseudomonas fluorescens* against early blight disease in tomato. *Appl. Microbiol. Biotechnol.*, 107, 19, 6071–6083, 2023, https://doi.org/10.1007/s00253-023-12642-w.
215. Guzmán-Guzmán, P., Kumar, A., de los Santos-Villalobos, S., Parra-Cota, F.I., Orozco-Mosqueda, M., d., C., Fadiji, A.E., Hyder, S., Babalola, O.O.,

Santoyo, G., *Trichoderma* Species: Our Best Fungal Allies in the Biocontrol of Plant Diseases—A Review. *Plants*, 12, 3, 432, 2023, https://doi.org/10.3390/plants12030432.
216. Dutta, P., *et al.*, Molecular interaction between plants and *Trichoderma* species against soil-borne plant pathogens. *Front. Plant Sci.*, 14, 2023, https://doi.org/10.3389/fpls.2023.1145715.
217. Rahman, M., *et al.*, Deciphering the antimicrobial activity of multifaceted rhizospheric biocontrol agents of solanaceous crops viz., *Trichoderma harzianum* MC2, and *Trichoderma harzianum* NBG. *Front. Plant Sci.*, 14, 2023, https://doi.org/10.3389/fpls.2023.1141506.
218. de Freitas, A.D.L., da Silva, J.C., de Souza Vandenberghe, L.P., Rogez, H., Moura, F.G., Bioactive coating containing harzianic acid: Control of *Fusarium oxysporum* and preservation of tomato quality. *Food Control*, 162, 110445, 2024, https://doi.org/10.1016/j.foodcont.2024.110445.
219. Rezaei, E.E., *et al.*, Climate change impacts on crop yields. *Nat. Rev. Earth Environ.*, 4, 12, 831–846, 2023, https://doi.org/10.1038/s43017-023-00491-0.
220. A. K, P., *et al.*, Impact of climate change and anthropogenic activities on aquatic ecosystem – A review. *Environ. Res.*, 238, 117233, 2023, https://doi.org/10.1016/j.envres.2023.117233.
221. Yang, Y., *et al.*, Climate change exacerbates the environmental impacts of agriculture. *Science*, 385, 6713, eadn3747, 2024, https://doi.org/doi:10.1126/science.adn3747.
222. Gomathy, M. and Kalaiselvi, K., Climate change and its impact on agriculture, in: *Advanced Technologies for Smart Agriculture*, pp. 259–282, River Publishers, New York, 2023.
223. Yuan, X., Li, S., Chen, J., Yu, H., Yang, T., Wang, C., Huang, S., Chen, H., Ao, X., Impacts of Global Climate Change on Agricultural Production: A Comprehensive Review. *Agronomy*, 14, 7, 1360, 2024, https://doi.org/10.3390/agronomy14071360.
224. Omotoso, A.B. and Omotayo, A.O., The interplay between agriculture, greenhouse gases, and climate change in Sub-Saharan Africa. *Reg. Environ. Change*, 24, 1, 1, 2023, https://doi.org/10.1007/s10113-023-02159-3.
225. Ackerl, T., Weldemariam, L.F., Nyasimi, M., Ayanlade, A., Climate change risk, resilience, and adaptation among rural farmers in East Africa: A literature review. *Reg. Sustain.*, 4, 2, 185–193, 2023, https://doi.org/10.1016/j.regsus.2023.05.004.
226. Yang, Q., *et al.*, Bibliometric Analysis on the Impact of Climate Change on Crop Pest and Disease. *Agronomy*, 13, 3, 920, 2023, https://doi.org/10.3390/agronomy13030920.
227. Guégan, J.-F., de Thoisy, B., Gomez-Gallego, M., Jactel, H., World forests, global change, and emerging pests and pathogens. *Curr. Opin. Environ. Sustainability*, 61, 101266, 2023, https://doi.org/10.1016/j.cosust.2023.101266.
228. Mutengwa, C.S., Mnkeni, P., Kondwakwenda, A., Climate-Smart Agriculture and Food Security in Southern Africa: A Review of the Vulnerability of

Smallholder Agriculture and Food Security to Climate Change. *Sustainability*, 15, 4, 2882, 2023, https://doi.org/10.3390/su15042882.
229. Arslan, A., *et al.*, Climate Smart Agriculture? Assessing the Adaptation Implications in Zambia. *J. Agric. Econ.*, 66, 3, 753–780, 2015, https://doi.org/10.1111/1477-9552.12107.
230. Zizinga, A., *et al.*, Climate change and maize productivity in Uganda: Simulating the impacts and alleviation with climate smart agriculture practices. *Agric. Syst.*, 199, 103407, 2022, https://doi.org/10.1016/j.agsy.2022.103407.
231. Zhao, J., Liu, D., Huang, R., A Review of Climate-Smart Agriculture: Recent Advancements, Challenges, and Future Directions. *Sustainability*, 15, 4, 3404, 2023, https://doi.org/10.3390/su15043404.
232. Vishnoi, S. and Goel, R.K., Climate smart agriculture for sustainable productivity and healthy landscapes. *Environ. Sci. Policy*, 151, 103600, 2024, https://doi.org/10.1016/j.envsci.2023.103600.
233. Oyetunde-Usman, Z. and Shee, A., Adoption of drought-tolerant maize varieties and interrelated climate smart agricultural practices in Nigeria. *Agric. Food Secur.*, 12, 1, 43, 2023, https://doi.org/10.1186/s40066-023-00429-1.
234. Erekalo, K.T. and Yadda, T.A., Climate-smart agriculture in Ethiopia: Adoption of multiple crop production practices as a sustainable adaptation and mitigation strategies. *World Dev. Sustain.*, 3, 100099, 2023, https://doi.org/10.1016/j.wds.2023.100099.
235. Frimpong, F., *et al.*, Water-smart farming: review of strategies, technologies, and practices for sustainable agricultural water management in a changing climate in West Africa. *Front. Sustain. Food Syst.*, 7, 2023, https://doi.org/10.3389/fsufs.2023.1110179.
236. Dissanayaka, D.M.N.S., Dissanayake, D.K.R.P.L., Udumann, S.S., Nuwarapaksha, T.D., Atapattu, A.J., Agroforestry—a key tool in the climate-smart agriculture context: a review on coconut cultivation in Sri Lanka. *Front. Agron.*, 5, 2023, https://doi.org/10.3389/fagro.2023.1162750.
237. Adam, B. and Abdulai, A., Minimum tillage as climate-smart agriculture practice and its impact on food and nutrition security. *PLoS One*, 18, 12, e0287441, 2023, https://doi.org/10.1371/journal.pone.0287441.
238. Swarnam, T.P., *et al.*, Climate smart crop-livestock integrated farming as a sustainable agricultural strategy for humid tropical islands. *Int. J. Agric. Sustain.*, 22, 1, 2298189, 2024, https://doi.org/10.1080/14735903.2023.2298189.
239. Ashfaq, M., Mushtaq, I., Mehmood, M.A., Kayani, S.B., Rauf, A., Climate-Smart Strategies for Integrated Pest Management, in: *Plant Quarantine Challenges under Climate Change Anxiety*, K. A. Abd-Elsalam, S. M. Abdel-Momen (Eds.), pp. 407–434, Springer International Publishing, Cham, Switzerland, 2024, https://doi.org/10.1007/978-3-031-56011-8_13.
240. Bi, Y., *et al.*, Effects of *Bacillus subtilis* on cotton physiology and growth under water and salt stress. *Agric. Water Manage.*, 303, 109038, 2024, https://doi.org/10.1016/j.agwat.2024.109038.

241. Armada, E., Roldán, A., Azcon, R., Differential Activity of Autochthonous Bacteria in Controlling Drought Stress in Native Lavandula and Salvia Plants Species Under Drought Conditions in Natural Arid Soil. *Microb. Ecol.*, 67, 2, 410–420, 2014, https://doi.org/10.1007/s00248-013-0326-9.
242. Pingali, P.L., Hossain, M., Pandey, S., Price, L.L., Economics of nutrient management in Asian rice systems: Towards increasing knowledge intensity. *Field Crops Res.*, 56, 1, 157–176, 1998, https://doi.org/10.1016/S0378-4290(97)00126-3.
243. Tao, H., *et al.*, Growth and yield formation of rice (*Oryza sativa* L.) in the water-saving ground cover rice production system (GCRPS). *Field Crops Res.*, 95, 1, 1–12, 2006, https://doi.org/10.1016/j.fcr.2005.01.019.
244. Bouman, B.A.M., Peng, S., Castañeda, A.R., Visperas, R.M., Yield and water use of irrigated tropical aerobic rice systems. *Agric. Water Manage.*, 74, 2, 87–105, 2005, https://doi.org/10.1016/j.agwat.2004.11.007.
245. Panda, D., Mishra, S.S., Behera, P.K., Drought Tolerance in Rice: Focus on Recent Mechanisms and Approaches. *Rice Sci.*, 28, 2, 119–132, 2021, https://doi.org/10.1016/j.rsci.2021.01.002.
246. Singh, P., Singh, R., Madhu, G.S., Singh, V.P., Seed Biopriming with *Trichoderma Harzianum* for Growth Promotion and Drought Tolerance in Rice (*Oryza sativus*). *Agric. Res.*, 12, 2, 154–162, 2023, https://doi.org/10.1007/s40003-022-00641-8.
247. Quazi, S.A.J., ferdous, J., Shozib, H.B., Khaton, A., Zaidi, N.W., Role of *Trichoderma asperelloides* and *Trichoderma brevicompactum* in improving drought tolerance in rice. *Plant Stress*, 12, 100457, 2024, https://doi.org/10.1016/j.stress.2024.100457.

Index

Also of Interest

Check out these other related titles from Scrivener Publishing

Climate Smart Agriculture, Edited by Anitha Velu, Prasanth Aruchamy, Raghu Ramamoorthy, Rajesh Kumar Dhanaraj, and Seifdine Kadry, ISBN: 9781394336333. Transform the future of sustainable farming with this guide to mastering the deep reinforcement learning architectures and algorithms that turn complex environmental data into precise, high-yield decisions for climate-smart agriculture.

Sustainable Agriculture Production Using Blockchain Technology, Edited by Rekh Ram Janghel, Rajesh Doriya, Jaykumar Lachure, and Yogesh Kumar Rathore, ISBN: 9781394248674. Revolutionize the agricultural supply chain with this essential guide, which provides the practical knowledge to leverage blockchain technology for transparency, traceability, and trust, alongside AI for overcoming modern farming challenges.

Mathematical Modeling in Agriculture, Edited by Sabyasachi Pramanik, Niranjanamurthy M., Ankur Gupta, and Ahmed J. Obaid, ISBN: 9781394233717. The main goal of the book is to explore the idea behind data modeling in smart agriculture using information and communication technologies and tools to make agricultural practices more functional, fruitful and profitable.

Agroforestry, Edited by Abhishek Raj, Manoj Kumar Jhariya, Arnab Banerjee, Ramesh Kumar Jha, and Krishan Pal Singh, ISBN: 9781394231133. This book offers comprehensive insights into the management of agroforestry for livelihood security and sustainable development in the tropics, addressing ecological interactions, productivity, and the monetization of carbon credits, while also outlining a future roadmap and policy challenges.

Smart Agritech: Robotics, AI, and Internet of Things (IoT) in Agriculture, Edited by Santosh Kumar Srivastava, Durgesh Srivastava, Korhan Cengiz, and Pramod Gaur, ISBN: 9781394302959. The main goal of *Smart Agritech: Robotics, AI, and the Internet of Things (IoT) in Agriculture* is to explore how emerging technologies such as robotics, artificial intelligence (AI), and IoT can be leveraged to improve efficiency, sustainability, and productivity in agriculture.

DIGITAL AGRICULTURAL ECOSYSTEM: Revolutionary Advancements in Agriculture, Edited by Kuldeep Singh and Prasanna Kolar, ISBN: 9781394242931. The book comprehensively explores the dynamic synergy between modern technology and agriculture, showcasing how advancements such as artificial intelligence, data analytics, and smart farming practices are reshaping the landscape to ensure food security in the era of climate change, as well as bridging the gap between cutting-edge research and practical implementation.

THE DIGITAL AGRICULTURAL REVOLUTION: Innovations and Challenges in Agriculture through Technology Disruptions, Edited by Roheet Bhatnagar, Nitin Kumar Tripathi, Nitu Bhatnagar and Chandan Kumar Panda, ISBN: 9781119823339. The book integrates computational intelligence, applied artificial intelligence, and modern agricultural practices and will appeal to scientists, agriculturists, and those in plant and crop science and management.

HANDBOOK OF AGRICULTURAL BIOTECHNOLOGY: Volume I Nanopesticides, Edited by Charles Oluwaseun Adetunji and Julius Kola Oloke, ISBN: 9781119836148. The book provides detailed information on the application of nanopesticides for the management of numerous agricultural pests and pathogens, as well as to foster an innovative pathway toward future sustainable agriculture and food systems.

HANDBOOK OF AGRICULTURAL BIOTECHNOLOGY: Volume II Nanobioherbicides, Edited by Charles Oluwaseun Adetunji, Julius Kola Oloke, ISBN: 9781119836155. The book provides a detailed examination of the application of nanobioherbicides that come from plants including information on the different metabolites derived from numerous plants that could become bioherbicides.

HANDBOOK OF AGRICULTURAL BIOTECHNOLOGY: Volume III Nanofungicides, Edited by Charles Oluwaseun Adetunji and Julius Kola Oloke, ISBN: 9781119836162. This book provides state-of-the-art information about recent advancements in the application of nanobiofungicides for effective management of post-harvest diseases and agricultural fungal diseases, including blights, mildews, molds, and rusts.

HANDBOOK OF AGRICULTURAL BIOTECHNOLOGY: Volume IV Nanoinsecticides, Edited by Charles Oluwaseun Adetunji and Julius Kola Oloke, ISBN: 9781119836179. The book provides detailed information about the application of repellent products that contain plant-based ingredients known as nanobioinsecticides, including the pesticide evaluation scheme guidelines for repellent testing, relevant information about the procedures to evaluate several repellent compounds, the development of new products that offer high repellency, and guidelines for consumer safety.

ECORESTORATION FOR SUSTAINABILITY, Edited by Arnab Banerjee, Manoj Kumar Jhariya, Surendra Singh Bargali, and Debnath Palit, ISBN: 9781119879718. With climate change and land degradation advancing at an alarming rate, the state of the global ecosystem and how it can be restored in our march toward sustainability is incredibly important, and this new, groundbreaking volume tackles these important issues head on.

FUNCTIONAL POLYMERS IN FOOD SCIENCE, Part 1: Food Packaging, Edited by Giuseppe Cirillo, Umile Gianfranco Spizzirri, and Francesca Iemma, ISBN: 9781118594896. Provides an extensive overview of the link between polymer properties and food quality. *NOW AVAILABLE!*

FUNCTIONAL POLYMERS IN FOOD SCIENCE, Part 2: Food Processing, Edited by Giuseppe Cirillo, Umile Gianfranco Spizzirri, and Francesca Iemma, ISBN: 9781118595183. Provides an extensive overview of the link between polymer properties and food quality. *NOW AVAILABLE!*

ADVANCES IN FOOD SCIENCE AND TECHNOLOGY Volume 1, Edited by Visakh, P.M., Sabu Thomas, Laura B. Iturriaga, and Pablo Daniel Ribotta, ISBN: 9781118121023. Written in a systematic and comprehensive manner the book reports recent advances in the development of food science and food technology areas. *NOW AVAILABLE!*

ADVANCES IN FOOD SCIENCE AND NUTRITION Volume 2, Edited by Visakh P.M., Laura B. Iturriaga, and Pablo Daniel Ribotta, ISBN: 9781118137093. This important book comprehensively reviews research on new developments in all areas of food chemistry/science and nutrition. *NOW AVAILABLE!*

Applied Water Science Volume 1: Fundamentals and Applications, Edited by Inamuddin, Mohd Imran Ahamed, Rajender Boddula and Tauseef Ahmad Rangreez, ISBN: 9781119724766. Edited by one of the most well-respected and prolific engineers in the world and his team, this is the first volume in a two-volume set that is the most thorough, up-to-date, and comprehensive volume on applied water science available today. *COMING IN SUMMER 2021*

Sustainable Water Purification, by Safiur Rahman and M. R. Islam, ISBN: 9781119650997. This is the only book that takes a zero-waste approach to propose 100% sustainable water purification techniques. *NOW AVAILABLE!*

Books in the series, "Bioprocessing in Food Science"

NUTRACEUTICALS FROM FRUIT AND VEGETABLE WASTE, Edited by Vidisha Tomer, Navnidhi Chhikara, Ashwani Kumar, and Anil Panghal, ISBN: 9781119803508. Nutraceuticals from Fruit and Vegetable Waste, the latest volume in the series, "Bioprocessing in Food Science," is an essential reference on the valorization of fruit and vegetable waste and the state of the art for fruit and vegetable processing.

NUTRITIONAL SCIENCE AND TECHNOLOGY: Concept to Application, Edited by Tejpal Dhewa, Anil Kumar Puniya, and Anil Panghal, ISBN: 97811198089. This new volume in the groundbreaking series, "Bioprocessing in Food Science," Nutritional Science and Technology, is a comprehensive resource covering topics such as human nutrition, food safety concerns, and emerging technologies like nanotechnology, and it also addresses the need for innovative food products, contamination, and regulatory challenges in the global food market.

HARVESTING FOOD FROM WEEDS, Edited by Prerna Gupta, Navnidhi Chhikara, and Anil Panghal, ISBN: 9781119791973. An exciting new volume in the groundbreaking new series, "Bioprocessing in Food Science," Harvesting Food from Weeds focuses on the latest processes, industrial applications, and leading research on this important, evolving new area, for engineers, scientists, students, and other industry professionals.

MICROBES IN THE FOOD INDUSTRY, Edited by Navnidhi Chhikara, Anil Panghal, and Gaurav Chaudhary, ISBN: 9781119775584. This newest volume in the groundbreaking new series, "Bioprocessing in Food Science," focuses on the latest processes, industrial applications, and leading research on microbes in the food industry, for engineers, scientists, students, and other industry professionals.

NOVEL TECHNOLOGIES IN FOOD SCIENCE, Edited by Navnidhi Chhikara, Anil Panghal, and Gaurav Chaudhary, ISBN: 9781119775584. The third book in the series, "Bioprocessing in Food Science," covers the latest and greatest new technologies in food science.

FUNCTIONAL FOODS, Edited by Navnidhi Chhikara, Anil Panghal, and Gaurav Chaudhary, ISBN: 9781119775560. Presenting cutting-edge information on new and emerging food engineering processes, *Functional Foods*, the second volume in the groundbreaking new series, "Bioprocessing in Food Science," is an essential reference on the modeling, quality, safety, and technologies associated with food processing operations today.

THERMAL FOOD ENGINEERING OPERATIONS, Edited by Nitin Kumar, Anil Panghal, and M. K. Garg, ISBN: 9781119775591. Presenting cutting-edge information on new and emerging food engineering processes, Thermal Food Engineering Operations, the first volume in the new series, "Bioprocessing in Food Science," is an essential reference on the modeling, quality, safety, and technologies associated with food processing operations today.

Printed and bound by CPI Group (UK) Ltd, Croydon, CR0 4YY
17/09/2026
14964282-0006